CHINA IN THE 21ST CENTURY

AGRICULTURAL TRADE IN CHINA

China in the 21st Century

Additional books in this series can be found on Nova's website under the Series tab.

Additional E-books in this series can be found on Nova's website under the E-book tab.

Agriculture Issues and Policies

Additional books in this series can be found on Nova's website under the Series tab.

Additional E-books in this series can be found on Nova's website under the E-book tab.

CHINA IN THE 21ST CENTURY

AGRICULTURAL TRADE IN CHINA

MARK T. DEVINE
EDITOR

Nova Science Publishers, Inc.
New York

Library of Congress Cataloging-in-Publication Data

Agricultural trade in China / editor: Mark T. Devine.
p. cm.
Includes index.
ISBN 978-1-62100-602-2 (hardcover)
1. Produce trade--China. 2. Exports--United States. I. Devine, Mark T.
HD9016.C62A326 2012
382'.410951--dc23
2011035662

Published by Nova Science Publishers, Inc. † New York

CONTENTS

PREFACE

China is the world's largest agricultural economy and the leading producer and consumer of many agricultural commodities. In recent years, its massive population and tremendous income growth have fueled a rapid increase in both the quantity and quality of food and fiber consumed. While China has met much of its needs by increasing domestic production, it has also emerged as a leading global importer of several agricultural commodities, including cotton, soybeans, vegetable oils, and hides and skins. China's increase in imports has benefited its trade partners significantly, but only for a narrow range of products. This book provides information and analysis regarding the conditions of competition in China's agricultural market and trade and their effects on U.S. agricultural exports.

Chapter 1- This report describes and analyzes government policies and other factors that affect the conditions of competition in China's agricultural market and trade, as well as their effects on U.S. agricultural exports. It provides an overview of (a) China's agricultural imports, exports, consumption, and production during 2005–09; (b) Chinese government regulations relating to the agricultural market; (c) competitive factors affecting the Chinese agricultural sector, with case studies to analyze these factors; and (d) Chinese tariffs, nontariff measures (NTMs), and free trade agreements (FTAs). The study also uses economic modeling to provide an analysis of the effects of Chinese tariffs, FTAs, and certain NTMs on U.S. agricultural exports. Results suggest that the elimination of Chinese tariffs and nontariffs measures could lead to an additional $3.9 billion to $5.2 billion in U.S. agricultural exports to China.

China's agricultural trade has grown rapidly, particularly since its market liberalization leading up to its 2001 accession to the World Trade Organization. China's chief agricultural exports consist of labor-intensive horticultural products shipped mainly to regional markets, while imports are limited to a small number of commodities, including soybeans, vegetable oils, poultry, cotton, and hides and skins. Chinese consumption patterns are changing because of urbanization, rising incomes, and the importance of food safety and quality to consumers.

China is a major global producer of agricultural products, especially fruits, vegetables, rice, cotton, and pork. Overall, China is largely self-sufficient, with the exception of a few key commodities which it imports. Government support for China's farm sector has grown significantly since 2004 when, in an important shift in economic policy, the government began supporting agriculture instead of taxing the sector to support industrial development. Most government programs for agriculture fall into four categories: direct payments, price support programs, agricultural infrastructure projects, and regulatory reforms (e.g., food safety and standards).

Chapter 2- With China becoming the second largest U.S. market in fiscal year 2010, China's emergence as a major agricultural importer is well-known. While China is a large net food importer, it is also becoming a formidable competitor in the export market with shipments nearly tripling over the past 10 years and market share increasing. While some of China's exports are bound for the United States, others directly compete with U.S. products in foreign markets. Exports of consumer-oriented high-value products (HVPs) have shown particular growth, especially to nearby markets in Japan and Southeast Asia. Although China faces production constraints and booming domestic consumption, future exports, particularly of high value products, have room for expansion.

Chapter 3- Agricultural exports to China finished fiscal year 2010 at $15.1 billion as the country surpassed Mexico to become the number two market for U.S. agricultural exports. Record soybean shipments along with strong Chinese demand for cotton and distillers dried grains (DDG's) contributed to the overall record in U.S. exports to China. Moving forward, exports in FY 2011 are forecast to reach another record at $17.5 billion.

Chapter 4- India and China share many similarities. The world's only two population billionaires are both among the global leaders in GDP growth over the past decade and both face similar challenges of producing enough food to meet the demands of an increasingly affluent populace. However, when it comes to agricultural trade, China and India appear to have taken divergent paths. China has emerged as a major market for agricultural products whereas India (though somewhat liberalizing trade) remains a largely closed market. This is particularly true for U.S. agricultural exports, as China is the fourth largest and a fast growing market, while India is not among the top 20 markets. However, both markets hold significant potential for U.S. agricultural exports, albeit with a different mix of products.

In: Agricultural Trade in China
Editor: Mark T. Devine

ISBN: 978-1-62100-602-2

Chapter 1

China's Agricultural Trade: Competitive Conditions and Effects on United States Exports*

United States International Trade Commission

Commissioners

Deanna Tanner Okun, Chairman
Irving A. Williamson, Vice Chairman
Charlotte R. Lane
Daniel R. Pearson
Shara L. Aranoff
Dean A. Pinkert

Karen Laney
Acting Director of Operations
Michael Anderson
Acting Director, Office of Industries

Project Leader
Joanna Bonarriva
Joanna.Bonarriva@usitc.gov
Deputy Project Leader
Marin Weaver
Marin.Weaver@usitc.gov
Office of Industries
Laura Bloodgood, David Coffin, John Fry, John Giamalva,
Brendan Lynch, Douglas Newman, and Alison Rozema

* This is an edited, reformatted and augmented version of the United States International Trade Commission publication, Investigation No. 332-518, USITC Publication 4219, dated March 2011.

Office of Economics
Jose Signoret, Danielle Trachtenberg, and Marinos Tsigas
Primary Reviewers
Kimberlie Freund and Alexander Hammer
Administrative Support
Phyllis Boone, Monica Reed, and Wanda Tolson
Under the Direction of
Jonathan R. Coleman, *Chief*
Agriculture and Fisheries Division

ABSTRACT

This report describes and analyzes government policies and other factors that affect the conditions of competition in China's agricultural market and trade, as well as their effects on U.S. agricultural exports. It provides an overview of (a) China's agricultural imports, exports, consumption, and production during 2005–09; (b) Chinese government regulations relating to the agricultural market; (c) competitive factors affecting the Chinese agricultural sector, with case studies to analyze these factors; and (d) Chinese tariffs, nontariff measures (NTMs), and free trade agreements (FTAs). The study also uses economic modeling to provide an analysis of the effects of Chinese tariffs, FTAs, and certain NTMs on U.S. agricultural exports. Results suggest that the elimination of Chinese tariffs and nontariffs measures could lead to an additional $3.9 billion to $5.2 billion in U.S. agricultural exports to China.

China's agricultural trade has grown rapidly, particularly since its market liberalization leading up to its 2001 accession to the World Trade Organization. China's chief agricultural exports consist of labor-intensive horticultural products shipped mainly to regional markets, while imports are limited to a small number of commodities, including soybeans, vegetable oils, poultry, cotton, and hides and skins. Chinese consumption patterns are changing because of urbanization, rising incomes, and the importance of food safety and quality to consumers.

China is a major global producer of agricultural products, especially fruits, vegetables, rice, cotton, and pork. Overall, China is largely self-sufficient, with the exception of a few key commodities which it imports. Government support for China's farm sector has grown significantly since 2004 when, in an important shift in economic policy, the government began supporting agriculture instead of taxing the sector to support industrial development. Most government programs for agriculture fall into four categories: direct payments, price support programs, agricultural infrastructure projects, and regulatory reforms (e.g., food safety and standards).

Several factors of production—such as low labor costs and government support—enhance the competitiveness of Chinese agricultural products, while other factors—such as its land tenure system and its fragmented transportation and cold storage infrastructure—weaken its competitiveness. Many of these agricultural competitive factors are illustrated in case studies of China's fresh apple, pork, processed foods, and wheat sectors.

Chinese average agricultural tariffs are relatively low, but remain high for several import-sensitive products, including beverages, tobacco, and nuts. China also maintains restrictive tariff-rate quotas for certain commodities, such as wheat, cotton, and sugar. The Commission's quantitative estimates indicate that China's NTMs particularly its sanitary and phytosanitary (SPS) measures have a larger effect on U.S. exports to China than its tariffs: SPS measures substantially limit or effectively prohibit certain U.S. agricultural products. China's FTAs and preferential trade agreements are largely with trade partners in East Asia and Oceania. Their provisions vary in scope, but overall have had a negligible effect on total U.S. agricultural exports because for several food and agricultural products, the level of U.S. exports to China is either very large or very small relative to the level of China's imports from its PTA partners.

ACRONYMS AND TERMS

ACFTA	ASEAN-China Free Trade Agreement
AD	Antidumping
ADBC	Agricultural Development Bank of China
ADOGA	American Dehydrated Onion and Garlic Association
AI	Avian Influenza
APHIS	Animal and Plant Health Inspection Services
AQSIQ	Administration of Quality Supervision, Inspection, and Quarantine
ARF	Automatic Registration Form
ASA	American Soybean Association
ASEAN	Association of Southeast Asian Nations
BRIC	Brazil, Russia, India, and China
BSE	Bovine Spongiform Encephalopathy
Bt	Bacillus thuringiensis
CCFTA	China-Chile Free Trade Agreement
CEPAs	Closer Economic Partnership Agreements
CES	Constant elasticity of substitution
CET	Constant elasticity of transformation
CIA	Central Intelligence Agency
C.I.F.	Cost insurance freight
CIQ	Chinese Inspection and Quarantine
CNCA	Certification and Accreditation Administration of China
CODEX	Codex Alimentarius
COFCO	Cereals, Oils and Foodstuffs Import and Export Company
Commission	U.S. International Trade Commission
CVD	Countervailing duties
DDGS	Distiller's dried grains with solubles
DETs	Differential export taxes
DOC	Department of Commerce
DSE	Macao Economic Services
EHP	Early Harvest Program
EIU	Economist Intelligence Unit
ERS	Economic Research Service
EU	European Union
FAO	Food and Agricultural Organization of the United Nations
FAS	Foreign Agricultural Service
FCR	Feed conversion ratios

FDI	Foreign direct investment
FFF	Frozen french fries
F.O.B.	Free on board
FTA	Free trade agreement
GDP	Gross domestic product
GE	General equilibrium
GM	Genetically modified
GMOs	Genetically modified organisms
GNI	Gross National Income
GOC	Government of Chile
GTAP	Global Trade Analysis Project
GTIS	Global Trade Information Service, Inc.
ha	Hectares. One hectare is equal to 2.47 acres.
HKSAR	The Government of Hong Kong Special Administrative Region
HPAI	Highly Pathogenic Avian Influenza
HS	World Customs Organization harmonized system
HTS	Harmonized Tariff Schedule
IMF	International Monetary Fund
IPPC	International Plant Protection Convention
IPR	Intellectual property rights
JECFA	Joint FAO/WHO Export Committee on Food Additives
KFC	Kentucky Fried Chicken
LDCs	Lesser developed countries
LPAI	Low Pathogenic Avian Influenza
MEA	Ministry of Economic Affairs
mmt	Million metric ton
mt	Metric ton
MFN	Most favored nation
MHKCEPA	The Mainland and Hong Kong Closer Economic Partnership Arrangement
MMCEPA	The Mainland and Macau Closer Economic Partnership Arrangement
MOA	Ministry of Agriculture
MOFCOM	Ministry of Commerce
MOH	Ministry of Health
MOU	Memorandum of Understanding
MRL	Maximum Residue Limits
mu	A traditional Chinese land unit of measure equal to about 1/15 of a hectare or 1/6 of an acre
MY	Marketing year
NCS	National Customs Service
NDRC	National Development and Reform Commission
NMPF	National Milk Producers Federation
NPC	National Potato Council
NTMs	Nontariff measures
OECD	Organisation for Economic Co-operation and Development
OIE	World Organization for Animal Health
PE	Partial equilibrium

PFI	Pet Food Institute
PRA	Pest Risk Assessment
PRRS	Porcine reproductive and respiratory syndrome
PSD	Production, Supply, and Distribution
PSRs	Product Specific Rules
PSE	Producer Support Estimate
PTAs	Preferential trading arrangements
QIP	Quarantine Import Permit
R-CALF	Ranchers Cattlemen Action Legal Fund, United Stockgrowers of America
RMB	Renminbi
ROC	Republic of China
ROO	Rules of origin
RoW	Rest of the world
SAC	Standardization Administration of China
SAG	State Administration of Grain
SARs	Special Administrative Regions
SAT	State Administration of Taxation
SPS	Sanitary and Phyto-sanitary
STE	State Trading Enterprise
TBT	Technical barriers to trade
TCK	*Tilletia controversa Kuhn*
TID	Trade and Industry Department
TRQs	Tariff-rate quotas
UAE	United Arab Emirates
UNCTAD	United Nations Conference on Trade and Development
UNESCAP	United Nations Economic and Social Commission for Asia and the Pacific
USAID	United States Agency for International Development
USDA	U.S. Department of Agriculture
USDEC	U.S. Dairy Export Council
USITC	U.S. International Trade Commission
VAT	Value-added tax
WHO	World Health Organization
WPA	Western Pistachio Association
WTO	World Trade Organization

Executive Summary

China is the world's largest agricultural economy and the leading producer and consumer of many agricultural commodities. In recent years, its massive population and tremendous income growth have fueled a rapid increase in both the quantity and quality of food and fiber consumed. While China has met much of its needs by increasing domestic production, it has also emerged as a leading global importer of several agricultural commodities, including cotton, soybeans, vegetable oils, and hides and skins. China's increase in imports has benefited its trade partners significantly, but only for a narrow range of products. At the same time, domestic policies to promote agricultural production and maintain self-sufficiency in staple foods increase the competitiveness of domestically produced goods over imported products and, in some cases, keep imports out altogether. As the Chinese agricultural sector has modernized and become more productive, China has also become an important global exporter of several horticultural products, including mandarin oranges, apples, apple juice, certain vegetables, and garlic.

This report responds to a request by the Senate Committee on Finance (Committee) for information and analysis regarding the conditions of competition in China's agricultural market and trade and their effects on U.S. agricultural exports. The Committee asked that the report include the following:

- an overview of China's agricultural market, including recent trends in production, consumption, and trade;
- a description of the competitive factors affecting the agricultural sector in China, in such areas as costs of production, technology, domestic support and government programs related to agricultural markets, foreign direct investment policies, and pricing and marketing regimes;
- an overview of China's participation in global agricultural export markets, particularly in the Asia-Pacific region and in those markets with which China has negotiated trade agreements;
- a description of the principal measures affecting China's agricultural imports, including tariffs and nontariff measures such as sanitary and phytosanitary measures and technical barriers to trade; and
- a quantitative analysis of the economic effects of China's most favored nation (MFN) tariffs, preferential tariffs negotiated under China's free trade agreements, and China's nontariff measures on U.S. agricultural exports to China and on imports from the rest of the world.

Major Findings and Observations

China's Agricultural Trade

China's agricultural trade has grown rapidly, particularly since China liberalized some of its markets in the lead-up to its 2001 accession to the World Trade Organization (WTO). During 2005–10, imported food and fiber became increasingly important in light of rising consumer incomes.

China ranked as the world's second-largest agricultural importing country behind the United States in 2009. China became a net importer of agricultural products in 2003, and this trade deficit is likely to persist as future growth in food demand, driven by rapidly rising per capita income, is expected to outpace increases in domestic production. China's agricultural imports from the world grew at an average annual rate of 24 percent during 2005–10, reaching $66.4 billion in 2010 (table ES.1). In 2010, China's main agricultural imports were soybeans and cotton, primarily from the United States, which together accounted for about one-half of all imported agricultural products in that year. After soybeans and cotton, leading imports were palm oil (8 percent), hides and skins (4 percent), dairy (4 percent), and wool (3 percent). During 2005–10, China imported very small volumes of beef, pork, and grains.

China's chief agricultural exports consist of labor-intensive horticultural products shipped mainly to regional markets, while agricultural imports are concentrated in a small number of products.

Consistent with its natural resource endowments of abundant rural labor and limited agricultural land on a per capita basis, China's agricultural exports are concentrated in labor-intensive products (compared with the United States and its other main trading partners), such as fresh and processed fruits and vegetables. In 2009, China was the fourth leading global agricultural exporting country (behind the United States, Brazil, and Canada). In third-country markets, the United States and China compete only in a narrow range of products—mainly a few fresh fruits and processed vegetables in a small number of primarily Asian markets.

Table ES.1. Chinese and U.S. agricultural production and trade: Comparative statistics

	China	United States
Population (2009)	1,338 million	307 million
Cropland (2009)	122 million hectares	164 million hectares
Cropland per agricultural worker (2007)	0.4 hectares	78.5 hectares
Value of farm production (2007)	$537 billion	$329 billion
Agriculture as share of GDP (2009)	10.6 percent	1.2 percent
Value of agricultural imports (2010)	$66 billion	$86 billion
Top five agricultural imports by value	Soybeans, cotton, palm oil, dairy products, and hides and skins	Coffee, cocoa products, wine, malt beverages, beef
Top five agricultural import suppliers	United States, Brazil, Argentina, European Union-27 (EU-27), Australia	Canada, EU-27, Mexico, China, Brazil
Value of agricultural exports (2010)	$36 billion	$119 billion
Top five agricultural exports by value	Processed vegetables, fresh vegetables, miscellaneous processed foods, fresh fruit, and animal feed	Soybeans, corn, wheat, cotton, and animal feed
Top five agricultural export markets	Japan, EU-27, Hong Kong, United States, Korea	Canada, China, Mexico, Japan, EU-27

Source: GTIS, Global Trade Atlas database (accessed February 18, 2011); USDA, ERS, "Foreign Agricultural Trade of the United States," 2011; CIA, *The World Factbook: China*; World Bank, Data: China; USDA, NASS, *Agricultural Statistics 2009*, 2009, Table 9-44; National Bureau of Statistics of China, *China Statistical Yearbook 2009*, 2009, Table 12-4.

Consistent with its relatively small endowment of arable land per capita, China's agricultural imports include many land-intensive products, such as cotton and soybeans. However, China imports very few land-intensive grains, such as rice, wheat, and corn, owing to government policies that promote self-sufficiency in grains and other staple foods. In 2010, the United States was the leading supplier of agricultural products to China, largely because U.S. soybeans accounted for 17 percent of all China's agricultural imports that year. China's agricultural imports from the United States reached $17.8 billion in 2010, accounting for 27 percent of China's total agricultural imports. Between 2005 and 2010, China's imports from the United States almost tripled because of growth in four products—soybeans, cotton, processed animal feed, and hides and skins—which together accounted for 84 percent of all U.S. agricultural products imported into China in 2010. At the same time, China's purchases of U.S. animal products, grains, and vegetable oils—sectors where the United States is highly competitive internationally—remained negligible.

Chinese Domestic Consumption

Consumption of agricultural products is growing in China as its large population, with incomes rising, demands a higher quantity and quality of food.

With the largest population in the world, China has a huge consumer base and is the largest consuming country for many agricultural products; it consumes one-third of all rice, one-quarter of all corn, and one-half of all pork and cotton. Incomes have been rising quickly in China; between 2000 and 2006 (the latest year for available data), rural and urban incomes rose 59 percent and 87 percent, respectively. High-income provinces, generally on China's east coast, contain a concentration of affluent consumers who are demanding a greater variety of food products, as well as more processed and convenience foods. Middle-class Chinese benefit from higher wages in urban areas, while rural Chinese incomes are being boosted by government policies aimed at closing the wide rural-urban income disparities. Demand for nonfood agricultural products, particularly cotton and hides and skins, will continue to grow in tandem with China's textile, apparel, and footwear industries.

Chinese consumption patterns are changing because of urbanization, rising incomes, and the importance to consumers of food quality and safety.

As per capita income in China grows, traditional mainstays of the Chinese diet, like grains and tubers, are giving way to rising consumption of non-traditional items like meat and fruit. Urban Chinese, with much higher incomes than their rural counterparts, consume more food overall, and in particular consume more fresh vegetables, fruit, and edible oils. They are also more likely to buy chilled, frozen, and perishable foods, as they are more likely to have refrigerators. Rural workers who migrate to cities are exposed to certain foods that are largely unique to urban areas, such as snacks and processed foods, and adopt new consumption patterns. Food quality and safety are important factors affecting Chinese food preferences, particularly in light of recent publicity about adulterated food and poisonings from Chinese food products. As a result, high-income urbanites focus their spending on high-quality products and also seek out products offering assurance of more safety, including organic foods.

Agricultural Sector Characteristics

Agriculture Accounts for 11 Percent of the Chinese Economy

In 2009, China's agricultural production, valued at $543 billion, accounted for 11 percent of gross domestic product (GDP). Employing just under half of the population, the sector is dominated by millions of farmers with small plots of land, which average just 0.6 hectares per household. In contrast, agricultural production in the United States accounted for 1 percent of GDP and employed less than 2 percent of the population, with an average farm size of 619 hectares.

China is a major global producer of agricultural products and is largely self-sufficient, with the exception of a few key commodities.

Since the late 1970s, Chinese agricultural production has grown, such that today China ranks as the leading global producer of many agricultural commodities. With limited good cropland and with scarce water resources in some areas, agricultural sector growth has been achieved largely through substantial increases in productivity, a result of both market-based policy reforms and the adoption of modern agricultural technology and farming practices. Historically, government policies have been geared toward ensuring that enough staple food is produced domestically to feed the population. While this goal is still important, the increased share in total production of labor-intensive products, such as horticulture, meat, and dairy, and the drop in the relative importance of traditional staples, such as grains and tubers, are notable current trends.

Chinese Agricultural Policy Objectives

China's support for the farm sector has grown substantially since 2004 when, in an important shift in economic policy, the government began supporting agriculture instead of taxing it.

In spite of the significant market-based agricultural policy reforms of the last decade, the Chinese government continues to play a central role in the sector. Elimination of agricultural taxes and increased funding for agriculture in the 2000s marked a renewed policy focus on agriculture and the rural economy by policymakers. Self-sufficiency in domestic grain production, raising farmers' incomes, and rural development have become major policy objectives of China's government; the Chinese government wants to shrink the gap between urban and rural incomes and to promote social harmony. Government programs for agriculture can be largely divided into four categories: direct payments, price support programs, agricultural infrastructure projects, and regulatory reforms (e.g., food safety and standards).

China's central, provincial, and local governments formulate agricultural policies in response to a policy environment shaped by historical events and current social factors.

Examining the roots of China's approach to implementing its agricultural policies is important to understanding why government funding of the farm sector varies significantly from province to province, and why Chinese agricultural policies often lack transparency. For both historical and practical reasons, provincial and local governments have significant autonomy over implementing agricultural programs and allocating funds. Regulations to protect farmer and consumer rights and to ensure fairness in the marketplace are often interpreted differently among provincial and local governments or simply ignored. Inconsistent regional implementation of regulations make it difficult for foreign companies to produce, trade, and sell in China's agricultural sector. Local government autonomy may also undercut the programs' efficacy in achieving central government objectives. Like those of other countries, China's agricultural policies reflect conflicting government objectives, including self-sufficiency, boosting farmer incomes, and ensuring affordable food for consumers, and to a lesser extent energy independence, conserving natural resources, and reducing harmful emissions.

Competitiveness in the Chinese Agricultural Sector

Several factors enhance the competitiveness of Chinese agricultural products, including low labor costs, government support, trade policies, and significant investments to modernize agricultural production.

Low labor costs remain the primary competitive advantage of the Chinese agricultural sector over the United States, even though the supply of abundant rural labor is shrinking and farm wages are rising. Government support for certain sectors, including direct payments, input subsidies, and tax exemptions, lowers production costs and boosts farmer incomes. Further, investments to modernize production and improve food safety, including investments by both the Chinese government and foreign companies, have played a key role in boosting the competitiveness of several agricultural sectors in China, such as pork and processed foods. In addition to domestic policies, trade policies, such as tariff-rate quotas (TRQs) and bans, limit import access for products such as grains and sugar whose continued domestic production is considered of strategic importance to the agricultural sector and the economy as a whole.

Several other factors weaken the competitiveness of the Chinese agriculture sector. These include the land tenure system, fragmented transportation and cold storage infrastructure, limited access to credit, and degradation of land and water resources.

China's land tenure system does not permit farmers to own, buy, or sell land. Farmers typically operate small plots (many less than 0.2 hectare), and because they are unable to consolidate land holdings, they cannot benefit from scale economies in producing and marketing their products. Unless China reforms its land tenure system it will likely face continuing difficulties in boosting farmers' incomes, encouraging investment, and increasing agricultural productivity. China's agricultural competitiveness is also weakened by the lack of an efficient, nationwide cold storage transportation network. The result is high spoilage rates, poor quality for those goods finding their way to market, and increased uncertainty for buyers. The development of cold storage chains in China is thwarted by a fragmented farm structure, comprising thousands of small-scale household farmers and small traders scattered throughout China.

Limited access to credit further weakens Chinese agricultural competitiveness. Farmers' inability to own land deprives them of collateral necessary for obtaining loans. This means that many small-scale farmers are unable to invest in mechanization and other technology to improve quality and productivity. Similarly, problems limiting the availability and quality of other factors of production, such as land and water, lessen the competitiveness of the sector. The already limited amount of land available for agriculture is being eaten away by ongoing urbanization and industrialization. At the same time, water resources are under strain, not only because of a lack of supply, but also from both overuse and pollution.

Several factors affecting competitiveness of Chinese agricultural products are illustrated by the fresh apple, pork, processed foods, and wheat sectors.

Fresh Apples

China is the world's largest producer and exporter of apples, by volume. Apple production in China is labor-intensive, requiring significant labor resources to plant, tend, and harvest, and China's low labor costs are key to the industry's competitiveness. The U.S. apple industry, also one of the world's largest producers and exporters, competes with Chinese apples in the high-quality segment of export markets and has been increasing exports to China to take advantage of China's growing wealth and consumption of fruit.

Pork

In the Chinese pork sector, government policies have encouraged domestic production and insulated many small-scale producers of low-quality pork from global market conditions. However, rising labor and feed costs, as well as increasing demand among Chinese consumers for higher quality pork, may lessen Chinese pork competitiveness in the future, creating opportunities for global exporters from the United States and Europe.

Processed Foods

China is one of the world's largest producers and exporters of processed foods. In recent years its production has increased, primarily because of rising disposable income and urbanization, which are driving demand growth in the domestic market. Processed food production is capital-intensive throughout most of the world, but in China there is great variation in the level of mechanization, and the industry relies more heavily on labor than other major producers. As labor costs continue to rise in China, however, the food-processing sector is becoming more mechanized and capital-intensive.

Wheat

Wheat is a staple and is viewed in China as important to food security. Despite limited land and water resources for wheat production, the central government has put in place domestic and trade policies designed to maintain domestic production and control import supplies.

Chinese Tariffs and Nontariff Measures

Chinese average agricultural tariffs are relatively low, but they peak for several import-sensitive products, including beverages, tobacco, and nuts. China also maintains restrictive TRQs for certain commodities, such as certain grains, cotton, and sugar.

China substantially reduced its tariffs and replaced absolute quotas with TRQs for various agricultural products in preparation for its WTO accession in December 2001. China's simple average tariff (not weighted by trade) for agricultural products fell from 42 percent ad valorem in 1992 to 24 percent ad valorem in 1998. To fulfill its WTO commitments, China committed to reduce its agricultural tariffs to a simple average of 15 percent ad valorem (table ES.2). The trade-weighted average tariff for agricultural products was 12 percent ad valorem in 2007. China's applied tariffs are highest for corn, wheat, and rice (65 percent ad valorem for over-quota imports); tobacco (57 percent); and raw cane and refined sugar (50 percent for over-quota imports). Other high-tariff products include cotton (40 percent for over-quota imports), certain fermented beverages (40 percent), beverage bases (35 percent), and a variety of nuts (30 percent). Although average agricultural tariffs are relatively low, the Chinese market's price sensitivity means that even a low tariff, added to the 13 to 17 percent value-added tax (VAT) assessed at the border, can raise the delivered cost of imports prohibitively.

China maintains TRQs for wheat, corn, rice, cotton, sugar, and wool; those for wheat, corn, rice, and sugar are administered largely by state-owned enterprises. Over-quota tariff rates are generally prohibitive and TRQ fill rates are low for grains. China's TRQs for agricultural products represent about 9 percent of domestic consumption of wheat, 5 percent of corn, 4 percent of rice, 13 percent of sugar, and 2 percent of cotton. These relatively small shares, as well as often nontransparent TRQ administration, indicate the importance the Chinese government attaches to self-sufficiency in these products.

Table ES.2. China's final bound WTO tariffs, MFN applied tariffs, and share of imports entering duty free, by agricultural product group

		Final bound tariffs (percent ad valorem)		MFN applied tariffs (percent ad valorem)		Share of imports entering duty free (percent)
Product group	Simple average	Share that is duty free[a] (percent)	Maximum	Simple average	Share that is duty free[a] (percent)	
Animal products	14.8	9.4	25	14.7	10.1	4.1
Dairy products	12.2	0	20	12.0	0	0
Fruit, vegetables, plants	15.0	4.8	30	14.8	5.9	3.6
Coffee, tea	14.9	0	32	14.7	0	0
Cereals and preparations	23.7	2.6	65	23.9	3.4	0
Oilseeds, fats, and oils	11.6	6.2	30	10.6	5.4	0.1
Sugars and confectionery	27.4	0	50	27.4	0	0
Beverages and tobacco	23.9	2.4	65	22.9	2.2	1.8
Cotton	22.0	0	40	22.0	0	0
Other agricultural products	11.9	10.2	38	11.5	9.4	2.3

Source: WTO and ITC, *World Tariff Profiles 2009*, 60. [a]Share of total HTS 6-digit subheadings that is duty-free.

A wide array of Chinese non-tariff measures (NTMs) substantially limit, or effectively prohibit, imports of certain U.S. agricultural products.

China's NTMs can be broadly classified into three categories: restrictions that keep products out of China, measures that raise costs for traders, and conditions that increase uncertainty and therefore risk. Chinese NTMs effectively prohibit imports of U.S. beef, pears, fresh potatoes, pet food, and strawberries, and significantly restrict imports of U.S. apples and pork (table ES.3). The VAT exemption for Chinese primary agricultural producers significantly disadvantages imported agricultural products by raising their cost relative to domestically produced ones, especially because the VAT on imports is assessed on the import's cost plus the import tariff paid. NTMs affect virtually all products by increasing costs or creating uncertainty, and can make U.S. products uncompetitive or dissuade U.S. exporters from entering the Chinese market. In some instances, China's NTMs are applied inconsistently; in other instances, they are relaxed or removed altogether when China's demand for imports increases. China also links NTMs to its own market access for certain products abroad.

China's Trade Agreements

China's free trade agreements (FTAs) were first negotiated with China's special administrative regions (SARs) and then with Southeast Asian and more distant countries.

The first FTAs were Closer Economic Partnership Agreements (CEPAs) with Hong Kong and Macau, two SARs in China. These were followed by FTAs with members of the Association of Southeast Asian Nations (ASEAN) and with Pakistan, Chile, New Zealand, Singapore, Peru, and Costa Rica. Apart from the FTAs with New Zealand, Hong Kong, and Singapore, the foregoing agreements are with developing countries. In addition, an FTA was implemented with Bangladesh, India, Korea, Laos, and Sri Lanka (the China-Asia Pacific Trade Agreement); this FTA essentially solidified and extended a prior agreement. Several of China's FTAs account for a large amount of trade, including those with Hong Kong, Macau, and the ASEAN countries.

The provisions of China's FTAs vary in scope.

The most comprehensive FTAs in terms of product coverage have been negotiated with areas and countries with relatively small agricultural sectors; there are no product exclusions for tariff reductions under China's FTAs with Hong Kong, Macau, ASEAN countries, and Singapore. For other FTAs, agricultural product coverage is generally limited to products that are complementary or offer only limited competition to domestic Chinese agricultural producers. China generally has not reduced or eliminated its high tariffs on sensitive products (such as wheat, corn, rice, and sugar) in FTAs with countries that are significant producers of such products.

Table ES.3. Reported Chinese NTMs affecting imports of U.S. agricultural products

NTM	Description
H1N1 influenza restriction	U.S. pork has been denied access due to fears related to "swine flu." The World Organization for Animal Health (OIE) has reported that there is no risk of influenza infection from consuming pork.
Ractopamine ban	China has a zero tolerance for ractopamine, a commonly used feed additive, in pork. This limits opportunities for farmers producing pork for other markets that could otherwise profitably export some cuts to China.
Zero tolerance for pathogens	Zero tolerance is unsupported by a scientific risk assessment. This policy can serve to limit imports of meat and poultry.
Bovine spongiform encephalopathy (BSE) restrictions	China stopped imports of U.S. beef following the discovery of BSE in the U.S. cattle herd in December 2003. This is contrary to OIE guidelines.
	Also related to BSE, China prohibits use of protein-free tallow ingredients derived from ruminants and imported ingredients in U.S. pet food exported to China, including ingredients that are themselves approved for import in China.

Table ES.3. (Continued)

NTM	Description
Low pathogenic avian influenza (LPAI) restrictions	China bans imports of poultry products from certain U.S. states in which LPAI has been detected. This is contrary to OIE guidelines.
Fire blight restrictions	Only two varieties of apples from three U.S. states are approved for import, and no pears. There is no known research demonstrating a risk of fire blight from symptomless commercial varieties of apples or pears.
Potato pest risk assessment	A Chinese pest risk assessment has been forthcoming for U.S. potatoes for a decade. Movement on this issue has reportedly been tied to U.S. movement on a variety of SPS issues affecting Chinese exports to the United States.
Strawberry ban	China does not allow imports of fresh strawberries, although there were no reported problems when this restriction was temporarily lifted during the Beijing Olympic Games.
Biotechnology regulations	All products containing genetically modified organisms must be labeled, the registration process cannot begin in China until registration is completed in the exporting country, and registrations must be renewed every three years.
VAT policies	VAT policies provide a cost advantage to Chinese domestic agricultural producers and processers that purchase domestic agricultural products rather than imports.
Labeling requirements	Some products reportedly must be labeled entirely in Chinese or must have non-Chinese characters on their labels covered with a sticker.
Customs measures	Some imports are subject to reference pricing, classification is not consistent, and clearance may be delayed.
Multiplicity and duplication	Multiple ministries and agencies are involved in licensing, certification, and inspection and do not share information among themselves.
Provincial and local variation	Regulations, standards, and enforcement can vary by location.
TRQ administration	Large allocations are reserved for state trading enterprises; only small allocations are available for private traders, and there is little reallocation.
Lack of transparency	Many Chinese ministries and regulatory agencies fail to follow agreed-upon comment and notification procedures. TRQ allocations and the identity of import license holders are not made public.

Source*:* Compiled by Commission staff.

Quantitative Findings

Chinese tariffs and TRQs are estimated to have reduced U.S. agricultural exports by as much as $2.1 billion in 2009.

Model simulation results prepared by Commission staff, suggest that China's food and agricultural tariffs and TRQs reduced U.S. food and agricultural exports to China in 2009 by between $1.3 billion and $2.1 billion (table ES.4). The tariff simulation captures the effects of the removal of Chinese tariffs and TRQs on agricultural imports from all sources. Among U.S. products most affected by China's agricultural tariffs were wheat (U.S. exports to China are estimated to have been reduced by between $489 million and $1.2 billion), poultry ($358–$363 million), pork offal ($51–$84 million), cotton ($28– $71 million), and alcoholic beverages ($32–$43 million). Absent tariffs, in the span of a few years, U.S. exports could expand more rapidly than modeling simulations indicate because of the possible additional effects of economic growth in China and of market development by U.S. exporters, two factors not included in the simulation.

Table ES.4. China: China's tariffs and simulated effects on U.S. exports to China in the absence of Chinese tariffs for selected agricultural products, 2009

Product	Chinese tariffs on U.S. products	Range of simulated change in U.S. exports to China absent Chinese tariffs	
	Percent		Million $
Wheat	68	489	1,192
Poultry	13	358	363
Pork offal	13	51	84
Cotton	5	28	71
Alcoholic beverages	29	32	43
All other	([a])	101	112
Total	6	1,251	2,090

Source: Commission staff calculations.
[a] Not applicable.

China's FTAs have a negligible effect on total U.S. agricultural exports.

Model simulation results prepared by Commission staff, presented in ranges to account for the statistical uncertainty in key economic parameters, suggest that the effects of China's FTAs on U.S. food and agricultural exports to China in 2009 may have ranged from a contraction of up to $21 million to an expansion of up to $48 million. The simulated effects range from negative (a contraction of U.S. exports) to positive (an expansion of U.S. exports) because they are small; as a result, the range for the aggregate, as well as for some individual products, straddles zero. This simulation assumes the full implementation of tariff and market access provisions for manufactured and agricultural goods in China's ratified FTAs; U.S. tariffs remain unchanged while China and its FTA partners experience tariff elimination or reductions. Among U.S. agricultural exports shown to be negatively affected by China's FTAs were wheat (U.S. exports to China were reduced by as much as $37 million), whey (a reduction of between $9 million and $12 million), grapes (a reduction of $9–$11 million), and apples (a reduction of $2– $3 million). Among U.S. agricultural exports positively affected by China's FTAs were poultry (an expansion of $63–$68 million), cotton (an expansion of $18–$24 million), and soybeans (an expansion of $15 million).

Quantitative estimates indicate that NTMs restrict U.S. exports more than tariffs.

In the absence of certain Chinese NTMs, it is estimated that total U.S. agricultural exports to China would have been $2.6–$3.1 billion higher in 2009 (table ES.5). Economic simulations were conducted on 12 U.S. agricultural product sectors for which (1) Chinese import prices were higher than world prices and (2) Commission research indicated that specific NTMs were impeding U.S. agricultural exports. Unlike the tariff simulation, this simulation estimates the impacts of the removal of all known and unknown NTMs specific to these products, not the elimination of a specific policy or set of policies. The sectors included in this simulation were wheat, several horticultural products (potatoes, apples, and stone fruits), cotton, and meat products (beef, pork, and poultry). The products for which the model indicated the greatest change in trade flows (and therefore considered to be most affected by Chinese NTMs) were wheat, cotton, and pork.

Table ES.5. China: Simulated effects of removal of China's NTMs for selected agricultural products, 2009

Product	Ranges of effects on U.S. exports to China			
	Million $		Percent change	
Wheat	1,452	1,704	1,722	2,022
Cotton	524	630	65	79
Pork offal	305	363	586	697
Frozen pork	49	56	215	245
Poultry	35	40	4	5
Apples	15	18	79	96
Stone fruits	1	1	12	16
Other products with NTMs	214	286	([a])	([a])
Total	2,595	3,098	146	174

Source: Commission staff calculations.
[a] Not applicable.

Figure.

1. Introduction

Overview

China is the world's largest agricultural economy. It is the leading producer of many agricultural commodities, supplying more than half of the world's pork; one-third of the world's horticultural products,[1] rice, and cotton; and close to 20 percent of the world's wheat, corn, and poultry.[2] With about one-fifth of the world's population, China is also the largest consumer of many agricultural products; its current share of global pork consumption is 50 percent, 40 percent for cotton, 30 percent for rice, and more than 25 percent for soybeans and soybean oil. While China generally has been successful in meeting its rapidly rising demand for food and fiber by increasing domestic production, it has emerged as a leading global importer of several agricultural commodities, including cotton, soybeans, vegetable oils, and animal hides. As its domestic agricultural production has grown, China has also become the largest exporter in global markets for several horticultural products, including mandarin oranges, apples, apple juice, and garlic and other vegetables.

China's increasingly important position in global agricultural markets followed decades of gradual growth in domestic food production and consumption. After the introduction of market-based reforms in 1978 that included the elimination of the collective production system and relaxation of government direction over certain farmer production and marketing decisions,[3] Chinese agricultural output grew significantly. For example, between 1978 and 2008, China almost doubled its production of grains (rice, wheat, and corn) and quadrupled its production of meats; production of fruit and milk was about 30 times greater in 2008 than in 1978.[4] At the same time, industrialization and urbanization led to unprecedented growth in the Chinese economy. During these three decades, population growth of about 1 percent annually, coupled with annual per capita income growth of 8 percent,[5] fueled a large increase in demand for more and higher-value agricultural products, especially by China's large and growing middle class.[6] China's rapid growth in food consumption was largely met by domestic production growth, enabling it to remain self-sufficient in most major commodities.

The U.S. agricultural sector is highly export-intensive.[7] In 2009, the United States exported $99 billion in agricultural goods, representing more than one-third of U.S. farm cash receipts.[8] With 95 percent of the world's population living outside U.S. borders,[9] U.S. agricultural exporters seek opportunities to increase sales in foreign markets, especially in countries with large populations, rapidly growing per capita incomes, and changing food consumption patterns increasingly oriented toward higher-value foods, such as meat, dairy products, horticultural products, and processed food.[10] China in particular is viewed as providing important agricultural trade opportunities in the future.[11]

Indeed, in the last decade, China has become a major export market for certain U.S. agricultural products.[12] Since joining the World Trade Organization (WTO) in December 2001, China has become significantly more open to agricultural goods from abroad; Chinese agricultural imports from the United States increased from $2.1 billion in 2002 to $17.8 billion in 2010, an average annual increase of about 31 percent.[13] In 2010, China overtook Mexico to rank second among leading export destinations for U.S. agricultural products, behind only Canada. Yet U.S. agricultural exports to China are highly concentrated in a few products, with soybeans, cotton, hides and skins, and processed animal feed accounting for 84 percent of the total in 2010. Many grain and meat products in which the United States is globally competitive[14] are not imported by China.

In statements submitted to the U.S. International Trade Commission (Commission or USITC), U.S. agricultural community members, business representatives, and policymakers expressed concern that exports of a wider variety of U.S. agricultural goods to China are deterred by tariffs and nontariff measures (NTMs) and that China's potential as a major destination for U.S. agricultural products is not being realized. There is also concern that Chinese government policies that have benefited Chinese agricultural production and exports may have weakened the U.S. competitive position vis-à-vis China in third-country markets, particularly in the Asia-Pacific region and in countries with which China has established free trade agreements (FTAs).[15] Comparative statistics for Chinese and U.S. agricultural production and trade are presented in table 1.1.

In its letter requesting this investigation,[16] the Senate Committee on Finance (Committee) asked the Commission to examine and report on the competitive factors affecting agricultural trade between China and the United States. The Committee asked that the report cover the period 2005 through 2009, or through the latest year for which data are available. Noting the importance of China as a market for U.S. agricultural exports for a small range of products, the Committee pointed out that the rapid rise of per capita income in China, along with resource constraints on Chinese domestic agricultural production, have the potential to increase U.S. exports in the future. However, certain Chinese government policies could weaken the U.S. competitive position in the Chinese market. Specifically, the Committee asked that the report include the following:

- an overview of China's agricultural market, including recent trends in production, consumption, and trade;
- a description of the competitive factors affecting the agricultural sector in China, in such areas as costs of production, technology, domestic support and government programs related to agricultural markets, foreign direct investment policies, and pricing and marketing regimes;
- an overview of China's participation in global agricultural export markets, particularly in the Asia-Pacific region and in those markets with which China has negotiated trade agreements;
- a description of the principal measures affecting China's agricultural imports, including tariffs and nontariff measures such as sanitary and phytosanitary measures and technical barriers to trade; and
- a quantitative analysis of the economic effects of China's most-favored-nation (MFN) tariffs, preferential tariffs negotiated under China's free trade agreements, and China's nontariff measures on U.S. agricultural exports to China and on imports from the rest of the world.

Scope of the Report

In response to the Committee's request, this report examines conditions of competition in China's agricultural market and trade, and their effects on U.S. agricultural exports. As requested by the Committee, the report provides three types of information: (1) background information on China's production, consumption, and trade in agricultural products; (2) information on the factors that affect the competitiveness of the Chinese agricultural sector and competition in the Chinese market for agricultural products; and (3) information on factors that directly affect U.S. exports and firms, including MFN and preferential tariffs, and NTMs.

Table 1.1. Chinese and U.S. agricultural production and trade: Comparative statistics

	China	United States
Population (2009)	1,338 million	307 million
Cropland (2009)	122 million hectares	164 million hectares
Cropland per agricultural worker (2007)	0.4 hectares	78.5 hectares
Value of farm production (2007)	\$537 billion	\$329 billion
Agriculture as share of GDP (2009)	10.6 percent	1.2 percent
Value of agricultural imports (2010)	\$66 billion	\$86 billion
Top five agricultural imports by value	Soybeans, cotton, palm oil, dairy products, and hides and skins	Coffee, cocoa products, wine, malt beverages, beef
Top five agricultural import suppliers	United States, Brazil, Argentina, European Union-27 (EU-27), Australia	Canada, EU-27, Mexico, China, Brazil
Value of agricultural exports (2010)	\$36 billion	\$119 billion
Top five agricultural exports by value	Processed vegetables, fresh vegetables, miscellaneous processed foods, fresh fruit, and animal feed	Soybeans, corn, wheat, cotton, and animal feed
Top five agricultural export markets	Japan, EU-27, Hong Kong, United States, Korea	Canada, China, Mexico, Japan, EU-27

Source: GTIS, Global Trade Atlas database (accessed February 18, 2011); USDA, ERS, "Foreign Agricultural Trade of the United States," 2011; CIA, *The World Factbook: China*; World Bank, Data: China; USDA, NASS, *Agricultural Statistics 2009*, 2009, Table 9-44; National Bureau of Statistics of China, *China Statistical Yearbook 2009*, 2009, Table 12-4.

The remainder of chapter 1 describes China's agricultural policies, including the government objectives driving the policies and the instruments used to achieve them. Chapter 2 surveys China's agricultural trade trends, focusing on trade with the United States, countries in the Asia-Pacific region, and China's FTA partners. This chapter also highlights China's trade with third-country markets where U.S. exports compete as well. China's agricultural consumption is discussed in chapter 3, which describes and analyzes China's consumption patterns, preferences, and trends—the ultimate drivers of current and potential demand for agricultural products, domestic and imported. Chinese farm-level production and domestic agricultural policies are described in chapter 4. Chapter 5 describes conditions of competition in the Chinese agriculture market, including factors that affect the cost of delivery, product differentiation, and the reliability of the supply of agricultural products in China. Chapter 6 presents case studies that highlight conditions of competition in the Chinese apple, pork, processed foods sectors, and wheat.

Information on the measures that directly affect U.S. agricultural exports and firms is covered in chapters 7–9.[17] Chapter 7 describes China's tariffs and gives quantitative estimates of the effect Chinese tariffs had on U.S. agricultural export levels to China in 2009. An overview of China's FTAs is presented in chapter 8. It includes quantitative estimates of the effects on U.S. agricultural exports of the elimination of tariffs between China and its FTA partners. Chapter 9 describes and analyzes China's agricultural NTMs and includes quantitative estimates of the effects that China's NTMs had on selected U.S. agricultural exports to China in 2009.

Agricultural products in this investigation are those that fall within the description of products covered by the WTO Agreement on Agriculture, part XIII, article 21. These products include 768 6-digit product codes classified in the World Customs Organization's Harmonized System (HS)—specifically, HS chapters 1 to 24, excluding fish and fish products (HS chapter 3)[18] —plus certain additional products in other HS chapters, such as milk proteins (HS chapter 35), hides, skins, and furs (HS chapters 41 and 43), wool (HS chapter 51), and cotton (HS chapter 52).

As requested by the Committee, certain information presented in this report, including trends production and consumption, focuses on the 2005–09 period. Shortly before this report was published, Chinese trade data for full-year 2010 became available. These data were incorporated where possible throughout the report, particularly in chapter 2 regarding Chinese agricultural trade. For some information with less timely statistics, the period from 2004 to the latest year for which data are available is used. Longer-term data are used to explain important long-term trends. The descriptive analysis on competitive factors and the quantitative analysis of the effect of Chinese trade measures is based on the latest available information and data.

Approach

As requested by the Committee, this report contains qualitative and quantitative analysis to evaluate the conditions of competition facing U.S. agricultural exports in China's agricultural market, including the effects of tariffs and NTMs on U.S. agricultural exports to China. The qualitative analysis consists of several parts: (1) a general discussion and analysis of the Chinese agricultural sector and factors affecting its competitiveness;

Chinese government policies and their effect on U.S. agricultural exports; and case studies to highlight the effect of specific Chinese policies and market conditions in particular sectors where U.S. agricultural products compete, both in China and in third-country markets.

The qualitative analysis was based on a review of existing literature, a public hearing, and interviews with U.S. government sources and agricultural sector representatives, including interviews with individual firms, trade associations, and exporters. Commission staff sought information from U.S. agricultural trade associations and U.S. firms with operations in China, contacting more than 80 commodity- and sector-specific agricultural trade associations and companies. Commission staff held extensive meetings with officials of the U.S. government, including the U.S. Department of Agriculture's (USDA) Economic Research Service (ERS), Foreign Agricultural Service (FAS), and Animal and Plant Health Inspection Service

(APHIS). Staff also traveled to China to meet with relevant U.S. and Chinese officials, USDA officials, academic researchers, importers, and market and logistics officials.[19]

Commission staff conducted extensive literature and data research on China's trade and domestic policies that affect U.S. agricultural products in the Chinese market. Relevant trade and production data were obtained from Global Trade Information Services (GTIS); the Commission's DataWeb; Chinese government Web sites, including those of the China National Bureau of Statistics and Ministry of Agriculture;[20] the United Nations' Food and Agriculture Organization; and the USDA. Information on China's tariffs and NTMs was obtained from the WTO, Organisation for Economic Co-operation and Development (OECD), World Bank, and USDA (ERS, FAS, and APHIS), as well as many private sector and academic sources.

To analyze the competitive factors affecting the Chinese agricultural sector, the Commission developed an economic framework that provides the assumptions, parameters, and structure shaping competitive conditions in agricultural trade. Competitive conditions in agriculture refer to the economic, institutional, and regulatory environment in which firms compete. Competitive factors are defined as direct and indirect determinants of the ability of suppliers to offer products with characteristics desired by buyers, who base their buying decisions on three main criteria: delivered cost, product differentiation, and reliability of supply. In this report, Commission analysis explores the relative importance of delivered cost, product differentiation, and reliability of supply in determining the competitiveness of the Chinese domestic agricultural industry vis-à-vis its competitors in China and in export markets.

In addition to descriptive information, the Committee requested that the Commission provide quantitative analysis of the economic effects of China's MFN tariffs, preferential tariffs negotiated under China's FTAs, and China's NTMs on U.S. agricultural exports to China and on imports from the rest of the world. The Commission's analysis was based on a simulation framework that consists of a partial equilibrium (PE) model and a general equilibrium (GE) model. The PE model focused on bilateral trade in food and agricultural products at the HS 6-digit level between the United States, China, and the rest of the world. The GE model used for the analysis was the Global Trade Analysis Project (GTAP) model, an economy-wide computable GE model of world trade specified at an aggregate product and sector level. The PE model was used to simulate the effects of Chinese tariffs and tariff-rate quotas on U.S. food and agricultural exports. The GE model was then used to simulate the economy-wide effects of those border measures. The two models were linked to provide consistent estimates of effects. A similar approach was applied in a recent USITC report on India.[21] Further, the GTAP model was used to simulate the effects of Chinese preferential tariffs for its FTA partners on U.S. agricultural exports.

Modeling simulations of the potential effects of NTMs were completed in a three-step process. First, price gap data were developed.[22] The existence of NTMs would likely raise the price of imports into China and restrict the quantities imported. Thus, the differences between the prices of goods imported by China and the export prices of countries that sell agricultural goods to China were estimated at a disaggregated level (HS 6-digit level) using unit values of imports from 2007–09. These were estimated separately for U.S. exports to China and third countries' exports to China taken as a group, adjusting for observable quality differences between exporters and for transportation costs. Second, a subset of products was identified for which available information indicated the presence of NTMs that may increase prices or restrict quantities. For these products, positive price gaps were treated as representing the economic effects of NTMs. Third, these price gaps were introduced into the simulation-modeling framework as being equivalent to tariffs, and the effects of their removal were estimated.

China's Policy Framework

The Government of China's Eleventh Five-Year Plan, covering the period 2006–10, identified China's primary domestic agricultural policy objectives as (1) developing modern agriculture; (2) increasing farmers' incomes; and (3) improving rural conditions.[23] The central government has also maintained its goal, outlined in other government documents, of reaching 95 percent food self-sufficiency in grains (corn, wheat, and rice).[24] The secondary objectives include energy independence, ensuring a safe food supply for consumers, and conserving natural resources. To meet these objectives, China's government actively regulates the agricultural sector using production and marketing policies, as well as closely monitoring international agricultural trade.[25]

As in many other countries, China's central, provincial, and local governments formulate agricultural policies in response to a policy environment linked to historical events and current social factors. As a result, officials articulate broad policy objectives and generate specific policy instruments influencing either macroeconomic factors (e.g., inflation rates, interest rates, and trade deficits) or microeconomic ones (e.g., industrial sectors and business investment). A simplified framework for Chinese domestic and trade policies in the agricultural sector appears in figure 1.1.

Policy Environment

China has suffered a series of droughts, famines,[26] political upheavals,[27] and other events over the past 100 years that the government is committed to avoid. In principle, the governmental structure consists of a one-party authoritarian system through which central government policies are presented to the populace in a "top-down" style of governance.[28] In practice, however, provincial and local governments often exercise significant autonomy in implementing national policies. Moreover, China possesses (1) a large rural population that depends on agriculture for its livelihood; (2) scarce productive farmland relative to its population; and (3) a consumer base that spends a large part of its income on food. In addition, China suffers from environmental degradation related to its rapid economic development, as well as a widespread lack of consumer confidence in the safety of the country's food supply. These factors often motivate the government to intervene and safeguard China's domestic agricultural production.

Today, nearly 40 percent of China's population of 1.3 billion is employed in the agricultural sector, and agriculture contributes about 11 percent to China's gross domestic product (GDP).[29] Over the past 30 years, China experienced strong economic growth, but laborers who did not migrate from rural areas to the cities typically have not benefited as much as their urban counterparts from better employment opportunities and rising incomes.[30] The Chinese agricultural sector has very low labor productivity, averaging only one-fifth the level of other sectors of the economy. The OECD identified low labor productivity as contributing to an income gap in 2007 in which per capita incomes were 3.3 times higher in urban areas of China than in rural ones.[31]

Policy Environment

- History of famines and food shortages
- History of political upheaval
- Large, poor, rural population employed in agriculture
- Limited agricultural land per capita
- One-party authoritarian system
- History of local autonomy vis-à-vis central government
- Environmental degradation
- Lack of confidence in safety of food supply

↓

Policy Objectives

- Address the interconnected needs of agriculture, rural communities, and farmers ("harmonious Socialist countryside")
- Improve the economic and social well-being of the rural population (mainly farmers)
- Maintain grain self-sufficiency and stable prices
- Ensure a safe food supply for all citizens
- Conserve valuable environmental resources
- Attain energy independence

↓

Policy Instruments

- Investments in rural infrastructure
- Investments in research and development for agriculture
- Minimum purchase prices for agricultural commodities
- Input subsidies, direct payments, and preferential credit
- Relief from agricultural taxes
- Food reserves
- Consumption subsidies (minimum living allowance)
- Strategic export and import controls
- Food safety law

Source: Compiled by Commission staff.

Figure 1.1. China's agricultural policy framework: Policy instruments are the outcome of policy environment and objectives.

Under the Chinese political system, it is common for policymakers to set up implementation procedures and enforcement mechanisms only after a regulation takes effect. Regulations to protect farmer or consumer rights and ensure fairness in the marketplace are sometimes interpreted differently among local governments, or simply ignored. This creates an environment of regulatory uncertainty for businesses, which suffer from the lack of administrative transparency and implementation details. Foreign companies in China tend to be relatively disadvantaged in this regard compared to domestic companies because foreign firms often lack the relationships that Chinese companies have with government agencies to receive timely regulatory clarifications. In addition, foreign food and agricultural companies operating in China are often surprised by the extent to which local governments have autonomy over the implementation of domestic agricultural support programs.[32]

In recent years, shortcomings within China's food safety regime in monitoring and enforcing the quality of products such as milk, eggs, and pet food have severely undermined consumer confidence in the system. As a result, the Chinese government has focused its attention on the issue of food safety, severely punishing key participants in several of the scandals, creating a politically powerful Food Safety Commission,[33] and enacting a sweeping new food safety law in 2009. However, as reports about unsafe food in China continue to be publicized, it is unclear how effectively government efforts will assuage consumer food safety fears.

Policy Objectives

With China's long history of famine and political revolutions, and in light of a rising number of "public order disturbances" by farmers and consumer groups in the mid- to late 2000s,[34] China's central government in Beijing began funneling additional resources to the farm sector and rural economy to meet its policy objectives. Raising incomes for farmers is important to the government because many of China's poor live in rural areas, and farming small plots is the primary livelihood for the majority of rural inhabitants. Boosting rural development by raising government funding for agricultural infrastructure and other spending is designed to promote social harmony and shrink the income gap between urban and rural workers. Self-sufficiency in grains, at least up to 95 percent of domestic demand, is considered by Chinese policymakers to be a key component of China's food security strategy.[35]

Food security is defined by the World Bank as "access by all people at all times to enough food for an active, healthy life";[36] the components include availability, access, and use.[37] Achieving food security requires that food be available in volumes that meet consumer needs and at prices they can afford. Food self-sufficiency relates to the productive capacity of the nation's agriculture sector, and is generally taken to mean the extent to which a country can satisfy its food needs from domestic production.[38] The two concepts differ in that self-sufficiency looks only to national food production for sources of supply, while food security takes into account the possibility of imports.[39] Largely because of China's history, food security and self-sufficiency, though distinct concepts, are often interconnected in Chinese policies.[40]

As in other countries whose governments intervene heavily in the agricultural sector, China's agricultural policies reflect conflicting government objectives. Efforts to raise farmers' incomes over the past 5–10 years have conflicted with the policy of 95 percent self-sufficiency in grain production because producing grains in China is less profitable than many other types of farming: for example, net cash returns earned by Chinese farmers from growing vegetables and fruits are several times higher than returns from growing rice, wheat, and corn.[41] Boosting farmers' incomes can also conflict with urban consumer welfare. For example, international food prices rose sharply during 2007 and 2008, but Chinese officials sought to limit domestic price increases by limiting exports of grains, cutting soybean tariffs, and importing vegetable oil. These actions had the effect of keeping consumer prices down, which benefited urban consumers but also denied additional income to Chinese farmers.

As noted, China's policies toward agriculture focus not only on food security through self-sufficiency and raising farmers' incomes, but also to a lesser extent on energy independence, ensuring a safe food supply for consumers, and conserving natural resources. Government programs promote domestic biofuel production and encourage water and soil conservation, but policies addressing environmental concerns and energy independence potentially conflict with grain self-sufficiency and increasing farmer incomes.[42] For example, producing biofuels affects China's mix of agricultural products and may end up increasing imports of feedstock sources.[43] China's policies toward the agricultural sector will evolve as the government continually rebalances conflicting policy objectives.

Policy Instruments

The tension between competing goals for the farm sector, consumers, and food security has caused the Chinese government to intervene heavily using multiple policy instruments. China's government agricultural programs can largely be subdivided into four categories: direct payments, price support programs, infrastructure, and regulatory reforms (e.g., food safety and standards). [44] Farmers' incomes are raised through minimum purchase prices, input subsidies, the elimination of agricultural taxes and fees, and government funding for capital equipment purchases to mechanize farming. The effects of these government payments on farmer income provide strong inducements to produce grains (rice, wheat, and corn) rather than more profitable livestock, horticulture, or other field crops.

With the exception of regulatory reforms, all of these programs are intended to boost farmers' incomes directly or to lower the delivery costs, potentially providing a competitive advantage over foreign goods. Improvements to agricultural infrastructure, such as rural transportation networks, the rural financial system, and marketing institutions, upgrade the environment for agriculture. Improved infrastructure lowers the cost of production and delivery, though sometimes over a longer time period than direct payments. Government-mandated regulatory reforms affect Chinese and foreign companies uniformly, provided that these policies are equitably implemented.

Chinese policymakers seek food security largely through domestic production, and imports are often viewed as a second-best source to achieve this objective. While USITC staff found little evidence of a coordinated effort across China's government agencies to broadly restrict agricultural imports, China bans or restricts entry of a number of agricultural products, or has threatened to do so. For example, U.S. apples, strawberries, potatoes, dairy products, and pork have all suffered from either delays in risk assessments or from risk assessments that are not based on internationally accepted scientific standards. China appears to link the application of NTMs, such as sanitary and phytosanitary requirements, to domestic policies by relaxing NTMs when policymakers determine that imports are needed to relieve food price inflation or shortages.[45]

Bibliography

Bloomberg Businessweek. "China Names Vice Premier Food Safety Commission Head," February 10, 2010. http://www.businessweek.com/news/2010-02-10/china-names-vice-premier-food-

Central Intelligence Agency (CIA). *CIA World Factbook: China*, updated May 20, 2010. https://www.cia.gov/library/publications/the-world-factbook/geos/ch.html (accessed October 27, 2010).

Central Intelligence Agency (CIA). *CIA World Factbook: Country Comparison; Population*, updated July 2010. https://www.cia.gov/library/publications/the-world-factbook/rankorder/2119rank.html (accessed November 2, 2010).

Food and Agricultural Organization of the United Nations (FAO). FAOSTAT database (accessed July 27, 2010).

Global Trade Information Service, Inc. (GTIS). Global Trade Atlas database (accessed various dates).

International Monetary Fund (IMF). *International Financial Statistics. www.imf.org* (accessed August 10, 2010).

Lohmar, Brian, Fred Gale, Francis Tuan, and Jim Hansen. *China's Ongoing Agricultural Modernization: Challenges Remain after 30 Years of Reform*. U.S. Department of Agriculture, Economic Research Service, April 2009. http://www.ers.usda.gov/Publications/EIB51/EIB51.pdf.

Lum, Thomas. "Social Unrest in China." Congressional Research Service, RL33416, May 8, 2006. National Bureau of Statistics of China. *China Statistical Yearbook 2009*. China Statistics Press, 2009.

Organisation for Economic Cooperation and Development (OECD). *Research and Markets: OECD Rural Policy Reviews: China 2009*. Paris: OECD, 2009.

———. *Agricultural Policies in Non-OECD Countries*. Paris: OECD, 2007.

———. *Agricultural Policies in Emerging Economies 2009*. Paris: OECD, 2009.

Owen, Elise. "Standards in China: Behind the Headlines." *China Business Review*, January–February 2010.

Promar International. *The Chinese Potato Industry in Transition*. Alexandria, VA: Promar, 2007. Porter, Michael E. *Competitive Strategy*. New York: The Free Press, 1980.

Thompson, Anne and Manfred Metz. *Implications of Economic Policy for Food Safety: A Training Manual*. Rome: Food and Agricultural Organization of the United Nations (FAO), 1999. http://www.fao.org/docrep/004/x3936e/X3936E03.htm (accessed July 23, 2009).

U.S. Agency for International Development (USAID). "Policy Determination: Definition of Food Security," April 13, 1992. http://www.usaid.gov/policy (accessed July 23, 2009).

U.S. Census Bureau. International Data Base. http://www.census.gov/ipc/www/idb/worldpopinfo.php (accessed December 2, 2010).

U.S. Department of Agriculture (USDA). Economic Research Service (ERS). "Briefing Room: China," March 12, 2009. http://www.ers.usda.gov/Briefing/China/data.htm (accessed December 3, 2010).

———. "Foreign Agricultural Trade of the United States," March 2010.

http://www.ers.usda.gov/Data/FATUS/DATA/McyTOP15.xls
http://www.ers.usda.gov/data/fatus/DATA/cytop25hvpimp.xls
http://www.ers.usda.gov/data/fatus/DATA/cytop25hvpexp.xls.
———. "Rural and Natural Resource Indicators." *Amber Waves*, June 2010.
U.S. Department of Agriculture (USDA). Economic Research Service (ERS) and Foreign Agricultural Service (FAS). *Outlook for U.S. Agricultural Trade,* August 31, 2010.
U.S. Department of Agriculture (USDA). Foreign Agricultural Service (FAS). *China Peoples Republic of: Bio-fuels, an Alternative Future for Agriculture,* by Kevin Latner, Caleb O'Kray, and Junyang Jiang. GAIN Report no. CH6049, August 8, 2006.
———. "India and China: Divergent Markets for U.S. Agricultural Exports." *International Agricultural Trade Report.* February 24, 2010.
http://www.fas.usda.gov/itp/china/India Chinamarket022010.pdf.
———. "National Export Initiative: Importance of U.S. Agriculture Exports," April 2010. http://www.fas.usda.gov/info/factsheets/tradevalue.pdf.
———. "U.S.-China Trade," August 4, 2010.
http://www.fas.usda.gov/itp/china/us-china.asp.
———. Production Supply and Distribution (PSD) Online.
http://www.fas.usda.gov/psdonline (accessed May 20, 2010).
U.S. Department of Agriculture (USDA). National Agricultural Statistics Service (NASS). *Agricultural Statistics 2009.* Washington, DC: GPO 2009.
http://www.nass.usda.gov/Publications/Ag Statistics/2009/index.asp
U.S. Government. President's Commission on Industrial Competitiveness. *Global Competition: New Reality.* Vol. 1. Washington, DC: GPO, January 1985.
U.S. International Trade Commission (USITC). *Guidelines for Developing an Economic Framework for an ITC Study.* Office of Economics. Unpublished, 2008.
———. *India: Effects of Tariffs and Nontariff Measures on U.S. Agricultural Exports.* USITC Publication 4107. Washington, DC: USITC, 2009.
———. *Sub-Saharan Africa: Effects of Infrastructure Conditions on Export Competitiveness.* USITC Publication 4071. Washington, DC: USITC, 2009.
World Bank. "Mid-Term Evaluation of China's Eleventh Five-Year Plan." Report no. 46355-CN, December 18, 2008.
World Bank. Data: China. http://data.worldbank.org/country/china (accessed Oct 21, 2010).
World Health Organization (WHO). "Glossary."
http://www.who.int/trade (accessed July 23, 2009).
Xiao, Yunlai, and Fengying Nie. *A Report on the Status of China's Food Security.* United Nations. Beijing: China Agricultural Science and Technology Press, 2009.

2. Chinese Agricultural Trade

Overview

China's agricultural trade is globally significant. In 2009, Chinese agricultural imports and exports accounted for 9 percent and 5 percent, respectively, of global agricultural trade.[1] China ranked as the world's second-largest agricultural importing country behind the United

States in that year, and for many agricultural products, including soybeans, vegetable oils, cotton, wool, and hides and skins, China was the top global importer. China was the world's fourth leading agricultural exporting country (behind the United States, Brazil, and Canada) in 2009 and the largest exporter of many types of fresh and processed horticultural products. However, despite the agricultural sector's large size, it plays a relatively small role in China's overall trade, with agricultural imports and exports accounting for about 3 percent of both imports and exports of merchandise during 2005–09.[2]

During 2005–10, China recorded an agricultural trade deficit that rose from $5.3 billion in 2005 to $30.8 billion in 2010 (figure 2.1).[3] China has been a net importer of agricultural products since 2003,[4] and the trade gap is likely to continue given that future growth in food demand, driven by rapidly rising per capita income, is expected to outpace increases in domestic production.[5] Between 2005 and 2010, China's agricultural imports rose at an annual average rate of about 24 percent, reaching a record $66.4 billion in 2010. This growth can be attributed to many factors, including increased consumer demand for higher-valued food products such as meat, horticultural goods, and processed food products, as well as strong demand by China's textile and apparel industry for fiber and for hides and skins. Lower trade barriers following China's entry into the World Trade Organization (WTO) also may have spurred growth in agricultural imports.[6] The United States is China's leading agricultural import supplier, accounting for 27 percent of its total agricultural imports during 2005–10. China's agricultural exports increased from $19.6 billion in 2005 to $35.7 billion in 2010, representing annual average growth of 13 percent. The United States ranks fourth among major markets for Chinese agricultural exports (behind Japan, the European Union [EU-27], and Hong Kong), accounting for about 9 percent of its agricultural exports during 2005–10. In 2010, China accounted for about 14 percent of total U.S. agricultural exports, overtaking Mexico to become the second largest export market for U.S. agricultural products behind Canada.[7]

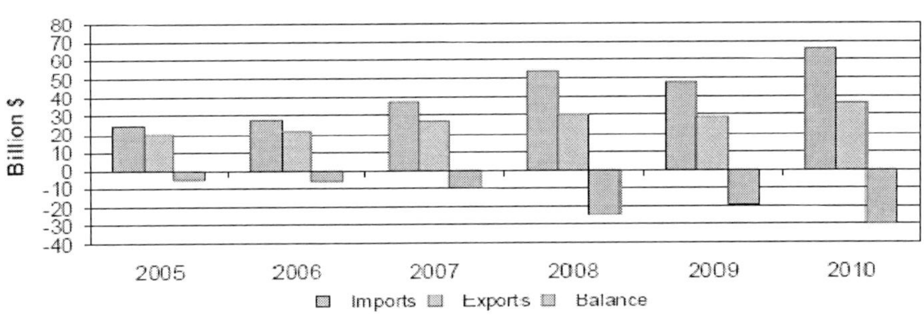

Source: GTIS, Global Trade Atlas database.

Figure 2.1. China agricultural trade balance: China maintained an agricultural trade deficit during 2005–10.

Relative to the United States, China has an abundance of rural labor and a scarcity of agricultural land.[8] Consistent with these natural resource endowments, China's agricultural exports are more concentrated in labor-intensive products such as fruits and vegetables, while its imports include many land-intensive products, like cotton and soybeans. However, China is not a major importer of land-intensive grains, such as rice, wheat, and corn. These are products in which the United States, as a land-abundant country, is highly competitive in international markets. The low level of grain imports is largely due to China's self-sufficiency policy, which seeks to avoid dependence on the international market for grains.[9] The Chinese central government encourages domestic production and discourages certain imports through a variety of programs, ranging from increased funding of infrastructure projects and guaranteed minimum prices to subsidies and import controls.[10]

Imports

Imports by Product

During 2005–10, Chinese agricultural imports more than doubled, increasing from $25 billion to $66.4 billion (table 2.1). Agricultural imports are highly concentrated in a few major products. In 2010, two products, soybeans (38 percent) and cotton (9 percent), accounted for almost one-half of all Chinese agricultural imports, while the top 10 import categories made up nearly 75 percent of the total. After soybeans and cotton, leading imports were palm oil (8 percent), dairy (4 percent), hides and skins (4 percent), and wool (3 percent). Among the fastest-growing imports during 2005–10 were several high-value products, including meat and dairy products, alcoholic beverages, fresh fruit, and miscellaneous processed foods. Major Chinese import flows are depicted in figure 2.2.[11]

The United States was China's leading agricultural import supplier during 2005–10, a period in which U.S. imports to China increased almost threefold to reach $17.8 billion in 2010 (table 2.1). China's imports from the United States are more concentrated than those from most of its other trading partners, with the top four imports—soybeans, cotton, hides and skins, and processed animal feed—accounting for 84 percent of the total in 2010. Soybeans accounted for 63 percent of all agricultural imports from the United States in 2010 and were responsible for most of the overall growth during 2005– 10. Although the United States is among the world's largest exporters of grains, vegetable oils, beef, and pork, these products amounted to less than 3 percent of Chinese imports from the United States between 2005 and 2010.

Table 2.1. China: Agricultural imports from the world and the United States by product, 2005–10 (million $)

Product	Source	2005	2006	2007	2008	2009	2010
			Animal products				
Live animals	World	109	63	74	104	140	269
	United States	17	18	19	42	33	34
Beef	World	9	8	14	18	44	84
	United States	0	0	0	0	0	0
Sheep and lamb meat	World	29	21	123	523	136	209
	United States	5	1	70	313	24	30
Sheep and lamb meat	World	55	50	78	106	139	157
	United States	0	0	0	0	0	0
Poultry	World	334	462	945	1,087	984	963
	United States	178	309	613	811	827	168
Dairy	World	630	823	1,129	1,411	1,784	2,874
	United States	78	106	159	196	145	243
Eggs	World	1	1	0	1	1	3
	United States	0	0	0	0	0	2

Table 2.1. (Continued)

Product	Source	2005	2006	2007	2008	2009	2010
Processed meats	World	4	8	8	6	4	7
	United States	2	2	2	1	1	3
			Grains				
Wheat	World	762	108	21	7	205	309
	United States	104	33	5	0	91	31
Rice	World	196	288	217	183	201	253
	United States	0	0	0	0	0	0
Barley	World	429	406	267	484	435	536
	United States	0	0	0	0	0	0
Corn	World	1	12	7	12	20	367
	United States	0	10	1	3	4	348
			Oilseeds and products				
Soybeans	World	7,779	7,489	11,472	21,814	18,787	25,081
	United States	3,160	2,719	4,234	8,443	9,333	11,319
Soybean oil	World	908	800	2,146	3,334	1,842	1,203
	United States	0	14	119	206	41	255
Palm oil	World	1,925	2,435	3,980	5,616	4,557	5,222
	United States	0	0	0	0	0	0
Coconut oil	World	78	93	106	193	103	294
	United States	0	0	0	1	1	2
Rapeseed oil	World	104	28	305	355	377	921
	United States	0	3	0	0	0	2
Other vegetable oil	World	148	392	757	881	589	899
	United States	6	7	13	47	16	18
			Horticulture				
Vegetables, fresh	World	3	7	1	3	6	8
	United States	0	0	0	0	0	0
Vegetables, processed	World	567	799	861	651	1,108	1,610
	United States	43	56	66	69	60	103
Nuts	World	108	129	169	286	363	510
	United States	20	17	28	67	110	151
Fruit, fresh	World	499	528	610	826	1,204	1,514
	United States	86	92	83	120	174	236
Fruit, processed	World	88	134	200	241	286	323
Juice	World	73	99	146	126	141	166
	United States	4	4	7	10	17	18
			Beverages				
Tea/coffee	World	30	44	61	83	68	118
	United States	5	9	11	14	8	10
Alcoholic beverages	World	389	553	815	1,082	1,052	1,570
	United States	14	17	18	29	35	45
Non-alcoholic beverages	World	18	30	48	52	59	72
	United States	3	3	3	5	4	2
		Sugar, sweeteners, confectionery					
Sugar	World	383	549	380	319	378	906
	United States	0	0	0	0	0	0
Other sweeteners	World	45	47	53	67	70	90

Product	Source	2005	2006	2007	2008	2009	2010
	United States	4	5	7	6	5	6
Product	Source	2005	2006	2007	2008	2009	2010
Cocoa products	World	177	183	211	313	262	437
	United States	11	15	17	16	27	33
			Processed foods				
Miscellaneous processed foods	World	435	495	563	771	874	1,194
	United States	103	127	121	147	121	158
Milled grain products	World	186	246	221	234	301	452
	United States	4	4	6	5	5	6
Processed animal feed	World	222	357	266	466	534	1,566
	United States	35	37	60	90	201	821
			Other				
Cotton	World	3,221	4,922	3,535	3,525	2,149	5,706
	United States	1,470	2,295	1,608	1,663	872	2,000
Wool	World	1,249	1,289	1,812	1,728	1,508	2,021
	United States	9	16	20	11	11	13
Hides and skins	World	1,483	1,605	1,822	2,067	1,701	2,415
	United States	678	826	864	930	642	822
Tobacco and products	World	384	463	541	788	844	791
	United States	12	63	69	103	112	155
All other	World	1,928	2,208	2,883	3,942	4,460	5,319
	United States	273	290	361	431	415	713
Total	World	24,991	28,178	36,850	53,707	47,717	66,439
	United States	6,342	7,128	8,620	13,826	13,407	17,834

Source: GTIS, Global Trade Atlas database.

Soybeans

Soybeans are China's largest agricultural import. During 2005–10, the value of soybean imports increased an average of 32 percent annually, and further growth is expected over the next five years. [12] Imports rose from $7.8 billion in 2005 to peak at $25.1 billion in 2010. [13] In 2009, China purchased more than one-half of total global soybean exports. [14] Most of China's soybean imports are processed into soybean meal for animal feed and cooking oil for human consumption. Large margins between the prices paid for imported soybeans and the prices received for processed meal and oil have made local processing highly profitable and have contributed to the recent growth of soybean imports.[15]

Soybeans are by far the single most important agricultural product traded between the United States and China (table 2.1). During 2005–10, China received almost one-half of total U.S. soybean exports (58 percent in 2010). The large quantities traded are consistent with the United States' position as the world's largest soybean producer and China's as the world's largest consumer. According to the American Soybean Association (ASA), U.S. soybean exports to China do not face trade barriers.[16] However, ASA has expressed concern that imports of genetically modified (GM) soybeans from the United States could become constrained by the cost of complying with Chinese GM regulations and the difficulty of gaining approval for new cultivars of soybeans.[17] In this regard, ASA notes that the Chinese government requires separate registration for soybeans with stacks[18] of two or more biotech traits, even if all traits have been separately approved.[19]

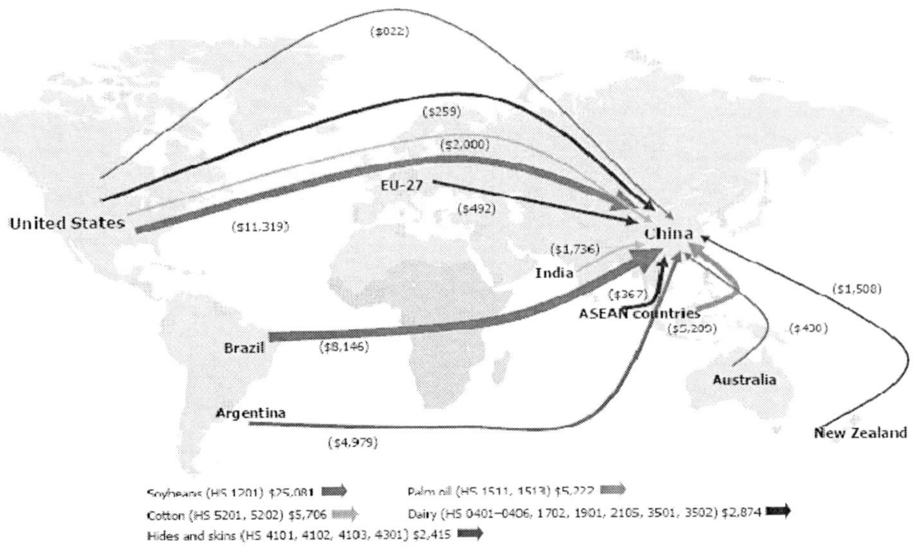

Figure 2.2. Chinese agricultural imports were highly concentrated in a few major products in 2010 (million $).

Vegetable Oils

Vegetable oils are China's second-largest agricultural import after soybeans. During 2005–10, 75–90 percent of China's vegetable oil imports consisted of palm oil and soybean oil. These two oils accounted for close to 23 percent and 30 percent of global imports, respectively, during 2005–09. Between 2005 and 2010, China's vegetable oil imports more than doubled in value, accounting for about 13 percent of all agricultural imports by China in 2010. Used mainly for cooking, vegetable oil imports have risen because of growth in per capita income and the increasing availability of low-cost palm oil from Malaysia and Indonesia. Chinese imports of soybean oil are relatively low compared to soybeans, reflecting, in part, the difference in tariff treatment between soybeans (duty free) and soybean oil (9 percent tariff).[20] Most of China's soybean oil imports are sourced from Brazil (67 percent in 2010), followed by the United States (21 percent), and Argentina (11 percent).

Cotton

Cotton is heavily used in China's large textile and apparel industry, which accounted for one-third of the value of world apparel exports in 2008.[21] As a result, China is the world's largest producer, consumer, and importer of cotton.[22] In 2009, China accounted for about one-third of global cotton imports, in spite of potential duties of 40 percent.[23] Because so much cotton is used in textile and apparel production, China's demand for cotton imports reflects both changes in domestic cotton production levels and global and domestic demand for Chinese cotton apparel and textiles. Cotton imports decreased each year during 2007–09 and by 2009 had declined 56 percent from 2006. The sharpest decline was between 2008 and 2009, when the global economic crisis caused a decrease in orders for Chinese apparel products.[24] However, in 2010, cotton regained its standing as China's second most important agricultural import, reaching $5.7 billion, when China's textile and apparel industry increased production.[25]

Between 2005 and 2010, cotton was China's second-largest agricultural import from the United States. U.S. cotton represented over one-third of China's cotton imports in 2010, and the United States was China's largest supplier during the period, followed by India, Uzbekistan, and Australia. Chinese imports of U.S. cotton peaked in 2006, dropped sharply in 2009, but then more than doubled the 2009 level in 2010. In addition to the weaker cotton demand following the global recession, a reduction in planted cotton area in the United States that began in 2006 had led to lower supplies available for export, including exports to China.[26] However, in 2010 U.S. production levels were the highest in three years, which meant more U.S. cotton was available for export at the same time global demand was recovering.[27]

Despite increased shipments that year, the United States has been losing market share in China to competitors, particularly India, which also accounted for about a one-third of China's cotton imports in 2010.[28] Indian cotton exports grew in line with increasing production as a result of India's adoption of Bt cotton.[29]

Hides and Skins

In addition to cotton, China's apparel, footwear, and accessories industries use imported hides and skins as inputs. During 2005–09, China accounted for about 30 percent of world hides and skins imports and was by far the world's leading importer. In 2010, China's hides and skins imports were $2.4 billion, the sixth highest category of agricultural imports that year. Applied tariffs on imported hides and skins are low at about 5 percent.[30] Imports grew each year between 2005 and 2010, except for 2009, when they dropped 18 percent. Like cotton, imports of hides and skins fell substantially because of the global economic crisis and the resulting drop in orders for Chinese apparel, footwear, and accessories worldwide, but rose again when the industry recovered in 2010. During 2005–08, close to one-half of Chinese hides and skins imports came from the United States, but this share fell to about one-third during 2009–10. These shipments consisted mostly of cattle hides.

Meat

During 2005–10, Chinese meat imports rose from $431 million to $1.4 billion, an average annual growth of about 35 percent. The increase is attributable principally to a rise in per capita meat consumption, which grew about 20 percent annually between 2005 and 2009.[31] During this period, poultry made up three-quarters of China's meat imports, followed by pork, with a 13 percent share, and sheep and lamb meat, with a 9 percent share. Growth in Chinese imports reflects a larger trend of rising poultry consumption in China that is outpacing domestic production. Poultry consumption in China rose because of health concerns about pork products, higher incomes,[32] and more frequent dining out by Chinese urbanites.[33] Poultry was one of China's top four agricultural imports from the United States during 2005–09, growing at an average annual rate of over 50 percent. The United States share of China's poultry imports rose from 53 percent to 80 percent during 2005–09, gradually edging out poultry imports from Argentina and Brazil. However, in 2010 U.S. market share fell significantly, with imports down 80 percent from 2009 due to China's countervailing duty (CVD) and antidumping (AD) investigations and subsequent imposition of duties on U.S. poultry products.[34]

Pork was the fastest-growing meat import by China during 2005–10, driven mostly by a fourfold increase between 2007 and 2008. Higher demand for foreign pork followed a domestic outbreak of blue ear disease in 2007, as well as poor weather that reduced the availability of domestic supplies.[35] This outbreak particularly encouraged imports of U.S. pork, which rose from $1 million in 2006 to peak at $313 million in 2008. However, once China recovered from the outbreak, its pork imports from the United States declined to only $30 million by 2010.

Grains

During 2005–10, Chinese wheat imports fluctuated widely. Imports dropped from $762 million in 2005 to $7 million in 2008 before rebounding to $309 million in 2010. Government policy largely drove these fluctuations.[36] In order to maintain production in the face of low grain prices in 2005, the government instituted a number of programs, which, together with higher international prices, led to expanded domestic wheat acreage beginning in 2006.[37] Higher domestic production lessened the need for wheat imports in 2007 and 2008.[38] When international wheat prices rose in 2008, the central government banned wheat exports, auctioned off grain reserves, and instituted retail price controls in an effort to slow food price inflation.[39] This domestic price ceiling further lowered wheat imports, as the price of wheat imports was not competitive with the price of domestic wheat. These actions were consistent with the government's goal of 95 percent self-sufficiency in grains.

China is the second-largest corn-producing country in the world,[40] using the grain primarily for animal feed and ethanol production. While China is among the world's leading corn-consuming countries, it imported very little until 2010, when imports increased eighteenfold from the year before to $367 million, virtually all of it from the United States. Like wheat, the government views corn as important for national food security[41] and provides support for domestic corn growers by guaranteeing prices for domestic corn from state-owned enterprises and by providing subsidized seed, while also controlling exports to insure that corn is available for domestic use.[42] Strong demand, coupled with poor production in the 2009/10 crop year (down by 5 percent from the previous year), led China to purchase about 1.5 million metric tons (mt) of U.S. corn in 2009/10.[43]

Other Products

Chinese imports of several high-valued processed agricultural products grew significantly between 2005 and 2010, most notably dairy products, alcoholic beverages, and fresh fruit. During 2005–10, imports of dairy products, consisting mostly of whole milk powder, infant formula, whey, and nonfat dry milk, increased more than fourfold, reaching $2.9 billion in 2010. Major dairy suppliers were New Zealand and the EU-27, which accounted for about 70 percent of all China's dairy imports in 2010. The United States is an important supplier of whey, which is by far the most important U.S. dairy export to China. Imports of infant formula increased sixfold between 2005 and 2010. This growth is attributed to both stronger demand for dairy products as per capita income grew in China and the preference for foreign infant formula in response to concerns over the safety of domestically produced product from 2007 to 2010.[44]

Alcoholic beverage imports by China grew nearly 34 percent annually between 2005 and 2010. Higher incomes and changing tastes among domestic consumers contributed to this growth.[45] Two products—wine and grape brandy (primarily Cognac)—accounted for over 80 percent of alcoholic beverage imports in 2010. Major suppliers included the EU27 (mostly France and Italy), Australia, Chile, and the United States. Although U.S. wine exports to

China grew by more than 50 percent annually during 2005–10, even faster growth by the EU-27 and Australia lessened the U.S. share of the Chinese wine market during this period. Even though per capita beer consumption in China is high by international standards, growth in beer imports was slower than wine because many foreign beer companies produce their beer in China.[46]

Chinese fresh fruit imports grew by about 26 percent annually during 2005–10. Among the leading products were bananas from the Philippines, grapes from Chile and the United States, guavas and longans from Thailand, and dragon fruit from Vietnam.

Imports by Major Trading Partner

During 2005–10, China's agricultural imports were concentrated among a few major partners (table 2.2 and figure 2.3). On average, the top three suppliers (United States, Brazil, and Argentina) accounted for more than one-half of Chinese imports. Of the top supplier countries to China, India, Brazil and Argentina recorded the fastest growth in imports between 2005 and 2010, while imports from Australia grew the least.

Imports from Brazil and Argentina were highly concentrated, with soybeans and soybean oil accounting for over 90 percent of the trade. In contrast, Chinese imports from the EU-27 were relatively diverse, with the largest import category, alcoholic beverages, accounting for about one-quarter of the trade during this period. Other important imports from the EU-27 included meat products (mainly pork) and animal skins. From Australia, China primarily imported inputs for the textile and apparel industries, including wool, cotton, and sheepskins. In 2010, Chinese wine imports from Australia were second only to those from France, and Australian producers have begun branding wines specifically for the Chinese market.[47] Chinese agricultural imports from Canada increased at an average annual rate of 30 percent between 2005 and 2010; two-thirds of those imports consisted of erucic acid rapeseed[48] for processing into canola oil and meal, and rapeseed oil.

Table 2.2. China: Agricultural imports by major trading partner, 2005–10 (million $)

Partner	2005	2006	2007	2008	2009	2010
United States	6,342	7,128	8,620	13,826	13,407	17,834
Brazil	3,011	3,803	4,823	8,789	8,442	10,729
Argentina	2,967	2,360	5,141	8,346	3,447	5,677
EU-27	1,819	1,952	2,621	3,438	3,239	4,719
Australia	2,388	2,311	2,596	2,923	2,473	3,885
Malaysia	1,461	1,823	3,124	4,223	3,130	3,619
Indonesia	934	1,264	1,845	2,684	2,281	3,001
Canada	977	630	1,081	1,577	2,488	2,784
India	346	1,076	1,212	1,515	814	2,381
Thailand	922	1,208	1,212	1,080	1,685	2,268
All other	3,824	4,622	4,575	5,305	6,311	9,542
Total	24,991	28,178	36,850	53,707	47,717	66,439

Source: GTIS, Global Trade Atlas database.

To date China has completed and implemented trade agreements with 20 countries,[49] and combined imports from these partners exceed imports from all other countries except the United States. During 2005–10, average annual growth in agricultural imports from the trade partners with trade agreements was 28 percent, compared with 24 percent for imports from all countries (table 2.3). Since many of China's preferential trade agreements came into force recently, the full tariff reductions have not yet occurred. Therefore, the full effect of these agreements is not yet observable in trade statistics. One factor that could limit the impact of duty reductions with Association of Southeast Asian Nations (ASEAN) countries is that products such as fresh fruits and vegetables are imported into China out of season, when domestic production is limited. Malaysia and Indonesia accounted for close to one-half of the imports from all preferential trading partners, with most of the imports consisting of palm oil.

Table 2.3. China: Agricultural imports from FTA-partner countries, 2005–10 (million $)

Country	2005	2006	2007	2008	2009	2010
ASEAN						
Malaysia	1,461	1,823	3,124	4,223	3,130	3,619
Indonesia	934	1,264	1,845	2,684	2,281	3,001
Thailand	922	1,208	1,212	1,080	1,685	2,268
Vietnam	177	336	423	446	703	674
Singapore	85	145	210	266	429	508
Philippines	108	158	157	236	226	415
Myanmar	22	22	52	170	106	183
Laos	2	6	12	17	29	30
Cambodia	2	2	3	3	6	3
Brunei	0	0	0	0	0	0
Total ASEAN	3,713	4,963	7,037	9,124	8,595	10,702
India	346	1,076	1,212	1,515	814	2,381
New Zealand	620	656	755	922	1,248	2,079
Chile	93	103	143	228	314	449
Korea	146	155	160	175	195	268
Pakistan	20	26	27	32	49	122
Hong Kong	46	57	82	81	88	112
Peru	5	10	16	31	33	47
Bangladesh	0	0	1	3	6	17
Sri Lanka	2	3	5	6	5	16
Macau	0	0	0	0	1	0
Total	4,991	7,049	9,437	12,117	11,349	16,193

Source: GTIS, Global Trade Atlas database.

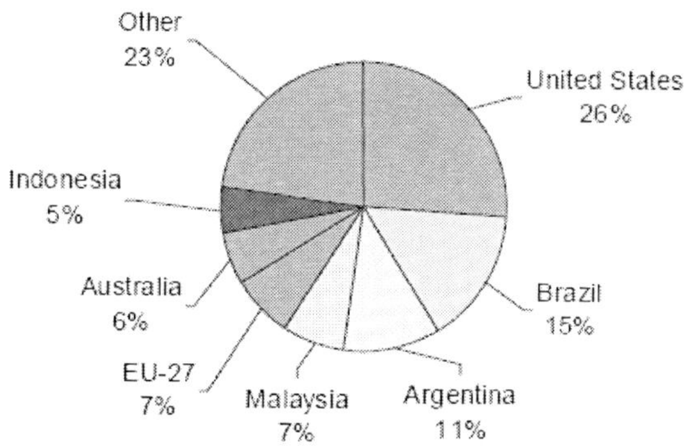

Source: GTIS, Global Trade Atlas database.

Figure 2.3. More than half of Chinese agricultural imports were supplied by three countries during 2005–10.

Exports

Exports by Product

During 2005–10, Chinese agricultural exports grew from $19.6 billion to $35.7 billion, an average of 13 percent annually. In 2010, close to one-half of China's agricultural exports were horticultural products, led by processed vegetables with shipments totaling $6.4 billion, followed by fresh vegetables ($3.7 billion) and fresh fruit ($2 billion). Miscellaneous processed foods, processed animal feed (mostly soybean oil cake and pet food), and meat products were the top non-horticultural products exported by China in 2010. Major Chinese export trade flows are depicted in figure 2.4.

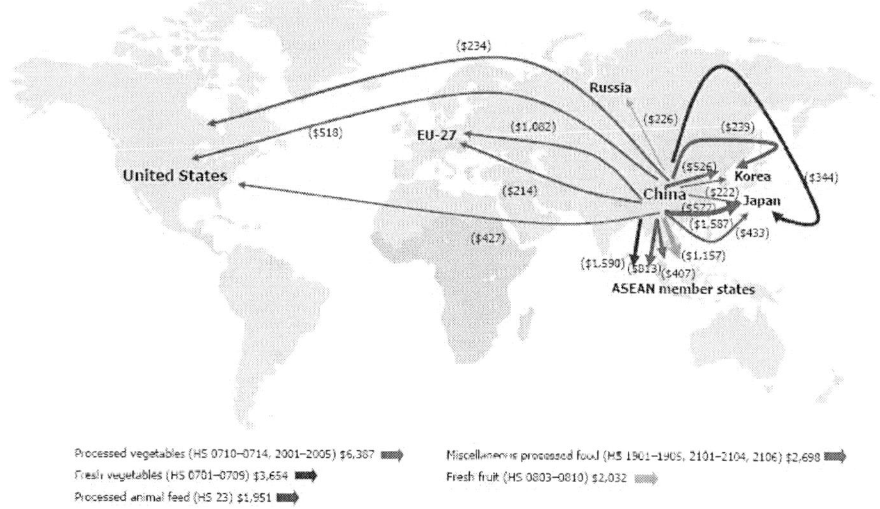

Source: GTIS, Global Trade Atlas database.

Figure 2.4. China's agricultural exports were highly concentrated in a few major products in 2010 (million $).

The United States was China's fourth-largest export market for agricultural products in 2010, behind Japan, the EU-27, and Hong Kong. In 2010, about one-half of Chinese agricultural exports to the United States were horticultural products, mainly processed vegetables (processed garlic and preserved mushrooms), fruit juice, and processed fruit (canned citrus and peaches) (table 2.4). Apple juice was another important export, as Chinese juice accounted for 72 percent of U.S. juice imports between 2005 and 2010, in spite of the antidumping duties that were levied on specific Chinese apple juice producers in 2000.[50] Between 2005 and 2010, Chinese exports to the United States grew by about $1.7 billion, with much of the increase from higher exports of pet food and processed horticultural products.

Table 2.4. China: Agricultural exports to the world and the United States by product, 2005–10 (million $)

Product	Source	2005	2006	2007	2008	2009	2010
			Animal products				
Live animals	World	329	333	375	507	442	454
	United States	15	18	20	27	19	20
Beef	World	42	64	79	96	61	109
	United States	0	0	0	0	0	0
Pork	World	406	401	284	276	263	332
	United States	0	0	0	0	0	0
Sheep and lamb meat	World	54	66	54	50	42	69
	United States	0	0	0	0	0	0
Product	Source	2005	2006	2007	2008	2009	2010
Poultry	World	194	167	261	324	336	428
	United States	0	0	0	0	0	0
Dairy	World	130	141	322	352	82	91
	United States	9	13	11	8	1	15
Eggs	World	79	78	94	132	122	143
	United States	2	2	3	5	4	4
Processed meats	World	1,187	1,276	1,342	1,119	1,157	1,462
	United States	27	19	10	10	15	24
Other products of animal origin	World	1,062	1,051	1,123	1,379	1,242	1,428
	United States	248	218	257	249	195	244
			Grains				
Wheat	World	37	161	481	31	2	0
	United States	0	0	1	0	0	0
Rice	World	225	409	479	482	524	416
	United States	0	2	2	1	0	0
Corn	World	1,097	412	874	73	32	33
	United States	0	0	0	0	0	0
Other grain	World	55	56	133	87	61	90
	United States	0	1	1	0	0	0
		Oilseeds and products					
Soybeans	World	170	146	196	351	237	118
	United States	16	16	18	36	30	10
Soybean oil	World	40	72	57	185	76	65

Product	Source	2005	2006	2007	2008	2009	2010
	United States	0	0	0	0	0	0
Rapeseed oil	World	21	90	17	11	13	5
	United States	0	0	0	0	0	0
Corn oil	World	71	76	58	151	21	17
	United States	0	0	0	0	0	0
Other vegetable oil	World	137	129	168	206	150	187
	United States	12	12	16	17	12	17
		Horticulture					
Vegetables, fresh	World	1,247	1,619	1,642	1,460	2,081	3,654
	United States	52	81	91	64	83	143
Vegetables, processed	World	3,266	3,871	4,753	5,300	4,988	6,387
	United States	249	341	412	433	409	517
Nuts	World	977	1,040	1,226	1,363	1,172	1,403
	United States	57	75	83	111	102	115
Fruit, fresh	World	658	789	1,067	1,529	1,815	2,032
	United States	4	13	24	17	15	14
Fruit, processed	World	861	1,031	1,352	1,567	1,382	1,604
	United States	174	235	326	419	381	398
Juice	World	526	671	1,354	1,259	762	864
	United States	182	218	483	595	352	412
Miscellaneous plants (ch 12)	World	621	690	844	954	1,005	1,228
	United States	32	43	44	45	33	41
		Beverages					
Tea/coffee	World	511	589	657	751	786	886
	United States	30	35	40	46	43	76
Alcoholic beverages	World	295	714	398	448	385	527
	United States	11	93	12	12	12	14
Non-alcoholic beverages	World	414	367	423	405	438	459
	United States	7	8	11	14	11	7
		Sugar, sweeteners, confectionery					
Sugar	World	111	61	48	28	34	64
	United States	0	1	1	2	2	3
Honey	World	88	105	94	147	126	183
	United States	24	27	14	13	0	1
Other sweeteners	World	307	401	517	649	737	995
	United States	66	73	113	130	145	145
Cocoa products	World	110	124	147	200	130	213
	United States	40	46	42	62	15	8
			Processed foods				
Miscellaneous processed foods	World	1,459	1,756	2,002	2,214	2,266	2,698
	United States	125	173	214	202	210	234
Milled grain products	World	200	236	509	540	466	551
	United States	1	6	5	7	9	13
Processed animal feed	World	474	505	994	1,615	1,758	1,951
	United States	47	95	210	321	325	427
Spices	World	417	400	434	564	617	771
	United States	46	37	44	56	64	92
				Other			

Table 2.4. (Continued)

Product	Source	2005	2006	2007	2008	2009	2010
Cotton	World	9	26	37	42	19	10
	United States	0	0	0	0	0	0
Wool	World	69	69	77	56	39	70
	United States	1	1	1	1	0	1
Tobacco and products	World	537	566	638	742	878	1,020
	United States	6	7	17	9	11	7
All other	World	1,170	1,311	1,743	2,423	2,159	2,686
	United States	9	125	123	197	175	227
Total	World	19,646	22,047	27,330	30,045	28,885	35,683
	United States	1,578	2,035	2,649	3,110	2,673	3,229

Source: GTIS, Global Trade Atlas database.

Horticultural Products

China is a large global exporter of horticultural products.[51] Between 2005 and 2010, about 45 percent of Chinese agricultural exports were horticultural products, consisting mostly of fresh and processed fruits and vegetables, and juice. During this period, exports rose from $8.2 billion to $17.2 billion, a rate of about 17 percent annually. Contributing to the rise in exports were low labor and input costs that make fruit and vegetable production and processing (both labor-intensive activities) profitable for Chinese exporters. Also, growth in domestic production in excess of domestic consumption growth provided surplus available for export.[52] The Chinese government has invested heavily in the food processing sector,[53] and large canneries (often designated as leading agribusinesses or "dragon head" enterprises[54] by the government) are also supported through loans from state banks and preferential legal treatment.[55]

Among China's horticultural exports, processed vegetables are by far the largest (table 2.4), reaching $6.4 billion in 2010. The top processed vegetable export was tomato paste (exported mainly to Russia, the EU-27, and Africa), followed by mushrooms (exported to the EU-27, the United States, and several Asian countries, including Japan, Thailand, and Hong Kong) and kidney beans (to the EU-27 and India). China also exported dried garlic, for which the United States is the leading market. Even though exports of fresh vegetables are constrained by poor cold storage facilities and an inability of many producers to meet foreign quality standards,[56] China's fresh vegetable exports exceeded $3.6 billion in 2010, compared to $1.2 billion in 2005. In 2010, fresh garlic accounted for almost two-thirds of these exports, with Indonesia by far the most important market, followed by Brazil, Vietnam, and the EU-27. The United States is the fifth largest market for Chinese fresh garlic, even though AD duties of up to 377 percent have been applied on Chinese product since 1994. [57] Other fresh vegetable exports important to China are carrots and onions, exported mostly to Japan and several markets in South Asia.

Fresh and processed fruit accounted for $3.6 billion of China's agricultural exports in 2010. Fresh apples were the leading fresh fruit export, followed by fresh pears and mandarin oranges. Many of these products were exported to Asia (mostly to Indonesia, followed by Thailand, Bangladesh, and the Philippines), although Russia is an important market for Chinese apple exports. China's processed fruit exports, mostly canned citrus, pears, and fruit mixtures, increased rapidly during 2005–10. The major market for these products was the

United States, where they were either consumed in the foodservice industry or repackaged into individual serving containers for retail sale.[58] Other major markets for Chinese processed fruit exports were Japan, the EU-27, and Canada.

Other Products

Miscellaneous processed foods and animal feed are also major agricultural exports for China. During 2005–10, miscellaneous processed food exports increased from $1.5 billion to $2.7 billion, making it China's third-largest agricultural export category after processed and fresh vegetables. Major exports in this category were noodles (HS190230), exported to Hong Kong, Japan, and the United States, and sauces (HS210390), exported to South Korea, Japan, Hong Kong, and the United States. China's processed animal feed exports increased about fourfold during 2005–10, reaching $2 billion in 2010. This category includes soybean oil cake (HS230400) and prepared feed additives for livestock (HS230990), as well as pet food for dogs and cats (HS230910). Major markets for soybean oil cake and feed additives are in Asia, including Japan, South Korea, and Vietnam. The United States was the largest purchaser of Chinese pet food, followed by Japan; sales to the United States surged from $12 million in 2005 to $343 million in 2010.

Exports by Major Trading Partner

Chinese export markets were highly concentrated among a few major trading partners. During 2005–10, about one-half of agricultural exports went to the top four trading partners, and three-quarters to the top 10 markets (table 2.5 and figure 2.5). Japan remained the top destination for Chinese agricultural exports, though its share fell from 26 percent in 2005 to 17 percent in 2010. The EU-27 was the second-largest market during this period, with a fairly stable share of about 14 percent. Hong Kong and the United States each accounted for about 10 percent of China's agricultural exports.

Table 2.5. China: Agricultural exports by major trading partner, 2005–10 (million $)

Partner	2005	2006	2007	2008	2009	2010
Japan	5,094	5,243	5,499	4,989	5,088	6,009
EU-27	2,485	3,006	3,938	4,672	4,000	4,797
Hong Kong	2,250	2,314	2,654	2,994	3,000	3,549
United States	1,578	2,035	2,649	3,110	2,673	3,229
Korea	1,871	1,771	2,484	2,086	1,832	2,213
Indonesia	387	560	880	800	962	1,603
Malaysia	560	668	868	887	973	1,324
Vietnam	262	330	461	692	915	1,267
Russia	611	710	923	1,083	908	1,162
Thailand	274	323	468	608	756	1,060
All other	4,275	5,087	6,504	8,124	7,777	9,469
Total	19,646	22,047	27,330	30,045	28,885	35,683

Source: GTIS, Global Trade Atlas database.

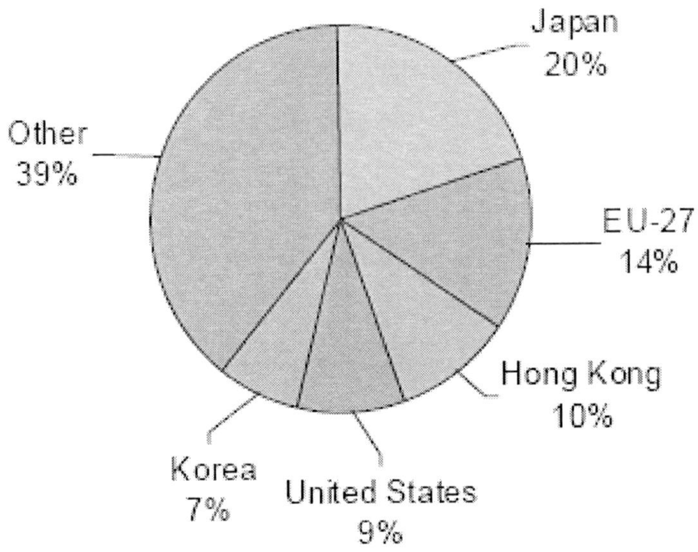

Source: GTIS, Global Trade Atlas database.

Figure 2.5. More than one-half of Chinese agricultural exports were shipped to four markets during 2005–10.

During 2005–09, Chinese agricultural exports to Japan were fairly stable at about $5 billion annually and then grew to $6 billion in 2010. Japan's proximity to China makes the two countries natural trading partners in many products. Many Japanese companies lease land in China and grow products for export back to Japan, though some of these companies also sell their products in China's domestic market. Japanese investment in Chinese agriculture is viewed positively in China because it increases the income of Chinese farmers and helps to develop China's agricultural sector.[59] Processed meats (chicken offal) are the largest agricultural export from China to Japan. Exports to the EU-27 market grew consistently between 2005 and 2010 with the exception of 2009, when they fell 14 percent. Processed fruits and vegetables were the single largest Chinese agricultural exports to EU-27 countries during 2005–10. During this period, Hong Kong was China's third-largest agricultural export market, and while some agricultural exports were likely consumed there, often traders in Hong Kong shipped products from mainland China for re-export to other markets.[60]

When grouped together, China's preferential trade partners accounted for more of China's agricultural exports than any individual trading partner. In 2010, China exported $13.8 billion in agricultural products to preferential trading partners, together accounting for 39 percent of its total agricultural exports (table 2.6). Fresh garlic was the largest export to these markets, with $1.2 billion shipped in 2010, accounting for 9 percent of all agricultural exports to preferential trade partners. Other major products exported to FTA partners were dried mushrooms, fresh apples, mandarin oranges, and live swine, but no other individual product accounted for more than 5 percent of exports in 2010. As with imports from preferential trade partners, the total effect of the agreements is not yet observable in the trade statistics, and the benefits of these agreements in terms of enhanced exports likely will not be fully realized for several years.

China-U.S. Competition in Third-Country Markets

To a large extent, agricultural exports of the United States and China do not compete in third-country markets. This is because the two countries have different factor endowments which influence their international competitiveness in broadly different products. As mentioned, relative to China, the United States is a land-abundant, labor-scarce country, a situation that gives it a comparative advantage in land-intensive agriculture, including grains, oilseeds, and livestock. In contrast, relative to the United States, China is a labor-abundant, land-scarce country; it therefore enjoys a comparative advantage in labor-intensive agricultural products, especially in horticulture.[61] In most cases, these resource endowments make China and the United States natural trading partners rather than competitors.

A partial exception is fresh fruit, of which both countries export high volumes. In the past, Chinese fresh fruit could not compete with that of the United States in global markets. This was mostly because of the poor quality of Chinese fruit, as scarce cold storage and poor rural infrastructure in China took its toll on perishable products traveling long distances for export.[62] However, in the last five years, the quality of many

Table 2.6. China: Agricultural exports to FTA-partner countries, 2005–10 (million $)

Country	2005	2006	2007	2008	2009	2010
Indonesia	387	560	880	800	962	1,603
Malaysia	560	668	868	887	973	1,324
Vietnam	262	330	461	692	915	1,267
Thailand	274	323	468	608	756	1,060
Philippines	289	409	490	475	550	637
Singapore	251	291	315	411	358	455
Myanmar	50	80	69	64	76	100
Cambodia	9	14	20	23	13	21
Laos	2	2	1	3	6	15
Brunei	4	4	5	7	7	10
Total ASEAN	2,087	2,680	3,578	3,970	4,617	6,491
Hong Kong	2,250	2,314	2,654	2,994	3,000	3,549
Korea	1,871	1,771	2,484	2,086	1,832	2,213
India	213	224	349	414	483	521
Pakistan	169	125	138	149	210	274
Bangladesh	47	67	107	101	186	242
Macau	123	139	163	187	188	201
New Zealand	36	48	67	77	68	85
Sri Lanka	26	38	54	49	58	82
Chile	12	18	40	65	51	66
Peru	5	9	17	26	31	34
Total	6,840	7,434	9,652	10,119	10,724	13,760

Source: GTIS, Global Trade Atlas database.

Chinese horticultural products has greatly improved, and competition with the United States, particularly in Asian markets, has grown.[63]

Nonetheless, competition between the United States and China in third countries appears to be limited to a few markets. Moreover, the United States and China currently tend to supply different market segments, differentiated by cost, quality, and/or variety preference.[64] For example, China has become the world's primary low-cost apple producer and, as a result, supplies markets where price is the main factor driving demand, while U.S. apples typically supply markets where consumers demand high-quality apples. As for pears, Chinese and U.S. products primarily supply distinct export markets because each country produces specific varieties that are preferred in different countries.[65]

As China becomes better able to supply high-quality fruit at a low cost to nearby Asian markets, competition with the United States could intensify.[66] The United States and China both export large quantities of apples to several Asian markets, including Indonesia, Malaysia, and India. Chinese apple exports to Vietnam, a low-cost apple market, averaged $53 million between 2008 and 2010, far outpacing average U.S. apple exports of $8 million to Vietnam during the same period.[67] In India, apple imports from all sources almost tripled between 2005 and 2009 to reach 90,800 mt.[68] During this period of tremendous growth, Chinese apples increased their market share relative to the United States as their quality improved while prices remained stable. China's share of Indian apple imports grew from 20 percent in 2005 to 40 percent in 2009, while that of the United States fell from 53 percent in 2005 to 42 percent in 2009.[69]

China and the United States both export large quantities of fresh and frozen vegetables, yet compete directly in only a few countries. For example, China and the United States are the top two exporters of carrots and turnips in the world, but have no markets in common among their top 10 export destinations. For onions and shallots, Japan is the only market where competition takes place, with China exporting $117 million in 2010 and the United States $35 million.[70] Competition is generally greater in frozen vegetables, which can be transported across long distances with little loss in quality. However, U.S. and Chinese producers tend to produce different varieties of frozen vegetables.[71]

Bibliography

American Soybean Association (ASA). Written submission to the U.S. International Trade Commission in connection with inv. no. 332-518, *China's Agricultural Trade: Competitive Conditions and Effects on U.S. Exports*, June 22, 2010.

California Cling Peach Board. Written submission to the U.S. International Trade Commission in connection with inv. no. 332-518, *China's Agricultural Trade: Competitive Conditions and Effects on U.S. Exports*, July 28, 2010.

Economist Intelligence Unit (EIU). *Business China: Main Report.* Beijing: March 2007.

———. *China: Food, Beverages and Tobacco Profile.* Beijing: 2008.

———. *China Food: Sub-Sector Update.* Beijing: January 2009.

———. "Off the Wagon." *Business China.* Beijing: March 1997.

Fry, John. "Update on the Contamination of Dairy Products in China." Executive Briefing on Trade. U.S. International Trade Commission, November 2008.

Gale, Fred. "A Tale of Two Commodities: China's Trade in Corn and Soybeans." Chapter 3 in *China's Agricultural Trade: Issues and Prospects*. U.S. Department of Agriculture, Economic Research Service, July 2007.

Gale, Fred, Bryan Lohmar, and Francis Tuan. "How Tightly Has China Embraced Market Reforms in Agriculture?" *Amber Waves*. U.S. Department of Agriculture, Economic Research Service, April 2009.

Gale, Fred, and Bryan Lohmar. "Who Will China Feed?" *Amber Waves*. U.S. Department of Agriculture, Economic Research Service, June 2008.

Gereffi, Gary, and Stacey Frederick. "The Global Apparel Value Chain, Trade and the Crisis: Challenges and Opportunities for Developing Countries." World Bank Development Research Group. Trade and Integration Team. Policy Research Working Paper 5281, April 2010. http://www.iadb.org/intal/intalcdi/PE/2010/05413.pdf.

Global Trade Information Services, Inc. (GTIS). Global Trade Atlas database (accessed February 16, 2011).

He, Jianwu, Li Shantong, and Sandra Polaski. "China's Economic Prospects: 2006–2020." Carnegie Papers. Washington, DC: Carnegie Endowment For International Peace, April 2007.

Huang, Sophia Wu. *Global Trade Patterns in Fruit and Vegetables*. U.S. Department of Agriculture, Economic Research Service, Agriculture and Trade Report no. WRS-04-06, June 2004.

International Monetary Fund (IMF). *International Financial Statistics. www.imf.org* (accessed October 7, 2010).

Lohmar, Bryan, Fred Gale, Francis Tuan, and Jim Hansen. *China's Ongoing Agricultural Modernization: Challenges Remain after 30 Years of Reform*. U.S. Department of Agriculture, Economic Research Service, Economic Information Bulletin no. 51, April 2009.

Shaosheng Jin, Suminori Tokunaga, Yuko Akune, and Masahiro Kageyama. "Agglomeration Effects and Japanese Food Industry Investment in China: Evidence from the Cities." Paper presented at International Association of Agricultural Economists Conference, Gold Coast, Australia, August 2006.
http://ageconsearch.umn.edu/bitstream/25307/1/cp060458.pdf.

U.S. Department of Agriculture (USDA). Economic Research Service (ERS). *China Is Using More Corn for Industrial Products*, by Fred Gale, Francis Tuan, Xiaohui Wang, and Zhi Cao. Electronic Outlook Report no. FDS-09K-01, December 2009.

———. *China's Rising Fruit and Vegetable Exports Challenge U.S. Industries,* by Sophia Huang and Fred Gale. Electronic Outlook Report no. FTS-320-01, February 2006.

———. "Cotton: Background." ERS Briefing Room, November 10, 2009. http://www.ers.usda.gov/Briefing/Cotton/background.htm.

———. *Cotton and Wool Situation and Outlook Yearbook*, by Leslie Meyer, Stephen MacDonald, and James Kiawu, November 2008.

———. *Cotton and Wool Situation and Outlook Yearbook* by Leslie Meyer, Stephen MacDonald, and James Kiawu, June 11, 2010.

———. *Cotton and Wool Situation and Outlook Yearbook*, by Leslie Meyer, Stephen MacDonald, and James Kiawu, December 13, 2010.

———. *Fiber Use for Textiles and China's Cotton Textile Exports*, by Stephen Macdonald and Sarah Whitley. Electronic Outlook Report no. CWS-08i-0, March 2009.

U.S. Department of Agriculture (USDA). Foreign Agricultural Service (FAS). *China, Peoples Republic of: Poultry and Products; Annual Report, 2008*, by Jianping Zhang and Chanda Beckman. GAIN Report no. CH8073, September 1, 2008.

———. *China, Peoples Republic of: Annual; Canned Deciduous Fruit,* by Joshua Emmanuel Lagos, Freddie Xu, and Wu Bugang. GAIN Report no. CH9085, October 14, 2009.

———. *China, Peoples Republic of: Grain and Feed, Annual*, by Kevin Latner and Jiang Junyang. GAIN Report no. CH6010, March 2006.

———. *China, People's Republic of: Grain and Feed, Annual*, by Chanda Beckman and Jiang Junyang. GAIN Report no. CH10008, March 1, 2010.

———. *India: Cotton and Products Annual, 2010*, by Santosh Singh. GAIN Report no. IN1029, April 12, 2010.

———.*India: Export of Cotton Allowed under License*, by Santosh Singh. GAIN Report no. IN1049. June 4, 2010.

———. *Livestock and Poultry: World Markets and Trade.* Circular Series DL&P 1-08, April 2008.

———. *National Plan for Expansion of Grain Production Capacity*, by Chanda Beckman and Jiang Junyang. GAIN Report CH10005, February 18, 2010.

———. Production Supply and Distribution (PSD) Online. http://www.fas.usda.gov/psdonline/psdResult.aspx (accessed various dates).

———. "Strong Processing Margins Support China's Expanding Soybean Import Market Demand." Oilseeds: World Market and Trade Circular Series FOP 01-10, January 2010. http://www.fas.usda.gov/oilseeds/circular/2010/January/oilseedsfull01-10.pdf.

U.S. Department of Agriculture (USDA). "Agriculture Secretary Vilsack's Statement on 2010 U.S. Farm Exports Record." News release, February 11, 2011.

U.S. International Trade Commission (USITC). *Canned Peaches, Pears, and Mixtures: Conditions of Competition between U.S. and Principal Foreign Supplier Industries.* USITC Publication 3972. Washington, DC: USITC, 2007.

———. *Fresh Garlic from China.* USITC Publication 3886. Washington, DC: USITC, 2006.

Waldron, Scott A., Colin G. Brown, and John William Longworth. *Rural Development in China: Insights from the Beef Industry*. Aldershot, England: Ashgate Publishing Limited, 2003.

World Trade Organization (WTO). Tariff Analysis Online (TAO). http://tariffanalysis.wto.org/report/tarifflines.aspx (accessed September 2, 2010).

Zappone, Chris. "How to Sell Wine to China." *Western Australia Today*, September 18, 2009. http://www.watoday.com.au/small-business/entrepreneur/how-to-sell-wine-to-china-20090918fuot.html.

3. Chinese Agricultural Consumption

Overview

China is a major global consumer of agricultural products. It consumes one-third of the world's rice, one-fourth of all corn, and one-half of all pork and cotton, and it is the largest consumer of oilseeds and most edible oils.[1] The traditional Chinese diet centers around staple foods (mainly grains and starches),[2] which account for nearly one-half of the daily caloric intake. Average Chinese per capita consumption recently stabilized at approximately 3,000 calories per day, one of the highest levels among Asian countries.[3] Over the past five years, per capita consumption of staple foods has been fairly stable in both rural and urban areas. However, food consumption patterns are changing with the Chinese increasing the portion of nonstaple food items, such as fruits, meat, and dairy products, in their diets.[4]

Chinese food consumption is influenced by factors such as population size and demographics, income, food prices, and general preferences. China has a population of over 1.3 billion people, ranging from poor rural farmers to wealthy urbanites.[5] Per capita income growth and urbanization are the two factors most responsible for altering recent consumption patterns in China. Rising income translates into higher per capita food consumption, while increasing urbanization is driving diversification of food choices because of greater availability and choice offered through increasingly diverse sales outlets.

Chinese consumers generally fall into one of three categories: rural consumers; urban low-income consumers; or urban high-income consumers. Although urban high-income consumers can afford to buy more and better-quality food, the ubiquity of food outlets[6] in cities means that nearly every urban resident, regardless of income, has available an increasingly diverse food selection. Compared to rural diets, urban diets contain less grain and more nonstaple items, including processed and convenience foods. Rural migrants to cities tend to adopt the urban diet.[7]

Along with more varied consumption, higher incomes are leading to changing food preferences, including the demand for better quality and safer foods.[8] Food preferences determine where urban Chinese purchase their foods, whether it be at local "wet markets," urban supermarkets, or restaurants. [9] Chinese value the diversity in food products that different shopping outlets offer.[10] In the future, analysts predict that further income growth and urbanization will continue to increase demand for a variety of higher-quality foods.[11]

Consumption Trends

Consumption Patterns

Between the 2005/06 and 2009/10 marketing years,[12] consumption of most agricultural products grew rapidly by volume, although growth rates differed among commodity groups (table 3.1).[13] During this period, consumption of oilseeds and oilseed products rose significantly, with that of soybeans, soybean meal and oil, and palm oil each increasing by 30 percent or more. Among the grains, corn consumption increased the most, rising 13 percent between 2005/06 and 2009/10. The higher consumption of corn, soybeans, and meals, which are ingredients for livestock feed, is a response to increased demand for meat. During 2005/06–09/10, consumption of pork (the most heavily consumed meat in China) increased by 8 percent, even though consumption fell in 2007/08. The drop in consumption that year was due to an outbreak of blue ear pig disease, which lowered production and caused pork prices to climb 85 percent.[14] Poultry consumption, although one-fourth the level of pork consumption, experienced the largest growth among meats, increasing almost 5 percent annually (or 21 percent overall) between 2005/06 and 2009/10. Consumption of milk powders (nonfat dry milk and whole milk powder) increased almost 20 percent during the same period, albeit from a low 2005/06 base. Greater consumption of milk powders, furnished mostly through increased imports, was in response to a drop in fluid milk consumption in 2008/09 and 2009/10 following a scandal involving melamine-contaminated powdered infant formula that led consumers to distrust domestic dairy products. [15] Consumption of cotton increased moderately between 2005/06 and 2009/10 in spite of a sharp dip in 2008/09, a result of the global economic recession that led to lower demand for textiles and apparel items manufactured from cotton.[16]

Table 3.1. China: Agricultural consumption, 2005/06–09/10

Commodity	2005/06	2006/07	2007/08	2008/09	2009/10	Annual growth	Growth 2005/06–09/10
Grains		Million metric tons (mt)				Percent	
Corn	137.0	145.0	149.0	152.0	155.0	3.1	13.1
Rice, milled	128.0	127.2	127.5	133.0	134.5	1.3	5.1
Wheat	101.5	102.0	106.0	105.5	105.0	0.9	3.4
Other grains	10.1	8.5	8.0	8.3	8.2	–4.7	–18.6
All grains	376.6	382.7	390.5	398.8	402.7	1.7	6.9
Oilseeds							
Soybeans	44.4	46.1	49.8	51.4	57.8	6.9	30.1
Rapeseed	13.7	11.9	11.4	13.7	15.1	3.2	9.8
Cottonseed	11.1	13.9	14.5	14.4	12.7	4.2	14.4
Peanuts	13.6	12.0	12.3	13.6	12.6	–1.4	–6.9
Sunflower seed	1.8	1.4	1.1	1.7	1.5	–0.9	–18.1
All oilseeds	84.7	85.4	89.1	94.8	99.7	4.2	17.8
Meat							
Pork[a]	45.1	46.1	42.7	46.4	48.7	2.1	8.0
Poultry	10.1	10.4	11.4	12.0	12.2	4.9	21.0
Beef and veal[a]	5.6	5.7	6.1	6.1	5.7	0.7	2.4
Meal							
Soybean meal	27.8	27.6	30.8	31.7	35.8	6.7	28.9
Rapeseed meal	8.3	7.5	7.1	8.3	9.1	3.0	9.8
Other meal	8.7	8.8	9.3	9.6	8.9	0.5	1.7
All meal	44.8	44.0	47.3	49.6	53.8	4.8	20.1
Dairy							
Fluid milk	28.6	33.0	36.3	35.4	29.6	1.6	3.5
Dry milk (nonfat and whole)	1.1	1.2	1.2	1.1	1.3	4.8	18.1
Edible oils							
Soybean oil	7.6	8.7	9.7	9.5	10.4	8.4	37.0
Palm oil	5.0	5.1	5.2	5.6	6.4	6.7	29.3
Rapeseed oil	4.5	4.3	4.1	4.9	5.3	4.1	15.7
Other	4.4	4.4	4.3	4.8	4.6	1.6	5.9
All edible oils	21.5	22.6	23.3	24.7	26.7	5.6	24.4
Cotton[b]	43.5	48.0	48.5	41.8	45.0	1.3	3.4

Source: USDA, FAS, PSD Online (accessed November 2, 2010).

[a] Carcass weight equivalent.

[b] 1,000 480-lb. bales.

Caloric Intake

The long-term trends in Chinese food consumption and nutritional status can be identified by looking at changes in the number of per capita daily calories consumed, as well as changes in the composition of the food from which those calories are derived. Since the beginning of the reform period in the late 1970s, Chinese caloric intake has increased significantly. Between 1977 and 2007, the daily intake of calories rose over 50 percent, with more than half of this increase occurring between 1977 and 1987.[17] Since then, China's caloric intake has continued to grow, but at a much slower rate, increasing only 5 percent between 1997 and 2007. On average, the Chinese consumed about 2,981 calories per capita per day in 2007. Although this was about four-fifths the level in the United States (3,748 calories per capita), China's daily caloric intake is above the global average and third among Asian countries, after South Korea and Kazakhstan; it is above that of India (2,352 calories per capita) but less than that of Brazil (3,113 calories per capita).[18]

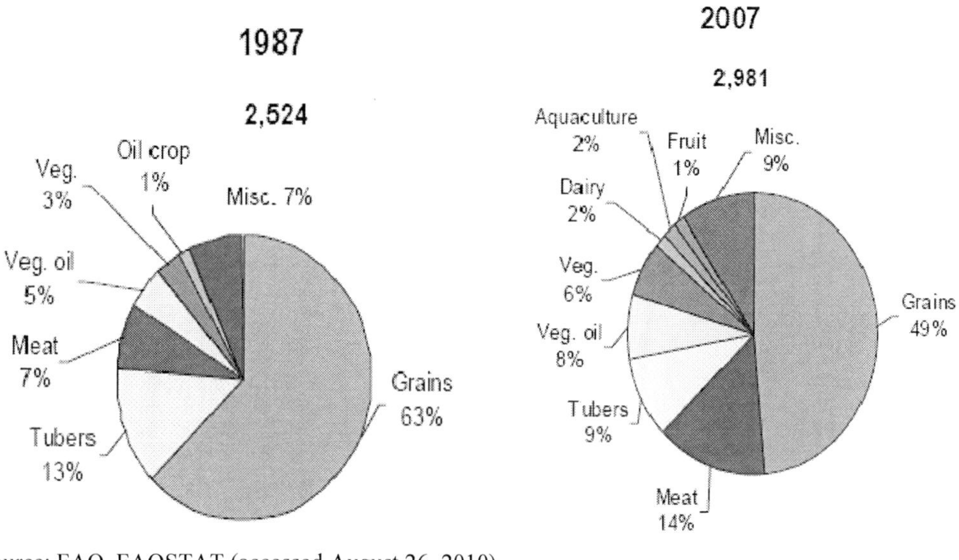

Source: FAO, FAOSTAT (accessed August 26, 2010).

Figure 3.1. Compared with 1987, Chinese daily caloric intake in 2007 had grown significantly and comprised more meat, fruits, and vegetables.

Over the past 20 years, the sources of calories in the Chinese diet have changed. In particular, the portion of calories derived from traditional staple foods, such as grains and tubers, has fallen, while the portion from traditionally nonstaple food products, such as meats, fruit, and dairy products, has increased (figure 3.1).[19] For example, calories from vegetables, meat, and fruit consumption increased by 108 percent, 122 percent, and 302 percent respectively between 1987 and 2007. Currently, grains and meats constitute the core of the Chinese diet, accounting for about two-thirds of calories consumed. The caloric importance of vegetables is growing both in absolute terms and as a share of overall caloric intake. Fruit remains a proportionally small part of the Chinese diet, even though calories derived from fruit consumption have increased over time.

Factors Affecting Consumption

Population

China is the world's most populous country, comprising one-fifth of the global population. Between 2005 and 2009, China's population grew by an estimated 3 percent (an annual average growth rate of 0.64 percent),[20] even though the population growth rate has been steadily declining since the 1990s (figure 3.2). China's population is projected to increase from its current 1.3 billion to 1.5 billion by 2030 and then decline.[21] The projected drop in population is primarily due to China's birth control policies limiting births, in principle, to one child per couple, although in rural areas two or three children households are not uncommon.[22] Because of such policies, China's population is older than that of other large developing countries and will continue to age.[23] For example, China's population has a median age of 34 years, compared with 29 in Brazil and 26 in India.[24] In 2008, the portion of China's population that was under 15 years old was three times higher than the portion over 65, but by 2030 these groups should be roughly equal in size.[25] In the future, as the Chinese population ages and declines, the growth in food consumption likely will decline as well. Nonetheless, given its massive population, China will remain a major global consumer of agricultural products.

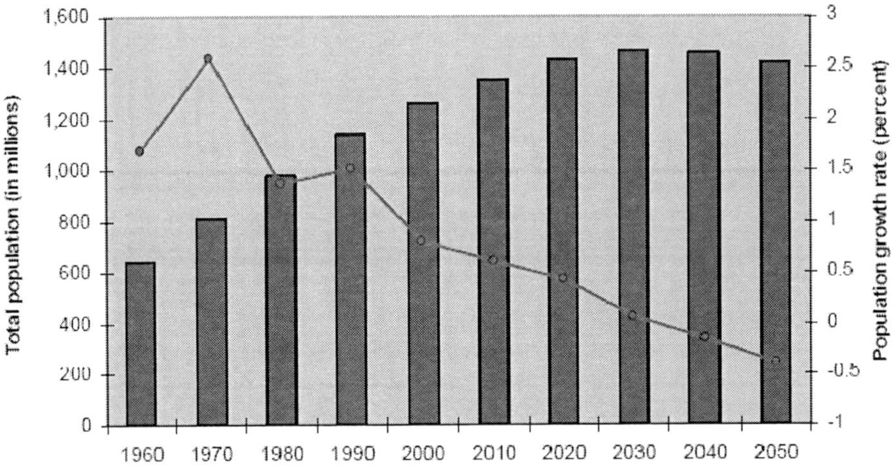

Source: OECD, OECD Factbook 2010: Economic, Environmental, and Social Statistics , 2010.

Figure 3.2. China's population is growing while its growth rate falls.

Urbanization

Food consumption patterns are also affected by the increased urbanization of China's population. During the 1960s and 1970s, just over 80 percent of China's population lived in rural areas (figure 3.3). However, beginning in the late 1970s the rural population began to decline as people moved to find employment in the cities, and by 2008 only 57 percent of the population was living in rural areas. Rural-urban migration continues to occur; between 2004 and 2008 the rural population fell by about 1 percent annually, while the urban population rose by 3 percent annually.[26] Today only 40 percent of Chinese people are rural farmers, whereas 20 years ago two-thirds were rural and produced a large portion of their own food.[27]

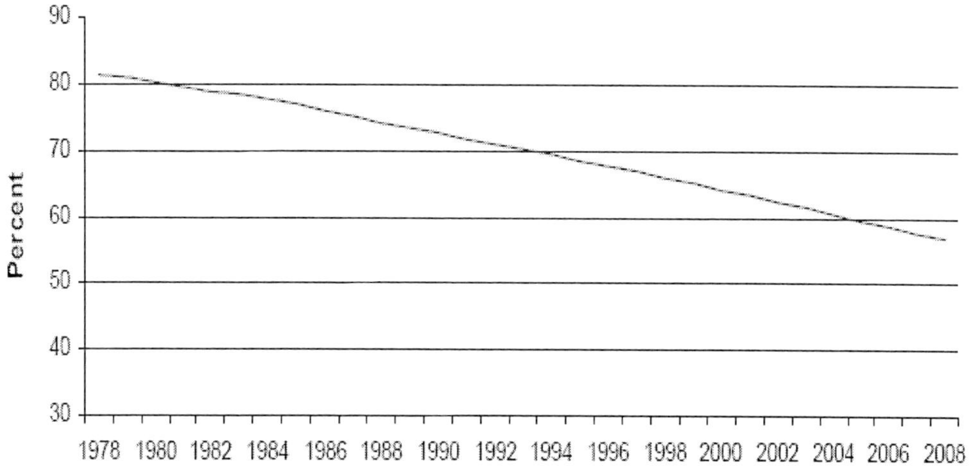

Source: The World Bank, Data: China (accessed July 26, 2010).

Figure 3.3. Rural population declined as a share of Chinese total population during 1978–2008.

Table 3.2. China: Per capita consumption of select commodities and population in rural and urban areas, 2004–08

Commodity	Area	2004	2005	2006	2007	2008
		Kilograms				
Grain[a]	Rural	219	209	206	199	199
	Urban	78	77	76	78	n/a
Vegetables	Rural	107	102	100	99	100
	Urban	122	119	118	118	123
Meat[b]	Rural	([c])	([c])	36	32	31
	Urban	31	33	32	32	31
Fruits[d]	Rural	17	17	19	19	19
	Urban	53	57	60	59	54
Edible oils	Rural	5	6	6	6	6
	Urban	9	9	9	10	10
Fresh eggs	Rural	5	5	5	5	5
	Urban	10	10	10	10	11
		Millions				
Population	Rural	784	777	770	762	754
	Urban	512	527	541	556	571

Source: USDA, ERS, *China Agricultural and Economic Data* (accessed date November 2, 2010).

[a] Grains includes tubers.

[b] Meat includes pork, beef, mutton, and poultry. [c]Data not available.

[d] Includes fruits and melons.

Urban and rural Chinese show different food consumption patterns, though the gap is narrowing. Urban Chinese consume more food overall, especially vegetables, fruit, and edible oils, but eat fewer grains than rural dwellers do (table 3.2).[28] Urban dwellers also consume significantly more fruit and fresh eggs than rural dwellers. In addition, they are more likely to buy chilled, frozen, or perishable foods than their rural counterparts, since they are more likely to own refrigerators.[29] Although urban per capita consumption remained fairly stable in terms of volume, urban consumers were purchasing foods of increasingly higher price and quality.[30] Meanwhile rural residents consumed less grain and vegetables, but slightly more fruit, during 2004–08.

Workers who migrate to cities normally adopt urban consumption patterns[31] and are exposed to certain foods that are largely unique to urban areas, such as snacks and processed foods.[32] Future rural-urban migration is therefore expected to continue to be a major cause of the changing pattern of Chinese consumption.[33] More information on migration can be found in box 3.1.

Income and Expenditure

Income Growth and Distribution

Historically, China has been a low-income country. In 1981, near the beginning of the reform period, 98 percent of the population lived on less than $2 a day, and 84 percent of the population on less than $1.25.[34] However, rising incomes over the past three decades have significantly lowered poverty levels. For example, between 1979 and 2002 income growth drew about 400 million people out of poverty,[35] and by 2005 only 36 percent of Chinese were living on less than $2 a day and 16 percent on less than $1.25.[36] More recently, even in the midst of a global recession, per capita incomes have continued to grow, rising 8 percent between 2008 and 2009.[37] In 2009, the average per capita gross national income (GNI) was $3,590, more than double the level in 2005.[38] Some Chinese earning well below this level receive a minimum living allowance (box 3.2).

BOX 3.1.

China's "Hukou" System Affects Food Consumption Patterns

Migration from rural to urban areas is changing food consumption patterns in China. This urbanization, in turn, is affected in numerous ways by the Chinese hukou system. This system, which is a way of assigning public services to citizens, was established in the 1950s as a household registration system that made public service distribution contingent on occupation and residency.

Urban dwellers received more benefits, such as education and healthcare, while rural citizens often received either land or a small allowance.

As urban centers grew, however, so did the population of rural migrants looking for work in cities. Starting at approximately 30 million in 1989, there are now an estimated 130 to 250 million Chinese migrants, accounting for 40 percent of the urban labor force.[a] Because the migrant workers are classified by hukou as rural, they are not covered by the same legal rights and protections as urban workers. Over 80 percent work seven days per week, fewer than 5 percent are covered by a pension, and only 21 percent are covered by employment contracts. They also work for minimal wages, which are sometimes not paid because employers know that the workers have no legal recourse. Migrants are without health benefits, and their children are also classified as rural dwellers no matter where they were born, meaning they have no right to an urban education.

Rural-urban migration alters consumption in the countryside in several ways. Migrant workers tend to adopt more urban consumption patterns while they are working in the cities and return home with the awareness of new foods and altered food preferences. Remittances from city-dwelling family members boost rural families' incomes, giving them the ability to increase food purchases. Further, over time, the decrease in available farm labor due to migration means the farm sector relies more on technology, causing an increasingly commercialized farm sector to emerge. This more agriculturally productive countryside generates more income, which, coupled with the recently acquired tastes of returned migrants, leads to changes in rural consumption.[b]

While the Chinese government is trying to institute reforms to the hukou system, including extending some urban services to rural areas, migrant workers still have little protection in cities. The prohibitive cost of offering education and health benefits to all urban dwellers keeps migrant workers out of the system. However, if the entire system were to be overhauled, migrant workers and their children would have access to all the same public services as their recognized urban counterparts; being legally covered by an enforceable contract, they would also receive higher incomes. This rise in public service coverage and income would provide incentives for increased urbanization, as well as lead to higher and more varied food consumption in both rural and urban areas.

a Scheineson, "China's Internal Migrants," May 14, 2009; industry officials, interviews by Commission staff, Beijing and Shandong province, China, September 6–16, 2010.
b Shapouri and Rosen, *Developing Countries*, June 2009, 36-37.

Income is distributed unevenly between urban and rural areas. In 2009, the average per capita disposable income in urban areas was $2,515 (8.8 percent higher than 2008) compared with just $754 (8.2 percent higher than 2008) in rural areas.[39] It is estimated that about 20 percent of the urban population is upper middle income (RMB 40,001– 100,000, or about $5,330–$13,326 per year[40]), and this figure could rise to 30 percent by 2015.[41] Although incomes of both rural and urban populations are rising, the gap between the two is widening. For example, between 2000 and 2006, rural incomes rose 59 percent, while urban incomes increased 87 percent.[42]

Per capita income is also unevenly distributed among the provinces. Generally, the coastal areas are more affluent than the interior of China.[43] In a 2008 survey of a selected group of workers, 7 out of the top 10 wage-earning provinces/municipalities were in coastal areas, [44] with Shanghai and Beijing having the highest per capita wages ($8,142 and $8,107 per annum, respectively), more than double the level of the majority of provinces. [45] Higher urban incomes play a role in these provincial income discrepancies, as most of the top wage-earning provinces or municipalities contain a major Chinese city. These high-income provinces contain a concentration of affluent Chinese consumers who are demanding a greater variety of food.[46]

Expenditure

Food is the largest class of household expenditure for all Chinese income groups; even housing takes a smaller share of average household income.[47] As income rises, the absolute amount of food expenditure increases, although the share of income spent on food falls (table

3.3). Urban residents spend substantially more on food than their rural counterparts. For example, urban lower-middle-income consumers spent RMB 3,640 ($533) on food annually, while their rural counterparts spent only RMB 1,317 ($193).[48] Higher incomes lead to an increase in both the quantity and quality of food demanded. However, while demand for higher quantities of food appears to level off in the top-income households, demand for higher-quality foods continues to rise with income.[49]

BOX 3.2. Urban Minimum Living Allowance Payments Increase Income for Poor Consumers

In 2006, the central government spent RMB 24.1 billion ($3.2 billion) on urban minimum living allowance payments.[a] The subsidy covers about 22 million urban Chinese, and these payments are designed to at least partially counteract increasing prices for basic food items. Local governments are permitted by the central government to increase urban minimum living allowance payments to their residents if they can afford it, but the central government requires that the basic payment be paid to eligible citizens.

The minimum living subsidy for each person varies in different cities. In Beijing, the unemployed, those unable to work, and elderly people who are financially dependent get RMB 330 ($44) a month. If a family's income is below RMB 330, they receive a subsidy to ensure their monthly income equals the minimum allowance.[b]

[a] United Nations, Food Prices Issues in the People's Republic of China, August 2008, 15.
[b] China Daily, "Urban Minimum Living Subsidy Increased," August 7, 2007.

Spending on food consumed outside the home is on the rise. In 2003, about 18 percent of urban household food expenditures and over 11 percent of rural household food expenditures were made outside the home.[50] In 2008, the average per capita annual expenditure on dining out was $127 among urban residents, up 26 percent from a year earlier.[51] Per capita expenditures on food consumed away from home vary among regions, with Shanghai spending the most ($300) and Tibet the least ($84).[52] Most such expenditures are made in restaurants, both independent establishments and fast-food chains like KFC and McDonald's. Although consumption away from the household is increasing, most foods are still eaten at home.[53] The exception is meat, with about half of all meat consumed outside the home.[54]

Food Prices

Along with increased income, changes in Chinese consumption trends are also spurred by shifting food prices. Through the 1980s and 1990s, after the reform period, the central government relinquished some control over food prices, and the price of grains rose sharply. This increase led urban consumers to shift their diets away from grains toward protein sources such as eggs for which price increases had been much more moderate.[55] Although pork is the most popular meat in China, chicken consumption has increased because of the efficiency of poultry farmers in producing a low-cost protein, as well as the fluctuating supply of pigs.[56] As mentioned, pork prices skyrocketed in 2007 in the face of lower supplies following an outbreak of a swine disease. In response, the government drastically increased pig farming subsidies, causing some farmers to shift into pig production. As a consequence, pork soon flooded the market and its price drastically fell, leaving pig farmers with little profit. Consumption of pork moved inversely with price, as consumers, especially low-income consumers, purchased 1 to 3 kilograms per capita less when prices were high.[57] This example illustrates the price sensitivity of Chinese consumers in making their food-purchasing decisions.

Table 3.3. China: Total expenditure[a] and share of expenditure on food by urban and rural populations, 2005–09

Income category	2005	2006	2007	2008	2009	Share of income spent on food
			Dollars[b]			5-year percent
			Urban			
Poor	324	370	453	556	623	48
Lowest income	380	429	531	652	717	47
Low income	524	598	741	892	987	44
Lower middle income	680	766	937	1,151	1,279	42
Middle income	892	992	1,196	1,489	1,656	39
Upper middle income	1,149	1,282	1,521	1,917	2,191	36
High income	1,477	1,652	2,011	2,575	2,820	34
Highest income	2,338	2,642	3,068	3,884	4,146	28
			Rural			
Poor	([c])	([c])	([c])	([c])	([c])	([c])
Lowest income	([c])	([c])	([c])	([c])	([c])	([c])
Low income	189	204	243	309	345	50
Lower middle income	233	256	310	382	420	48
Middle income	284	322	386	473	519	46
Upper middle income	351	405	484	603	672	43
High income	561	662	788	986	1,096	37
Highest income	([c])	([c])	([c])	([c])	([c])	([c])

Source: ISI Emerging Markets, CEIC database.

[a] Among other things, total expenditure includes spending on food, clothing, housing, medicine.

[b] Converted from renminbi to dollars based on the average-annual exchange rate. U.S. Federal Reserve System, Foreign Exchange Rates (Annual), January 2, 2008 and January 4, 2010.

[c] Not available.

Since food makes up a relatively large share of consumer spending in China, food price inflation has a significant impact on overall inflation. According to official data, food prices in December 2010 were up 9.6 percent, and consumer prices were up 4.6 percent, from a year earlier.[58] Official data on food prices may understate the impact on Chinese consumers. A survey in November 2010 reportedly found that average wholesale prices for a basket of 18 vegetables in 36 cities had increased 62.4 percent over the previous year. Factors contributing to recent food price inflation include poor weather that lowered the supply of some commodities, rising demand, increasing agricultural wages, and an expanded money supply.[59]

The government's efforts to curb inflation have included an increase in bank reserve requirements and interest rates and the elimination of road tolls on trucks carrying produce starting December 1, 2010. Government statements have targeted markets for cotton, grains, oils, and sugar, and raised the possibility of price controls on "important daily necessities."[60]

Food Preferences

Changing Preferences

Like that of other countries at similar stages of development, the traditional Chinese diet comprises mostly grains and other starches. As mentioned, higher incomes are allowing consumer preferences to change, particularly among the urban dwellers.[61] Consumption of nonstaple, higher-value foods such as meat (especially pork), dairy, fruits, vegetables, and processed food has grown significantly in the past three decades; in fact, 30 percent of the food currently consumed in China has been processed in some way.[62]

China's per capita expenditures for animal proteins[63] for 2008 averaged $184, up from $137 in the previous year. The Chinese consume about four times as much pork as poultry, the second most popular animal protein. Pork consumption has been encouraged by improved cold storage distribution, as the product can be transported greater distances to reach more customers.[64] Pork consumption levels are also high due to government support programs, including purchasing pork for reserves and occasionally subsidizing pork purchases for low-income consumers.

Chinese pork consumption is almost equal to that of China's more economically developed neighbor, Taiwan. Taiwan is a possible indicator of Chinese growth potential, due to its similar geography, preferences, and culture. Further potential expansion in Chinese meat consumption could come from poultry, given that per capita poultry consumption in Taiwan (26.16 kilograms) is almost three times the level in China (9.1 kilograms).

Dairy consumption also rises with income and is helped by government programs and wider access.[65] School milk programs have increased milk consumption by children, and at the same time the presence of large supermarkets and fast-food chains promote the availability of new dairy products.[66] From 2000 to 2005, fluid milk consumption grew 180 percent.[67]

Even the way in which traditional products, like grains, are consumed is changing. For example, wheat is being consumed less in traditional foods such as noodles and steamed bread and more in the form of Western products such as bread, cake, and processed food.[68] Moreover, growing meat consumption is resulting in increased consumption of oilseed meals and corn, both of which make up animal feed.

Demand for Food Safety

Food quality and safety are important factors affecting Chinese food preferences.[69] High-income urban groups that focus their expenditure on high-quality products also seek assurance that their food is safe.[70] Since late 2008, publicity about food poisonings has spurred increased demand for organic and safe products. Safety concerns can determine where certain foods are bought: fresh produce is usually purchased at a wet market because fresher produce is perceived to be safer, while meats are increasingly bought at a supermarket because of the availability of cold storage.[71] Some commodities have acquired a negative reputation following a health scare, leading Chinese consumers to distrust those products for several years afterwards. This is currently the case for domestically produced dairy products after melamine, an industrial chemical, was first found in Chinese milk powders in 2008.[72] Lacking a strong food safety system, Chinese consumers often look to prices as a proxy for food quality and safety, and 52 percent of consumers believe that low prices signify poor quality, compared to 16 percent of consumers in the United States.[73]

Chinese Shopping Preferences

In general, Chinese consumers prefer diversity in their shopping experience. Urban consumers tend to shop at multiple, heterogeneous venues for different products, ranging from local traditional wet markets to supermarkets (figures 3.4 and 3.5). Rural consumers still primarily shop at wet markets because supermarkets are not widely available outside cities. Reportedly, only one-half of Chinese consumers view themselves as a loyal customer of at least one store, splitting their purchases between supermarkets, wet markets, convenience stores, and roadside vendors. The Chinese have an appetite for branded foods and product diversity; the majority of Chinese report wanting to see brand- name products in their shopping outlets even if they cannot yet afford them, saying they would purchase them if they had more income.[74] Chinese consumers cite shopping as a favored leisure activity and devote much more time to it; they shop for 9.8 hours per week on average, compared to 7.2 in BRIC countries[75] and only 3.6 hours per week in the United States.[76]

Non-Food Consumption

Consumption of non-food agricultural items, including cotton and hides, is also increasing. China is the world's largest producer, importer, and consumer of cotton.[77] Most of the cotton is used as an input for finished products, such as textiles, apparel, and home furnishings.[78] While cotton consumption fell following the global economic crisis, by May 2010, China's total textile and apparel exports were growing again, up 19 percent from the previous year.[79] With the gradual recovery of the world economy and the return of domestic and global demand for apparel, consumption of both imported and domestic cotton is likely to increase.[80] Southern China, the country's largest manufacturing region for leather goods, has already experienced a turnaround of the economy and is expected to use more leather, especially bovine leather, and hides for products such as shoes, handbags, and luggage.[81]

Source: Commission staff.

Poultry, typically unrefrigerated, is cut to order in this traditional Chinese wet market.

Figure 3.4. Poultry for sale at a traditional Chinese wet market.

Source: Commission staff.

Unlike at Chinese wet markets, at supermarkets poultry and other meats are sold precut and kept chilled. Catering to consumer preferences, meat is usually unpackaged so that customers may select the pieces they want.

Figure 3.5. Poultry for sale at a modern Chinese supermarket.

Bibliography

Barboza, David. "Virus Spreading Alarm and Pig Disease in China." *New York Times*, August 16, 2007.

Bloomberg Businessweek. "China May Impose Temporary Price Controls as Inflation Surges," November 17, 2010. http://www.businessweek.com/news/2010-11-17/china-may-imposetemporary-price-controls-as-inflation

Central Intelligence Agency. *CIA World Factbook: Brazil,* updated August 19, 2010. https://www.cia.gov/library/publications/the-world-factbook.

———. *CIA World Factbook: China*, updated April 26, and May 20, 2010. https://www.cia.gov/library/publications/the-world-factbook.

———. *CIA World Factbook: India*, updated August 19, 2010. https://www.cia.gov/library/publications/the-world-factbook.

Chan, Wai-Chan, and Anne Tse. *The Consumer Trap: Retailers Need to Adapt to Entice Fickle Chinese Shoppers into Their Stores.* McKinsey & Company, May 12, 2007.

China Daily. "China's Urban, Rural Income Gap Widens," January 22, 2010.

———. "Urban Minimum Living Subsidy Increased," August 7, 2007. http://www.chinadaily.com.cn/china/2007-08/07/content_5449094.htm (accessed September 14, 2010).

Dean, Jason. "China Says it Can Subdue Prices." *Wall Street Journal Online, Asia News*, December 26, 2010. http://online.wsj.com/article/SB10001424052970203568004576043500174732200.html.

The Economist. "China Policy: Rethinking China's One-Child Policy." August 25, 2010.

Economist Intelligence Unit (EIU). *China: Food, Beverages, and Tobacco: Sub-sector Update.* Economist Intelligence Unit, January 5, 2009.

———. *China: Food, Beverages, and Tobacco: Sub-sector Update.* Economist Intelligence Unit, July 10, 2009.

———. *Country Profile 2007: China.* Economist Intelligence Unit, 2007.

European Commission. Directorate General for Agriculture and Rural Development. "China: Out of the Dragon's Den?" Monitoring Agri-trade Policy (MAP) no. 01-08, May 2008.

Federal Reserve System. "Foreign Exchange Rates (Annual)," January 2, 2008 and January 4, 2010. http://www.federalreserve.gov/releases/g5a/20080102/; http://www.federalreserve.gov/releases/g5a/current.

Food and Agriculture Organization of the United Nations (FAO). FAOSTAT Database. http://faostat.fao.org (accessed various dates).

Foster, Peter. "China Feels the Bite of Food Inflation." *The Telegraph*, November 22, 2010. http://www.telegraph.co.uk/finance/china-business/8149640/China-feels-the-bite-of-foodinflation.html.

Fuller, Frank, John Beghin, and Scott Rozelle. "Consumption of Dairy Products in Urban China: Results from Beijing, Shanghai, and Guangzhou." November 29, 2006. http://iisdb.stanford.edu/pubs/21640/dairydemand3citiesfinalmanuscript.pdf.

Gale, H. Frederick. "China's Growing Affluence: How Food Markets are Responding." *Amber Waves*, June 2003. http://www.ers.usda.gov/AmberWaves/June03/Features/Chinas GrowingAffluence.htm.

Gale, Fred, and Shida Henneberry. "Markets Adapt to China's Changing Diet." *Choices*, 2nd quarter 2009. http://www.choicesmagazine.org/magazine/article.php?article=75.

Gifford, Rob. "Food Fears Persist in China 2 Years After Milk Scare." *National Public Radio*, October 26, 2010.

Grant, Andrew, and Ian St-Maurice. "Winning the Hearts of Mainland Consumers." *South China Morning Post*, January 28, 2008. cited by McKinsey & Company. http://www.mckinsey.com/locations/greaterchina/mckonchina/industries s.aspx.

Hesketh, Therese, and Zhu Wei Xing. "The Effect of China's One-Child Family Policy after 25 Years." *The New England Journal of Medicine.* 353; 11. September 15, 2005.

ISI Emerging Markets. CEIC Database. http://www.ceicdata.com (accessed January 25, 2011).

Keidel, Albert. "The Causes and Impact of Chinese Regional Inequalities in Income and Well-Being." Carnegie Endowment for International Peace, December 2007. http://carnegieendowment.org/files/Keidel2.pdf.

Latner, Kevin. "Who Is Feeding China?" Paper presented at the Agricultural Outlook Forum. Washington, DC, February 19, 2010.

MacDonald, Stephen, and Sarah Whitley. "Fiber Use for Textiles and China's Cotton Textile Exports." *CWS-08i-01* U.S. Department of Agriculture, Economic Research Service, March 2009.

National Bureau of Statistics of China. *China Statistical Yearbook 2009*. China Statistics Press, 2009.

Organisation for Economic Co-operation and Development. *OECD Factbook 2010: Economic, Enviornmental, and Social Statistics*. http://www.oecd-ilibrary.org /economics/oecdfactbook 18147364 (accessed July 7, 2010).

Roberts, Ivan, and Niel Anders. *Developments in Chinese Agriculture*. Australia: ABARE eReport, July 2005.

Rozelle, Scott. *Vegetables in China: Final Report to the Western Growers Association*. March 2006.

Scheinson, Andrew. *China's Internal Migrants*. Council on Foreign Relations, May 14, 2009. http://www.cfr.org/publication/12943/chinas internal migrants.html.

Shapouri, Shahla, and Stacey Rosen. *Developing Countries Face Urbanization Growth, Food-Security Worries, and Food-Safety Challenges*. Food Security Assessment, 2008–2009. U.S. Department of Agriculture, Economic Research Service, June 2009. http://www.ers.usda.gov/Publications/GFA20/GFA20c.pdf.

Trading Economics. "China Inflation Slows in December, " January 20, 2011. http://www.tradingeconomics.com/Economics/Inflation-CPI

United Nations. *Food Prices Issues in the People's Republic of China*. Occasional Paper, August 2008.

U.S. Department of Agriculture (USDA). Economic Research Service (ERS). China Agricultural and Economic Data. http://www.ers.usda.gov/Data/China (accessed various dates).

———. *Demand for Food Quantity and Quality in China*, by Fred Gale and Kuo Huang. Washington, DC: USDA, January 2007.

U.S. Department of Agriculture (USDA). Foreign Agricultural Service (FAS). *China, Peoples Republic of: Food Price Inflation in 2010*, by Ryan R. Scott and Catherine Barker. GAIN Report No. CH10060, November 2, 2010.

———. *China, Peoples Republic of: Cotton and Products Annual; Annual Report*, by Chanda Beckman and Wu Xinping. GAIN Report no. CH10019, May 1, 2010.

———. *China, Peoples Republic of: Cotton Update*, by Wu Xinping. GAIN Report no. CH10033, June 29, 2010.

———. *China, Peoples Republic of: Guangdong Market for Leather Weathers the Storm, Spells Demand for American Hides and Finished Leather*, by Ursula Chen. GAIN Report no. CH9619, November 16, 2009.

———. *China, Peoples Republic of: Livestock and Products Annual; Annual Report*, by Mike Woolsey and Jianping Zhang. GAIN Report no. CH9069, September 14, 2009.

———. *China, People's Republic of: Oilseeds and Products Annual*, by Mark Petry and Wu Xinping. GAIN Report no. CH10006, February 23, 2010.

———. Production, Supply, and Distribution Online. http://www.fas.usda.gov/psdonline/psdhome.aspx (accessed various dates).

U.S. International Trade Commission (USITC). *Shifts in U.S. Merchandise Trade 2009*. USITC Publication 4179. Washington, DC: USITC, 2010.
World Bank. "Data: China." http://data.worldbank.org (accessed various dates).

4. Chinese Agricultural Production and Policies

Overview

Agriculture is important to the Chinese economy and employment. China's agricultural production accounted for 11 percent of gross domestic product (GDP),[1] and the sector employed just under one-half of the population in 2009. The sector is dominated by millions of farmers with small plots of land, averaging just 1.5 acres per household.[2]

Endowed with all of the world's major climates, China produces a wide range of agricultural products. Despite having limited good-quality cropland and very little water in some areas,[3] China has expanded its agricultural production since the late 1970s such that today China ranks as the leading global producer of many agricultural commodities. It achieved this expansion largely through substantial increases in productivity, a result of both market-based policy reform and the adoption of modern agricultural technology and farming practices. In addition to overall production growth, the composition of Chinese agricultural production has changed over time. Most notable is the increased share in total production of labor-intensive products, such as horticulture, meat, and dairy, and the drop in the relative importance of traditional products, such as grains and tubers.

In spite of agricultural policy reforms, the Chinese government continues to play a central role in the sector. China's support for the farm sector has grown significantly since 2004, and increases in funding for the sector indicate a renewed focus on agriculture and the rural economy by policymakers. To shrink the gap between urban and rural incomes and promote social harmony the major agricultural-related objectives of China's government are self-sufficiency in domestic grain[4] production, raising farmers' incomes, and rural development.

Production

General Production Patterns

China is the third-largest country in the world in area, and about 15 percent of its land is arable.[5] From the beginning of the reform period in 1978 until 2008, China's arable land area increased by about 4 percent annually; in the past decade it has been virtually unchanged.[6] Although the amount of land available for agriculture is being threatened by ongoing urbanization, industrialization, and desertification (box 4.1), so far there has not been a drastic decline in total agricultural land. Areas lost to nonagricultural uses have been offset by farms expanding into marginal areas, and land left fallow by farmers exiting the sector has been brought back into production by the farmers that have remained.[7] Under China's Eleventh Five-Year Plan (2006–10), the government continued its goal of maintaining a minimum of 120 million hectares (ha) in agricultural production in order "to ensure high levels of food self-sufficiency."[8] This goal is being addressed through laws and policies, such as the 2002

China's Rural Land Contracting Law and the Master Land Use Plan of 1997–2010 (box 4.2). Recent estimates of actual planted land range from about 120 to 155 million ha (compared with 164 million ha in the United States).[9]

With limited arable land, the significant growth in China's agricultural production has been achieved largely by making major improvements in productivity, rather than by expanding China's cropland.[10] For example, between 1978 and 2008, the area planted to cotton rose by 18 percent, while production increased by 246 percent; planted area of all oil-bearing crops rose 106 percent, while production increased 466 percent; and even though the area planted to tubers (root crops, such as potatoes and cassavas) fell by 29 percent, production dropped by only 6 percent.[11] For many agricultural products, China produces a large share of total global production (table 4.1).

As noted, two major factors drove the trend toward higher agricultural productivity in China—policy reforms (discussed later in this chapter) and adoption of improved agricultural technology beginning in the late 1970s. The adoption of yield-improving technologies was especially important to raising agricultural productivity. Use of biological technology, such as better seed varieties, contributed significantly to this trend,[12] beginning with the Green Revolution, which reached China in the 1980s. The Green Revolution also brought other advanced technologies such as irrigation and chemicals inputs.[13] Technological advances continued to be made into the 1990s and 2000s. For example, in the late 1990s, as the expansion of orchards slowed, farmers started upgrading them by replanting, grafting, and improving agronomic care.[14] Other yield improvements were made through the use of genetically modified (GM) seeds, which were approved by the Chinese government in the late 1990s. China has currently approved more than 200 GM seed varieties, including seed for cotton, sweet peppers, papaya, and tomatoes. It most recently approved GM seeds for corn and rice in 2009.[15] Since the late 1990s, China has been significantly increasing investments in plant biotechnology, and by the mid-2000s it had one of the largest public investments in plant biotechnology in the world.[16] China is likely to continue to see productivity gains from agricultural technology advancements.

BOX 4.1. China Has Made Relatively Minor Investments in Agricultural Land Abroad

Around the world, relatively wealthy, food-scarce countries are investing in agricultural land in Africa and South America. A 2008 jump in world food prices brought the issue of food security to the forefront, and net food importers have been purchasing farmland to grow crops in order to protect themselves from future high prices and food shortages.[a] Some countries that do not have enough arable land to grow crops are purchasing plots in land-abundant countries and raising food to send home for consumption. While many of these transactions are hidden from the public eye, the International Food Policy Research Institute (IFPRI) estimated in 2008 that 15 to 20 million hectares had been involved in purchases and transactions in recent years.[b] That area equates to the entirety of agricultural land in France.

China has engaged in this market through its "go-out policy,"[c] in response to the limits on its arable land and issues with water conservation, which make it advantageous to purchase fallow land in other countries.[d] Through its well-established relationship with Africa in the Forum on China-Africa Cooperation (FOCAC), investment in both land purchases and research has increased. For example, China has bought 2.8 million hectares to grow palm oil in the Democratic Republic of Congo and has set up at least 11 research stations through Africa with the aim of generating higher yields.[e] In 2009, there were an estimated 1 million Chinese farm laborers working in Africa.[f]

In addition to Africa, China is open to purchasing land in Southeast Asia and South America, and may be interested in harvesting soybeans, bananas, vegetables, and edible oil crops in those areas.[g]

Although China has a large population and growing food demand, it is still a relatively minor player in agricultural land acquisition.[h] Several countries are leading the way, including private companies in India and South Korea, and the governments of the Gulf states. One Indian individual leases over 1,200 square miles of land, an area larger than Rhode Island, just in Ethiopia[i], and a South Korean company attempted to secure a 1.3 million-hectare plot of land in Madagascar in 2008, though political issues blocked this purchase. Arab nations are energy-rich and food-poor, and they are trading oil for farmland.[k] Sudan has now leased a total of 2.5 million hectares to South Korea, Egypt, and the Gulf states.[l]

China may seek out more foreign land purchases if it faces shortages in certain crops because farmers have shifted production to higher-return commodities or due to decreased productivity.

[a] Kugelman and Levenstein, "Land Grab?" 2009, 2.
[b] Ibid., 1.
[c] Spencer, "China Looks Abroad," May 9, 2008.
[d] Bezlova, "Buying Farmland Abroad," May 9, 2008.
[e] Economist, "Buying Farmland Abroad: Outsourcing's Third Wave," May 21, 2009.
[f] Ibid.
[g] Thomson Financial News, "China Considers Buying Land," May 8, 2008.
[h] Spencer, "China Looks Abroad," May 9, 2008.
[i] Srivastava and Sharma, "Indian Co. Leases Land," November 12, 2010.
[j] Kugelman and Levenstein, "Land Grab?" 2009, 1.
[k] Ibid., 2.
[l] Ibid., 1.

BOX 4.2. Chinese Land Use Policies Support Food Security Goals

China's 2002 Rural Land Contracting Law addresses land use in rural areas. It states that rural land in China is owned by cooperatives and is reserved exclusively for agricultural purposes, such as for farming and for residences for farmers, together with essential buildings such as schools, hospitals, and agricultural support facilities. The law stipulates that all land to be used for commercial purposes or for non-farmer residences must first be transferred to the national government.[a]

In recent years, China's planned land use system has been carried out through the Master Land Use Plan of 1997– 2010 and annual land use plans which implement national goals. The first National Master Land Use Plan was introduced in 1997 when the central government became concerned about rapid urban expansion and resulting farmland losses. The Master Land Use Plan sets long-term regulations relating to both the area and the location of agricultural lands in a given region that are allowed to be converted to non-agricultural use. In effect, the plan requires provinces to balance agricultural land conversion to non-agricultural use and to seek new land for agricultural uses (known as the farmland supplement).[b]

China's land use regulations, which prescribe using land quotas and rules preserving basic cropland are considered necessary by the central government to achieve food security. For agricultural land to be converted into nonagricultural use, the conversion must be permitted under the Master Land Use Plan and located outside delineated areas of preserved cropland. However, these regulations are considered very difficult to implement. Without monitoring and accountability, and with significant local government flexibility to consider overall economic development goals, land use mandates can result in arbitrary policy implementations for farmers.[c]

[a] Dickinson, "China Real Estate Laws, Part I," n.d.
[b] Wang, Tao, and Tong, "Trading Land Development Rights," 2009, 3.
[c] Ibid.

Table 4.1. Chinese and U.S. rank in world production for selected agricultural products, 2008/09

Agricultural products	China's rank in world production	China's share of world production	U.S. rank in world production	U.S. share of world production
		Percent		Percent
Apples	1	43	2	6
Cabbages and other brassicas	1	53	9	2
Cucumbers and gherkins	1	64	5	2
Eggplants (aubergines)	1	56	21	(a)
Garlic	1	77	6	1
Onions, dry	1	31	3	5
Pears	1	65	2	4
Pig meat/pork	1	46	3	11
Potatoes	1	18	5	6
Rice, paddy	1	28	12	1
Sweet potatoes	1	77	12	1
Tomatoes	1	26	2	10
Watermelons	1	68	5	2
Wheat	1	16	3	10
Broilers	2	17	1	23
Maize (corn)	2	20	1	37
Cow milk, whole, fresh	3	6	1	15
Sugar cane	3	7	10	2
Beef	4	10	1	20
Soybeans	4	7	1	35

Source: FAO, FAOSTAT (accessed July 27, 2010).
[a] Less than 1 percent.

Recent Production Trends

In addition to the growth in agricultural output and productivity, trends show a gradual shift in the composition of agricultural production. Specifically, there has been a longterm trend away from the production of traditional staple foods, such as grains and tubers, toward increased production of nonstaple foods, such as fruits, dairy, and meat.[17] For example, in 1980, grains accounted for 80 percent of total sown area, while vegetables made up 2 percent and orchards, 1 percent.[18] By 2008, grain-sown area had fallen to 68 percent, while vegetables had risen to 11 percent and orchards to 7 percent. This trend was driven by changing consumer demand as a result of rising incomes and urbanization, and by the higher returns available to farmers from producing nonstaple rather than staple food products. However, government programs increase the returns on grains (compared to what they would earn on the open market), thus making them profitable enough for famers to produce.

During marketing years (MY) [19] 2005/06–09/10, production of most commodities grew (table 4.2).[20] Increases in production were large for horticultural products, especially fresh and processed fruit, which experienced double-digit annual growth during the five-year period. The production of certain commodities declined between 2005/06 and 2009/10, including tubers, peanuts, and soybeans.

Table 4.2. China: National agricultural production, 2005/06–09/10

Commodity	2005/06	2006/07	2007/08	2008/09	2009/10	Annual growth	Growth 2005/06–09/10[a]
Grains	Million metric tons (mt)					Percent	
Corn	139.4	151.6	152.3	165.9	155.0	2.9	11.2
Rice	180.6	182.6	186.0	191.9	([b])	2.0	6.3
Wheat	97.4	108.5	109.3	112.5	115.0	4.3	18.0
All grains	484.0	497.5	501.6	528.7	([b])	3.0	9.2
Oilseeds							
Soybeans	16.4	16.0	14.0	15.5	14.7	(2.3)	(10.1)
Peanuts	14.3	12.9	13.0	14.3	14.7	0.9	2.5
Rapeseed	13.1	11.0	10.6	12.1	13.7	2.0	5.0
Cottonseed	11.1	13.9	14.5	14.4	12.5	3.9	12.7
Vegetable oils							
Soybean	6.1	6.4	7.0	7.3	8.7	9.2	41.1
Rapeseed	4.6	4.1	3.9	4.7	5.2	3.5	11.3
Peanut	2.3	2.0	2.0	2.2	2.2	(0.8)	(4.0)
Cottonseed	1.3	1.6	1.6	1.6	1.5	3.0	10.9
Meals							
Soybean	27.3	28.5	31.3	32.5	38.6	9.2	41.4
Rapeseed	8.2	7.2	6.9	8.3	9.1	3.5	11.3
Cottonseed	4.0	4.7	4.9	4.8	4.4	2.9	10.2
Peanut	2.8	2.6	2.6	2.8	2.8	(0.3)	(2.0)
Horticulture							
Vegetables	564.5	539.5	564.5	592.4	([b])	1.7	4.9
Fruit	161.2	172.4	181.4	192.2	([b])	6.0	19.2
Tubers	34.7	27.0	28.1	29.8	([b])	(4.0)	(14.1)
Citrus fruit	15.9	17.9	20.6	23.3	([b])	13.6	46.4
Beans	21.6	21.0	17.2	20.4	([b])	(0.7)	(5.3)
Fresh fruit							
Apples	20.4	26.1	24.8	29.8	32.0	12.6	56.9

Table 4.2. (Continued)

Commodity	2005/06	2006/07	2007/08	2008/09	2009/10	Annual growth	Growth 2005/06–09/10[a]
Pears	11.3	12.0	12.9	13.5	13.8	5.1	21.9
Tangerines/ mandarins	8.1	9.0	11.0	12.7	13.3	13.5	65.2
Peaches and nectarines	7.5	8.2	9.0	9.6	9.8	6.9	30.6
Bananas	6.5	6.9	7.8	7.8	([b])	6.4	20.2
Oranges	4.5	4.8	5.5	6.0	6.4	9.3	42.7
Grains	Million metric tons (mt)					Percent	
Grapes	4.0	4.6	4.6	5.0	5.6	8.9	40.0
Grapefruit	1.9	2.0	2.2	2.5	2.9	11.7	55.3
Processed fruit							
Apple juice, concentrated	0.5	1.0	1.2	0.6	([b])	18.0	11.1
Peaches, canned[c]	0.2	0.2	0.3	0.4	0.3	13.9	61.8
Raisins	0.1	0.1	0.2	0.1	0.2	10.0	42.9
Pears, canned[c]	0.1	0.1	0.1	0.1	0.1	12.4	54.0
Meat							
Pork[d]	45.6	46.5	42.9	46.2	48.9	2.0	7.3
Poultry (broiler)	10.2	10.4	11.3	11.8	12.1	4.4	18.6
Beef and veal[d]	5.7	5.8	6.1	6.1	5.8	0.5	1.5
Dairy							
Fluid milk	28.6	33.0	36.3	35.5	29.6	1.6	3.4
Whole milk powder	0.9	1.0	1.2	1.1	1.0	2.1	6.4
Nonfat dry milk	0.1	0.1	0.1	0.1	0.1	(2.4)	(10.0)
Other							
Sugar cane	86.6	99.8	113.0	124.2	([b])	12.8	43.3
Sugar beets	7.9	7.5	8.9	10.0	([b])	8.9	27.4
Sugar, centrifugal	9.8	9.4	12.9	15.9	13.3	9.9	35.5
Cotton[e]	28.4	35.5	37.0	36.7	32.0	3.9	12.7
Walnuts, inshell basis	0.4	0.4	0.5	0.5	0.6	9.6	44.3

Sources: USDA, FAS, PSD Online (accessed May 20, 2010, and October 4, 2010); USDA, ERS, China Agricultural and Economic Data: National Data; ISI Emerging Markets, CEIC database.

Note: Parenthesis ( ) indicates a negative number.

[a] Growth rates are for 2005/06–2009/10 or until latest year available. [b]Not available.

[c] Net weight.

[d] Carcass weight equivalent.

[e] 1,000 480 lb. bales.

Grains

During 2005/06–2009/10, Chinese grain production (corn, wheat, and rice) increased 3 percent annually. According to the U.S. Department of Agriculture (USDA), corn production rose because increased plantings and good weather resulted in the highest yields in a decade in 2008/09.[21] Yields were also higher because of increased fertilizer use during this period.[22] In the case of rice, use of GM seeds helped increase yields and production. Even though China did not approve the use of GM rice seed until 2009, it has been planted in a number of provinces since 2004.[23] Of all the grains, the production of wheat increased the most between 2005/06 and 2009/10. Because of better crop management, wheat yields reached a 10-year high in 2008/09;[24] however, in 2009/10, yields dropped for both wheat[25] and corn because of a drought in the northern part of the country.[26]

Recent production trends in grains are heavily influenced by government support programs (see the section of China's agricultural policies later in this chapter).[27] In the Eleventh Five-Year Plan (covering 2006–10),[28] the Chinese government set a goal of maintaining grain production sufficient to supply 95 percent of domestic needs. Consequently, the government supported grain production between 2005/06 and 2009/10, making it profitable enough to entice farmers to produce grains instead of such other products as fruit and vegetables, which generally generate higher returns.[29]

Oilseeds and Products

Chinese soybean production fell from 16.4 million metric tons in 2005/06 to 14.7 million metric tons in 2009/10, a drop of over 10 percent. The decline in production resulted from fewer hectares planted to soybeans.[30] While soybeans enjoy some government support, [31] it is not as extensive as for grain production, suggesting that the government has chosen to allow most domestic soybean demand to be met through imports.[32] By 2009/10 domestic soybeans made up less than 20 percent of total domestic consumption, with the remainder furnished through imports.[33]

Chinese production of soybean oil and meal[34] grew 9 percent annually between 2005/06 and 2009/10 in response to rising demand for animal feed by China's growing livestock sector.[35] This growth could not have occurred without increased imports of soybeans.[36] Currently, the poultry industry uses the largest amount of commercial feed (derived from both corn and soy), but soybean meal is also used in the pork and aquaculture industries.[37] Increased soybean oil production has met China's rapidly rising demand for edible oils, accounting for just under one-half of all Chinese production of these products.[38]

Horticulture

During 2005/06–2009/10, rising per capita income generated increased demand for horticultural products, translating into higher producer prices and production; Chinese production of fruits and vegetables increased more than that of most other commodities, especially citrus fruit and apples, which increased by around 50 percent. Production of citrus fruit began to rise in 2007 because trees planted in an orchard expansion period beginning in 2002 started to yield fruit.[39] At the same time, the quality of citrus fruits improved because of better field management and increased use of pesticides, fertilizer, and other inputs. This was at least partly attributable to the positive influence of food processors, who provided farmers with chemicals and technologies to ensure the quality and safety of fruit they process.[40] Among deciduous fruits, apples had the strongest growth, as farmers responded to high apple prices and increased demand for fruit and juicing apples by increasing apple acreage between 2004 and 2008.[41]

Meats

Pork is the most heavily produced and consumed meat in China. However, production has fluctuated, increasing between 2005/06 and 2006/07, but declining 8 percent in 2007/08 following an outbreak of the deadly porcine reproductive and respiratory syndrome, commonly called blue ear disease. Pork production fell because deaths from blue ear[42] resulted in fewer pigs being slaughtered, and because animals were sent to market early, thereby lowering average slaughter weights. [43] After the outbreak, production rebounded as the government provided price supports and subsidies to farmers wishing to rebuild their sow herds.[44] Additionally, production was bolstered by an increase in the number of larger-scale hog farms (i.e., 50 head or more), because larger producers are able to buy better stock, have better disease management techniques, and ensure more predictable returns by entering into production contracts with slaughter facilities.[45]

The outbreak of blue ear in pigs gave a boost to poultry production, at least in the short run, as Chinese consumers substituted poultry for pork. Broiler production had the highest growth (19 percent) of any meat between 2005/06 and 2006/07, both because of its affordability compared to other meats and because of its popularity with food service firms, small restaurants, and popular quick-serve restaurants like KFC and McDonald's.[46]

Dairy

For most of 2005/06–07/08, milk production grew rapidly in response to several factors: strong dairy demand, especially among wealthy urban consumers;[47] increased foreign direct investment in the dairy industry; and better coordination of supply chains by large retailers.[48] This growth occurred in spite of high feed costs and a lack of cold storage.[49]

Starting in 2008, food safety scandals shook consumer confidence in Chinese dairy products, depressing demand and leading to a drop in domestic production. The first scandal involved illness and infant deaths from milk tainted with melamine, which was used to make infant formula.[50] Incidents of tainted milk continued to occur, including further findings of melamine and other dangerous chemicals in milk between 2009 and 2011.[51] In response to the poor publicity and because of a lack of qualified inspectors, the government closed a number of small milk collection points so it could improve inspection at higher-volume ones. Given that the average Chinese dairy herd is only about five cows, many farmers who lost their local collection points found it uneconomical to travel longer distances to deliver their milk.[52] As a result, many farmers exited commercial dairy production, reducing China's dairy production capacity and triggering a structural market deficit, meaning that domestic production of dairy products will be unable to meet Chinese demand in coming years.[53]

Cotton

Textile industry demand drove changes in Chinese cotton production during the five-year period. Until late 2008, rising domestic and global demand for textiles and apparel increased the demand for cotton by Chinese manufacturers. This boosted prices for cotton compared to alternative crops (such as grains and oilseeds), and farmers responded by increasing production.[54] However, global demand for textiles and apparel fell because of the global economic crisis beginning at the end of 2008,[55] which lowered the demand for cotton. As a result, by 2008/09, farmers' returns from cotton production dropped between 40 and 60 percent, and farmers responded by reducing planted area in 2009/10.[56] This reduction was partially offset by higher yields as Bt cotton became more widely used in many cotton-growing provinces.[57]

China's Agricultural Policies Related to Production

Introduction

China's support for the farm sector has grown significantly since 2004, when central government policymakers decided that rural development would be a key objective of the

Eleventh Five-Year Plan (2006–10).[58] Similar to the decision to enshrine private property rights in the constitution in 2004, encouraging rural development is a significant shift in economic policy for China (box 4.3). In the past, agriculture was taxed in order to fund investments and job creation in industrial sectors and urban areas in China's coastal region. But efforts to promote rural development by shrinking the gap between urban and rural per capita incomes and investing in agricultural infrastructure have caused government funds to flow in the opposite direction—away from fast-growing urban areas and back to the farms.

BOX 4.3. Chinese Property Use Rights Are Changing

Immediately after the Communist Revolution in 1949, peasants were granted full ownership of land as a way to break up large tracts held by wealthy landholders. But as China's Communist system became more entrenched during the 1950s and 1960s, farmers lost individual ownership rights to forced agricultural collectivization. Then, with the advent of the Reform Period, the government implemented the "household production responsibility system." This gave farmers "use rights" (temporary ownership and decision making) over certain pieces of land for short periods, with the goal of encouraging higher yields in food production. [a]

The central government passed a series of laws that further strengthened land use rights for farmers by establishing lease periods.[b] These systems were unevenly implemented by provincial and local governments, reflecting local preferences and party officials' attempts to balance the needs of farmer equality with productivity criteria.

In 2002, China's Rural Land Contracting Law (2002 Land Law) more clearly specified land use rights for farmers and assured the use of 30-year land use contracts.[c]

In 2004, the central government enshrined the concept of private ownership of real property in China's constitution. The constitution now distinguishes between land, which can only be owned by the state, and buildings and fixtures located on land, which can be privately owned. In 2007 China adopted a new Property Law which supersedes, but does not repeal, the 2002 Land Law. The 2007 Property Law clarified, where the constitution did not, that buildings and fixtures on land are owned separately from the land on which they sit.[d]

[a] Promar International, The Chinese Potato Industry in Transition, 2007, 10.
[b] In 1984, lease periods were set at 15 years. In 1993 30-year leases were permitted.
[c] Promar International, The Chinese Potato Industry in Transition, 2007, 10.
[d] Dickinson, "China Real Estate Laws, Part I," n.d.

China's economic growth and large foreign currency reserves give the central government significant capacity to meet its policy objectives. Government programs for agriculture can be largely subdivided into four categories: direct payments, price support programs, agricultural infrastructure, and regulatory reforms (e.g., food safety and standards). With the exception of regulatory reforms, all of these programs are intended to boost farmers' incomes directly or lower the cost of production.

According to the Organisation for Economic Co-operation and Development (OECD), Chinese government[59] support for agriculture is low compared to the support given by developed countries such as the United States, South Korea, Japan, Canada, and the European Union. However, it is in line with that of rapidly growing economies such as Brazil, Mexico, Russia, and South Africa.[60] Moreover, China's share of general services in total agricultural support is very high relative to the support programs in other countries, mostly due to China's large investments in modern research and extension services, food safety agencies, and agricultural price information services.

Understanding China's approach to implementing agricultural policies is important in discerning why government funding of the farm sector varies significantly from province to province and why Chinese agricultural policies often lack transparency. Particularly important is the fact that provincial and local governments have significant autonomy in implementing agricultural programs and allocating funds.[61] Regulations to protect farmer or consumer rights, and to ensure fairness in the marketplace, are often interpreted differently by different provincial and local governments, or simply ignored. These circumstances make it difficult for foreign companies to produce, trade, and sell in China's agricultural sector.[62] Local government autonomy also calls into question whether the programs are effective in achieving central government objectives. The central government has been accused of having too little power to enforce its regulations; this is often observed in a lack of enforcement of water, pollution, and food safety regulations.[63] Table 4.3 summarizes selected Chinese agricultural policies.

China's Current Policy Objectives Related to Agriculture

As noted at the beginning of the chapter, self-sufficiency in grain production, raising farmers' incomes, and rural development have been major policy objectives of China's government in recent years.[64] For three consecutive years, 2004–06, "No. 1 Documents," which are the top-priority documents jointly adopted each year by the Central Committee of the Communist Party and the central government, concentrated on agriculture and rural development. The documents for 2004 focused on "boosting growth in farmers' income," for 2005 on "strengthening comprehensive production capacity of agriculture," and for 2006 on the construction of the "new socialist countryside." All of these policy objectives were incorporated into the Eleventh Five-Year Plan for 2006–10 (box 4.4).

The Chinese government has also retained its long-term self-sufficiency objective for food grains. According to the Mid-Long-Term National Grain Security Plan (2008–20), issued in November 2008, China's agricultural sector will maintain its grain self sufficiency rate at above 95 percent through 2020. The Mid-Long-Term Plan also calls for the annual sown area for grains to be set above 100 million hectares and for an increase in average grain yields from 4.74 tons/hectare in 2008 to 5.25 tons/hectare in 2020.[65]

In the Eleventh Five-Year Plan, a full set of implementing measures were designed to meet each of China's policy objectives in agriculture (box 4.5). Central government funds were budgeted for rural infrastructure (roads, electricity and water supply, water conservancy, communication, rural schools and clinics, and sanitation systems), including 1.2 million kilometers of roads constructed or renovated by 2010. Rural incomes were targeted for expansion by encouraging the development of village and township enterprises, speeding up the migration of rural labor to urban areas (which increases rural incomes as rural labor supplies tighten), and improving capacity for grain production. Specific goals to be achieved by 2010 included grain output of 500 million metric tons (including soybeans).[66] For this purpose, the government was to enhance the direct subsidy policies for grain producers and reinforce the construction of farmland water conservancy, drainage, and irrigation systems.[67] Many of these measures will be described in further detail later in this chapter.

Table 4.3. China: Selected central government funding and regulations for agricultural production

Policy	Policy description	Policy effect/goal	Impact on U.S. exports
Elimination of agricultural taxes	In 2006, a wide range of formal and informal taxes, charges, and fees charged to farmers were eliminated at the national level.	Lowers the cost of production for farmers and increases their incomes.	Chinese farm products are more competitive with U.S. exports to China.
Exemption from the value-added tax (VAT) for farmers	For domestic production, the policy removes a tax of 13 percent on primary agriculture products and 17 percent on processed food.	Farmers benefit from lower taxes and higher demand for their goods. Purchasers benefit through lower prices.	Imports of agricultural goods are still charged the VAT, resulting in a higher tax burden on imports than on similar goods produced in China.
Minimum purchase prices	Sets minimum prices for certain staple crops, primarily grains.	To the extent that Chinese domestic prices are raised by the minimum purchase prices, they encourage farmers to produce higher volumes of those staple crops.	If minimum purchase prices are set above the world price, exports to China from all countries (including the U.S.) may be encouraged, absent trade barriers.
Area payments to grain producers and other direct payments (except input subsidies)	Area payments for grains (wheat, corn, and rice) and other government payments for improved dairy breeds, hog breeding farms, large poultry farms, and insurance subsidies.	Lowers the cost of production for farmers and increases their income. Payments for improved breeds or production efficiency also may encourage higher yields.	Imports are less competitive relative to domestic production of targeted products, and lower prices in China discourage imports.
Farm input subsidies (fuel, fertilizers, electricity, seeds)	These policies each differ in their implementation. See descriptions in the main text of this report.	Lowers the cost of production for farmers and increases their income. Encourages farmers to produce more. Keeps Chinese food prices low.	Imports are less competitive than domestic production, and lower prices in China discourage imports.
Water policies	Surface water: Managed by the Ministry of Water Resources and its local water resource offices, which oversee a network of irrigation districts. Surface water prices are set by local governments using national guidelines. Groundwater: At the national level, there are few water regulations, and local government regulations are weak and not enforced.	Water prices often fail to cover operating costs, leaving little revenue for water infrastructure improvements. Lowers the cost of production for farmers and increases their incomes.	Chinese farm products are more competitive with U.S. imports. Farmer incomes also rise, but farmers buy few imported goods.

Table 4.3. (Continued)

Policy	Policy description	Policy effect/goal	Impact on U.S. exports
Banking reforms and preferential lending practices	Since 2003, the central government has provided cash infusions to insolvent rural banks and credit unions. Local and provincial Chinese officials are also experi-menting with a variety of programs to drive higher production levels in the farm sector. Pilot projects include expanding the number of rural lenders, microlending schemes, and allowing more flexibility in setting interest rates.	Lowers production costs, increases production.	Lowers cost of domestic production relative to U.S. production costs.
Agricultural infrastructure	These policies include funding roads, telecommunications, power systems, and irrigation systems that directly impact agricultural production costs. In addition, they include projects which indirectly improve farm productivity, such as education systems, medical infrastructure, information networks, and agricultural research projects.	Promotes agricultural productivity in an attempt to shrink the income gap between rural and urban Chinese consumers.	Impact on U.S. exports undetermined, particularly in the long term. Lowers cost of domestic production and delivery relative to U.S. agricultural production and delivery, while also boosting income for rural regions of China.
Inspections and Food Safety	2009 Food Safety Law combined and expanded several of the central government's laws together into one comprehensive law.	Establishes and enforces food safety regulations for certain products.	In some cases, implementation places burdensome requirements on U.S. exporters (e.g., labeling); U.S. exports could benefit from uniform, predictable enforcement but also suffer when enforcement is tighter for imports than for domestic Chinese production.
Biofuels (ethanol and biodiesel)	Ethanol producers receive a refund of the VAT, exemption from a 5 percent consumption tax, a profit guarantee per metric ton, preferential supplies of feedstocks, and compensation for transportation or sales losses. Biodiesel producers receive funding for new production and use of alternative feedstocks.	Promotes China's energy independence.	Encourages U.S. exports to China for certain feedstocks.

Policy	Policy description	Policy effect/goal	Impact on U.S. exports
Food reserves	An extensive system of government-funded storage for grains, cotton, edible oils, and pork.	Intended to prevent mass starvation among China's populace and promote price stability.	Limited impact on U.S. exports because reserve volumes are typically domestically produced and because of trade policy mechanisms.
Foreign direct investment (FDI) regulations	2007 Foreign Investment Catalog.	China generally encourages FDI that enhances productive capacity or technology aimed at reducing pollution. Restrictions apply to conventional seed development, distribution and retail sales of agricultural production, oilseed processing, biofuels production, and some beverage production. The catalog prohibits FDI in the development and production of genetically modified plants and animals.	Generally encourages Chinese domestic production and indigenous innovation; discourages U.S. exports and U.S. FDI.

Source: Compiled by Commission staff.

BOX 4.4. The Chinese Government's Eleventh Five-Year Plan Increases Funding in Rural Areas

China's eleventh five-year plan, for the years 2006 through 2010, has the stated objective of creating a more harmonious socialist countryside, which includes advancing the economic welfare of the rural population. Income inequality between urban and rural areas is viewed by the government as a growing problem in Chinese society because of the concern that it will spark social unrest.[a] The Chinese government seeks to raise the welfare of rural provinces, move to a more service-oriented economy, and conserve valuable environmental resources.[b]

The plan contains a number of quantitative benchmarks,[c] several of which pertain to the rural population and agriculture:

Benchmark	2005 level	2010 goal
Efficiency coefficient of irrigation water	0.45	0.5
Coverage of new rural cooperative health system	75.7	80.0
Per capita net income rural households (RMB)	3,255	4,150

Source: World Bank, Information from Mid-Term Evaluation, December 18, 2008.

The stated goal of the Chinese government is to develop the countryside and close the rural-urban income gap while still maintaining 95 percent self-sufficiency in grains.[d] According to a midterm review conducted by the World Bank, China is on track to meet its benchmarks, primarily because of increased funding to rural areas. For example, in 2006, Premier Wen Jiabao promised a 14 percent increase from 2005 funding for rural areas, leading to a total expenditure of RMB 339.7 billion ($45.3 billion) in that year.[e] The Chinese government, via funding to provincial governments, is improving agricultural productivity through increased extension services and technological investment.[f] Over 90 percent of agricultural research funding is publicly financed,[g] accounting for approximately $2.6 billion in 2005.[h] Agricultural machinery per 100 square kilometer of agricultural land increased from 84 in 2004 to 147 in 2007.[i]

In 2005, China's central treasury spent RMB 297.5 billion ($39.6 billion) on agriculture and rural development.

Much of the funding supported rural taxation reform (RMB 66.2 billion or $8.8 billion) and improving rural infrastructure (RMB 98.9 billion or $13.2 billion), while smaller amounts went to compulsory education, direct grain subsidies, and general poverty reduction.[j] By 2008, the Chinese government reported spending nearly $87 billion on rural programs, a 37-percent increase from 2007 and more than double 2005 spending levels.

[a] BBC, "China to Tackle Rural Problems," February 22, 2006.
[b] Fan, "China's Eleventh Five-Year Plan (2006–10)," 2006, 709–710.
[c] World Bank, *Mid-Term Evaluation of China's 11th Five Year Plan*, 2008, 11, 29.
[d] World Bank, *Mid-Term Evaluation of China's 11th Five Year Plan*, 2008, 52, 63; Lohmar et al., *China's Ongoing Agricultural Modernization*, April 2009, 19–20.
[e] Xinhua News Agency, "Wen Hears Farmers' Concerns on 11th Five-Year Plan," March 20, 2006.

[f] World Bank, *Mid-Term Evaluation of China's 11th Five Year Plan*, 2008, 54, 71.
[g] Lohmar et al., *China's Ongoing Agricultural Modernization,* April 2009, 30.
[h] FAO, *How to Feed the World 2050*, October 12–13, 2009, 1.
[i] World Bank, WDI Online Database. The World Bank defines agricultural machinery per 100 sq. km. of agricultural land as the number of wheel and crawler tractors (excluding garden tractors) in use in agriculture at the end of the calendar year.
[i] Xinhua News Agency, "Investment in Farming Increased to $37 Bln," June 27, 2006.

BOX 4.5. China's Governing Structure Affects Agricultural Policy Implementation

China's political system is multilayered. The central government covers the entire country, while subnational governments include 31 provinces, more than 600 hundred cities, over 2,000 counties, nearly 100,000 townships, and nearly 1 million villages. For the most part, every government office in China has been assigned a bureaucratic rank and any territorial level of government (e.g., a province) contains offices with several different ranks. At the national level, the State Council (China's cabinet) is at the apex and commissions, such as the State Planning Commission and the Food Safety Commission, are just below the State Council in rank.

Ministries such as the Ministry of Agriculture or the Ministry of Health are one step below Commissions, and bureaus within ministries are a bureaucratic rank below ministries.[a]

A key feature of the Chinese system is that governmental units are unable to issue orders binding units of the same rank. The result is that ministries cannot issue binding orders on provinces because they hold an equivalent bureaucratic rank. This is operationally quite different from the governmental structure that binds the federal and state governments in the United States. China's system forces participants into consensus to operate effectively.[b]

China's central government has no interagency policy and administrative coordination process such as that in the United States. The result is that licensing paperwork and other administrative requirements are often duplicative. All companies based in China, foreign or domestic, must build relationships with each Chinese government agency that has regulatory authority over the company's operations.

For example, a U.S.-owned food processing company which imports certain food items and also sources domestically in China needs relationships with perhaps dozens of agencies, including the Administration of Quality Supervision, Inspection, and Quarantine, Ministry of Commerce, Ministry of Health, Ministry of Agriculture, Customs, and the State Food and Drug Administration.[c]

In the Chinese political system, it is common for policymakers to put in place implementation procedures and practical enforcement mechanisms only after a regulation officially takes effect. For example, a regulation may require that products be tested at approved facilities before any facilities have been approved by the Chinese government to do such testing.[d] The lack of explicit implementation details, or administrative transparency, can hamper foreign companies more than domestic Chinese companies because foreign firms often lack the relationships that Chinese companies have with government agencies to receive clarifications in a timely way.[e]

[a] Lieberthal, "China's Governing System and Its Impact on Environmental Policy Implementation," 3–4.
[b] Ibid.
[c] Industry official, interview by Commission staff, Shanghai, China, September 13, 2010.
[d] Owen, "Standards in China: Behind the Headlines," 42.
[e] Ibid.

China's Support for Agriculture Relative to other Countries

As noted, China's government support for agriculture is low compared to that of developed countries, such as the United States and European Union, but in line with that of other rapidly growing economies. As measured by the OECD's PSE,[68] the amount of support provided to Chinese farmers was low (and sometimes negative) during the 1990s, but gradually rose to 9 percent in 2007, the latest year for which Chinese data are available.[69] Compared with other countries at a similar level of development, including Brazil, Mexico, Russia, and South Africa, China's support for farmers falls in the middle of the range (table 4.4). China's PSE reflects changes in the central government's policy priorities from grain self-sufficiency and low consumer prices toward a stronger focus on raising farm household incomes.[70]

Government support to China's agricultural sector indicates that Chinese policymakers are placing a renewed emphasis on the rural economy.[71] Indirect support, in the form of general services, is very high relative to similar support programs in other countries, due largely to investments in agricultural infrastructure. General services include modern research and extension services, food safety agencies, and agricultural price information services, most of which provide benefits to producers and consumers throughout the economy. Compared with direct payments to farmers, general services support is less production-distorting to the sector.[72]

Table 4.4. Agricultural producer support estimates for selected countries, 2006–09 (percent of gross farm receipts)

Country	2006	2007	2008	2009
South Korea	59.5	58.3	46.2	51.7
Japan	51.6	46.4	47.5	47.8
EU	29.0	24.1	22.4	23.5
Canada	22.9	18.8	13.0	20.2
Mexico	13.2	12.9	12.0	12.5
Russia	17.5	10.8	([a])	([a])
United States	11.2	10.0	7.9	9.8
China	11.0	8.6	([a])	([a])
Australia	4.5	5.0	4.0	2.7
Brazil	6.1	5.0	([a])	([a])
South Africa	7.5	3.3	([a])	([a])

Sources: 2010 OECD PSE database; 2009 OECD Factbook, 2009, 227. [a]Not available.

Selected Government Programs Promoting Agriculture in China

Elimination of Certain Agricultural Taxes

Until the beginning of the 2000s, Chinese farmers faced a wide range of formal and informal taxes, charges, and fees, which varied among provinces. In 2000, officially recorded agriculture-related taxes included the animal husbandry tax, a tax on special agricultural products, a tax on the use of cultivated land, and a contract tax.[73] In total, the taxation burden shouldered by farmers in 2000 was estimated to be between RMB 180 billion ($24 billion) and RMB 220 billion ($29 billion), which was more than 10 percent of farmers' net annual income.[74] Rural taxation became a major source of dissatisfaction among farmers, and many local governments in China's rural regions faced protests in the late 1990s and early 2000s.[75]

China's rural tax reform, implemented by the central government in 2003, lowered the tax burden on farmers. The reform began by integrating most agricultural taxes, fees, and charges into one tax and then capping the tax at a maximum rate of 8.4 percent of annual agricultural output value.[76] Then, the central government announced in 2004 that the new agricultural tax would be phased out over five years.[77] But as the agricultural tax reform effort gathered steam, the government announced in March 2005 that reforms should be further accelerated, with the goal of phasing out all national farm taxes by 2006. In 2005, 28 provinces exempted farmers from agricultural taxes, and at the beginning of 2006, the central government eliminated all agricultural taxes.[78]

Value-Added Tax Exemption for Agricultural Products

The largest source of revenue for China's central government is the value-added tax (VAT),[79] a tax assessed on the value of products at each transaction point in the production and distribution chain.[80] China's State Council Decree No. 538 (Article 2.2.a) of 2008 sets the VAT rate for food grains and vegetable oils at 13 percent. Article 2.2.e of the decree also states that the VAT rate shall be 13 percent on "other goods prescribed by the State Council," which in principle applies to all agricultural products except certain processed foods taxed at 17 percent.

In fact, farmers selling their products are exempt from the VAT. Article 15 of Decree No. 538 provides a full exemption from payment of the VAT for "self produced agricultural products sold by agricultural producers."[81] Moreover, the Chinese government maintains several other VAT exemptions for agricultural products produced domestically, both to provide financial assistance to farmers and because it is considered impractical for uneducated Chinese farmers to keep track of purchase VAT (the tax charged to purchasers) and sales VAT (the tax charged to sellers).[82]

Exemptions to the VAT for China's agricultural sector fall into four categories. Based on the order they fall in the production chain, they are farm inputs, farm sales, processor-imputed VAT, and processor-exempted products.[83] The farm inputs exempt from the VAT include seeds, pesticides, herbicides, agricultural machinery, and some fertilizers. These inputs may add up to as much as one-third of the cost of the crop to the farmer.[84] Because all products from agricultural producers[85] are sold without paying the VAT, buyers (e.g., food processors) of goods from these producers are able to deduct 13 percent of these inputs' value when calculating the VAT they charge at the next point of sale (processor-imputed VAT). This exemption protects processors, when they source domestically produced food inputs, from paying the VAT that farmers are exempted from paying and also from double payment of the

VAT for certain farm inputs that were not VAT-free.[86] The practical effect of this policy, which allows buyers to deduct the VAT that they did not originally pay, is that domestically produced agricultural products gain a cost advantage over competing imported products.[87]

Sales of domestic feed products are also VAT-free under the "farm sales" rule. These products include mixed feed, compound feed, premix feed, feed concentrate, bran, distiller's dried grains with solubles, oilseed meal (excluding soybean meal), fishmeal, and bone meal. According to this policy, feed mills and oilseed crushers (excluding soybean crushers) are VAT exempt as well. All other grain processors, including flour mills, starch manufacturers, and distillers, are subject to the VAT on their sales. Crushers selling most oil meals, with the exception of soybean meal, are not required to charge the VAT or provide a VAT sales receipt.

Minimum Purchase Price

Minimum purchase prices, also known as floor prices, are applied to farm-level purchases of selected grains in China (table 4.5). The minimum prices were first announced in 2004 for early indica rice and japonica rice at RMB 1,400 ($187) and RMB 1,500 ($200) per metric ton (mt), respectively.[88] Coverage was extended to include middle and late indica rice in 2005. However, as market prices in both years remained largely above the minimum price levels, no government purchases at the floor price were undertaken in 2004, and in 2005 the intervention was limited to early indica rice. In 2006, new minimum prices were announced and the coverage was further extended to include wheat. As wheat market prices slid below the minimum levels, the government designated state-owned warehouses in major wheat-producing provinces to buy wheat at minimum prices. According to China's State Grain Administration, wheat purchases amounted to 41 million mt through September 2006, which accounted for about 40 percent of China's total production in 2006. Intervention purchases of early indica rice were much smaller, staying below 4 million mt, compared with total rice production of about 181 million mt in 2006.[89]

Table 4.5. China: Government floor price for grains, 2007–10

Items	2007	2008	2009	2010	Growth 2009–10	Purchase period
		RMB/mt			Percent	
Rice						
Early indica (unmilled)	1,440	1,540	1,800	1,860	3.3	July–September
Japonica (unmilled)	1,500	1,640	1,900	2,100	10.5	November–February
Wheat						
White wheat	1,440	1,540	1,740	1,800	3.4	May–September
Red wheat	1,380	1,440	1,660	1,720	3.6	May–September
Corn	1,400	1,500	1,500	([a])	([a])	December–April
Corn average floor price						

Source: USDA, FAS, China, Peoples Republic of: Grain and Feed; Annual 2010, March 1, 2010, 17.
Note: Floor prices in dollars range from RMB 1,380 ($184) to RMB 2,100 ($280).
[a] Not available.

In 2009, the provinces covered by the minimum price support program for wheat were Heilongjiang, Jilin, Liaoning, Inner Mongolia, Shandong, Hebei, Henan, Anhui, Jiangsu, Shanxi, Hunan, Hubei, and Jiangxi. These provinces produce approximately 80 percent of China's commercial sales of wheat.[90] During the previous three years, wheat procurement at the floor price averaged about 37 percent of China's total wheat production.[91]

For corn, the purchase program covers the northeastern provinces, including Heilongjiang, Inner Mongolia, Jilin, and Liaoning, in 2009/10. In addition to the floor price of $220.50 per mt, the government encourages end users (mainly feed mills) and state trading companies in southern China provinces to purchase corn from northeastern provinces. For purchases made from December 1, 2009, to April 30, 2010, the central government offered a subsidy of $10.30 per mt to traders or feed mills.[92]

Direct Payments

Since 2004, the Chinese central government has provided significant direct payments to farmers, primarily for growing grains (corn, rice, and wheat).[93] In addition to payments made based on the area used for grain production, seed subsidies are provided for grains (including barley on a trial basis in 2010) and oilseeds (soybeans, peanuts, and rapeseed). Payments to offset rising fertilizer and fuel prices were initiated in 2006, expanding significantly in 2008 and 2009 (table 4.6).[94] By 2009, the scope of government payments had widened to include other farm input costs, more government funds for machinery purchases and seeds, and targeted payments, which often varied by farm size, for breeding and raising dairy cattle and sows.[95] For example, to encourage larger-sized hog farms (50 or more slaughtered annually), the central and local governments made payments of RMB 2.5 billion ($0.3 billion) in 2009 to improve hog-raising conditions.[96]

Table 4.6. China: Selected government direct support programs, 2005–09 (million $)

Year	Area payments (grains only)	Seed subsidy	Machinery subsidy	Fuel/fertilizer subsidy
2005	1,941	574	44	0
2006	2,088	603	88	1,838
2007	2,221	979	294	4,059
2008	2,221	1,775	588	9,382
2009	2,221	([a])	1,471	10,529

Sources: USDA, FAS, China, Grain and Feed Annual 2010, March 1, 2010, 16; USDA, FAS, China, Grain and Feed Annual 2010, March 3, 2009, 13.

Note: In 2007 and 2008, the seed subsidy covered soybeans, rice, wheat, corn, rapeseed, and cotton. In 2009, the seed subsidy was expanded to also cover potatoes. In 2010, it was further expanded to cover hulless barley and peanuts on a trial basis. Exchange rate: $1 = 6.8 RMB.

[a] Not available.

In relation to China's overall agricultural production value, direct payments through 2007 were estimated to be minor; increases in rural incomes were largely attributed to increased product sales and nonfarm income sources.[97] But documented government funding more than doubled in size from 2007 to 2009, and by all accounts, has continued to grow in 2010. Consequently, the impact of these payments on China's farm sector is increasing significantly, and total payments to Chinese farmers from all levels of government are now estimated to be larger than direct national government payments to U.S. farmers.

Area Payments to Grain Producers

Grain producer payments in China were introduced in 2004, usually at the rate of RMB 10 ($1.33) per *mu* of area sown to rice, wheat or corn. In total, RMB 11.6 billion ($1.5 billion) was paid under this program in that year.[98] Total outlays for area payments increased from 2005 through 2007, when coverage became nationwide and payment amounts were capped by the central government at $2.2 billion (table 4.6). The implementation of grain producer payments varies from province to province because central government grants are issued to grain risk funds controlled by provincial authorities, which may supplement grain payments with local funds.[99] Typically, at least half of the grain risk fund is used for direct payments to farmers.

Input Subsidies

Fuel and Fertilizers

In 2006, the central government started a program intended to partially compensate farmers for price increases in fuel, fertilizer, and other agricultural inputs.[100] By 2007, the comprehensive subsidy on fuel and fertilizer for grain farmers totaled RMB 27.6 billion ($3.7 billion). According to the Ministry of Finance, the comprehensive subsidy averaged about $14.50 per farm household in that year. By 2009, fuel and fertilizer subsidies totaled $10.5 billion (table 4.6).

In the case of fertilizers, government support is part of several separate programs targeting fertilizer producers, with cost reductions being passed along to farmers purchasing the input. In December 2003, the government issued an electricity subsidy for small and large nitrogen fertilizer producers totaling RMB 6.3 billion (approximately $0.8 billion) in cost reductions every year. Transportation for fertilizer has been subsidized by a total of RMB 5.0 billion (approximately $0.7 billion) every year since 2003. Since 2004, the Chinese government has given a bank loan subsidy to qualified fertilizer distributors who store fertilizer during the winter season. This totaled $43.9 million to $73.3 million every year.[101]

Seeds

Since 2002, farmers have benefited from subsidies for purchasing improved soybean seeds. In 2004 and 2005, this scheme was extended to cover improved seeds of wheat, maize, and rice. The budget allocations from the Ministry of Agriculture for this program amounted to RMB 2.85 billion ($0.4 billion) in 2004. By 2008, the latest year for which data are available, seed rebate payments had increased to $1.8 billion (table 4.6). Implemented at the provincial level, the program is intended to lower the cost of production and boost farm yields. In some provinces, seed companies are provided the funding, and, in turn, make seed available at discounted prices. In other provinces, farmers are paid directly to purchase seed, so the subsidy is akin to direct income support.[102] According to one estimate, farmers who directly receive government seed funds only pay about 50 percent of their seed costs out of their own pockets.[103]

Farm Machinery

To further its goals of modernizing the farm sector and increasing agricultural productivity, the central government began paying a small subsidy in 2002 for the purchase of farm machinery. The subsidy primarily assisted larger farms with sufficient scale to warrant mechanization[104] and currently covers between 20 and 40 percent of the cost of new machinery, depending on the type of machinery and product being produced.[105] Funds may be paid directly to the farmer, business, or village cooperative which bought the machinery or to the machinery dealer. The largest incentives (as a percent of total purchase cost) are given for machinery used in harvesting and packing. Farm machinery payments have reportedly accounted for significant conversions from non-mechanized to mechanized production and packing in the last year.[106]

Water Policies

According to China's National Water Law, as revised in 2002, all property rights to surface water and groundwater belong to the state, including the right to use, sell, or charge fees.[107] Nevertheless, oversight and management of water use is generally poor. Surface water is managed by the Ministry of Water Resources and its local offices, which oversee a network of irrigation districts. Surface water prices are set by local governments using national guidelines, and these prices often fail to cover operating costs, leaving little revenue for water infrastructure improvements.[108] Farmers typically do not pay for surface water by volume, but instead pay a fixed amount based on the area they irrigate or for the use of a well.[109] China's water management policies provide little incentive to conserve because farmers are charged only a fraction of the water's value.[110]

Many farmers use groundwater reservoirs for irrigation in water-scarce areas of China, particularly in the north. The National Water Law does not allow groundwater extraction if pumping is harmful to the long-term sustainability of groundwater use. But at the national level, there are few water regulations that specifically address groundwater management, and local government regulations are weak and not enforced, in fact, in most regions of China, groundwater resources are almost completely unregulated. According to a 2009 survey, less than 10 percent of well owners in areas of northern China obtained a required well drilling permit before drilling. In addition, water extraction charges were rarely, if ever, imposed on users in these northern China communities, and no volume restrictions were imposed on well owners.[111] However, many agricultural processors, such as companies that process fruits and vegetables, often pay the local town for water by the ton and have seen their water charges increase significantly over the past few years.[112]

Any government policies in China that provide better incentives for users to conserve water, particularly through higher per volume rates, would almost certainly run counter to central government objectives of raising farmers' incomes and furthering rural development. Furthermore, water conservation conflicts with self-sufficiency goals in grain production, because grains require significantly more water per hectare than alternatives such as livestock and horticulture.[113]

Banking Reforms and Preferential Lending Practices

Until the end of the 1990s, preferential loans in agriculture were given mostly to state marketing organizations to fund the purchase and storage of certain agricultural products, primarily grains. In the 2000s, most of these programs were discontinued, and by February

2006, the Agricultural Development Bank of China, the policy bank implementing government programs, announced that commercial rates would also apply to grain marketing enterprises.[114] But since 2003, reform of China's rural financial system has been a priority for the central government, particularly providing cash infusions to insolvent rural banks and credit unions.[115] After the start of the Eleventh Five-Year Plan in 2006, which reinforced policies focused on rural development, preferential rates were applied to loans targeting rural development and poverty alleviation. By the end of 2006, rates for rural development loans were just above half the commercial rates. [116] However, loan funds were often diverted to supplement sub-provincial government budgets or given to industrial enterprises rather than benefiting farmers and the rural poor.[117]

A national survey conducted in 2009 indicated that less than 9 percent of China's farmers had loans from established institutions.[118] Rather than secure loans from these groups, farmers most often rely on loans from relatives or underground lenders at unsubsidized rates.[119] Farmers find it difficult to take advantage of government programs to encourage lending because under China's system of state ownership of land, farmers do not personally own the land on which they farm. Therefore, few assets are available to secure financing for agriculture, and this has restricted the flow of funds for farm investments to enhance productivity.[120]

For certain agricultural products, the central government continues to intervene with bank loan subsidies to assist companies with declining profits or large financial losses. For example, the Ministry of Finance reported subsidies of RMB 190 million ($25 million) to the nation's dairy companies in 2010, more than double the RMB 80 million ($11 million) in subsidies granted to dairy companies in 2009. According to Xinhua news reports, the government objective is to help dairy companies stabilize production and resume their purchases of raw milk. But the Ministry of Finance also noted that the funds encouraged an increase in cattle herds and protected farmers' interests.[121]

Given the constraints imposed by the current land tenure laws, local and provincial Chinese officials are also experimenting with a variety of programs and subsidies to drive consolidation, efficiency, and higher production levels in the farm sector. These pilot projects are expanding the number of rural lenders and allow more flexibility in setting interest rates, as well as village-wide credit evaluations, joint borrowing, and government loan guarantee schemes designed to increase microlending, normally in amounts of less than $1,000. Other village governments are testing arrangements that permit farmers to use their limited land-use rights (e.g., timber rights, fruit orchards, and marketing contracts) as loan collateral.[122]

Agricultural Infrastructure

Investment in agricultural infrastructure is a major budget expenditure item for China's central government to achieve development targets in the rural sector. Agricultural infrastructure includes not only roads, telecommunications, and power and irrigation systems that directly impact agricultural production costs, but also includes goods and services that indirectly improve farm productivity, such as education systems, medical infrastructure, information networks, agricultural research projects, and the development of financial services for the farm sector.[123] Under the Chinese budgetary accounting system, government expenditures for agriculture consist of four major items: direct expenditures on agricultural production, rural relief funds, rural capital construction, and agricultural science and technology promotion.[124] Agricultural infrastructure comprises many projects in the last three categories. In 2005, rural capital construction accounted for 21 percent of the Chinese government's agricultural spending, rural relief funds for 5 percent, and science and technology promotion funds for 1 percent.[125] Spending on agricultural infrastructure totaled RMB 117.5 billion ($15.7 billion) in 2005, increasing to RMB 140.8 billion ($18.8 billion) in 2007, the last year for which OECD data are available.[126]

Large government grants are currently being provided to local government officials for agricultural infrastructure and capital projects. The central government transferred RMB 17 billion ($2 billion) in 2009 to more than 1,000 grain-producing counties in China to promote grain infrastructure and industry development, including storage facilities, distribution networks, and processing upgrades.

The government continues to accept primary responsibility for funding pollution control, land rehabilitation, and the maintenance and development of transport and irrigation infrastructure.[127] A recent example is the October 2010 agreement between China's Ministries of Commerce and Agriculture to work together over the next five years to improve the distribution and marketing chain for agricultural goods. The ministries will fund the construction of large agricultural wholesale markets located between major production areas and transportation hubs, as well as promote business cooperation between large chain enterprises and agricultural regions. The two departments also agreed to improve efforts to standardize farm produce distribution and monitor the quality of vegetables, fruits, and meat to help ensure food safety.[128]

Increased funding for agricultural infrastructure projects indicates a renewed focus by China's central government on encouraging rural development, but analysts find it difficult to assess the actual level of support specifically earmarked for agricultural infrastructure. In particular, it is difficult to verify that all budgetary expenditures from various government bodies are included in the data.[129] Each ministry has its own system of channeling funds from the central government to the village level.[130] Another difficulty is that information on budgetary expenditures is aggregated. In many cases, the scope of a payment category is so broad that it is impossible to separate payments that are made to farmers directly, payments to services provided collectively to the village as general services, and payments to support the development of rural areas in general, including non-agricultural activities.[131]

Inspection Services and the 2009 Food Safety Law

While China has funded food inspection services for at least 20 years, food safety became a higher priority for policymakers after it was widely reported that animal feed, milk powders, and eggs tainted with melamine had killed several infants and made thousands of consumers sick in 2008.[132] Since then, not only has expenditure on inspection services increased, but Chinese government officials at all levels of government have worked to upgrade food safety standards.[133]

A recent effort by the Chinese government to upgrade food safety and standards is the 2009 Food Safety Law, which took effect on June 1, 2009. The law requires stronger food product supervision, safety standards, and penalties against violators, and its scope extends the full length of the production chain to encompass farmers, food processors, marketers, and distributors.[134] The Food Safety Law extends government regulatory authority to controlling the use of pesticides, fertilizers, veterinary drugs, and feed additives,[135] as well as the labeling of ingredients and additives on packaging.[136]

The Food Safety Law seeks to create a safer food production system in China, as well as boost confidence among Chinese consumers about the nation's food supply. At the present time, it is unclear whether implementation of the new law is furthering these goals. Farmers and food companies face efforts to implement regulations from overlapping government agencies, each requiring inspections. The Ministry of Health released a circular in June 2009 to instruct government agencies on how to apply the new law, and implementing regulations for the law were released in July 2009. But other agencies play a role in enforcement. The State Council coordinates national-level enforcement, the Ministry of Agriculture supervises the production of agricultural products, the State Administration for Industry and Commerce supervises the transport and distribution of food products, and the Administration of Quality Supervision, Inspection, and Quarantine supervises imports and exports of food products and regulates food production. Administrative and regulatory overlaps are still being worked out between the agencies.[137]

Products for which production standards are not set by the central government are controlled at the provincial level, in principle to supplement the national law.[138] For example, the Sichuan Provincial Department of Quality Supervision set standards for local pickle products in 2009 because the central government was silent on the issue.[139] Additionally, provincial governments carry out enforcement. In many instances, local authorities require that annual quality supervision reports be completed by food processing facilities and sent to the authorities, but unannounced safety inspections, food testing, and site visits to monitor quality standards are rare.[140]

Food Reserve Policies

Central, provincial, and local governments in China maintain an extensive system of food and fiber reserves for domestically produced grain, cotton, edible oils, and pork.[141] Under the Chinese system, these reserves appear to serve two government goals.[142] The first is a measure of China's food security status, since the reserves serve as a hedge against significant food shortages. Minimum government grain reserves must be at least 25 percent of consumption, and China's cultivated land area for grains is not permitted to fall below 120 million hectares (296.5 million acres).[143]

The second government use for the reserves is ensuring price stability. Inventories are sometimes increased in an attempt to raise or stabilize food prices in the face of global price declines. For example, in 2008–09, the Chinese government became a large-scale purchaser of grains and soybeans to prop up domestic prices as global prices began to fall.[144] Another example is an effort by the central government to boost hog prices in early 2010 by establishing reserves totaling 24,000 mt. However, these reserves are not likely to influence price stability in China to a significant extent. Typically, data on China's food reserves are considered secret government information, so inventory data are unable to transmit price signals to move consumer prices. For example, as grain prices were rising in 2008, few consumers and grain analysts were aware that China held large and rising inventories.[145]

China's Agricultural Laws and Policies for Foreign Direct Investment

Foreign direct investment (FDI) in China is primarily governed by the Foreign Investment Catalog, last updated in 2007. Through the catalog, China classifies industries into categories for which foreign investment is encouraged, restricted, or prohibited. In general, government incentives are available to projects which fall into the "encouraged" not eligible for special incentives but not facing particularly difficult approval requirements.[146]

BOX 4.6. Biofuel Policies May Conflict with Food Security Objectives

China's policies toward agriculture are not only related to food security and raising farmer's incomes, but also to a lesser extent on promoting energy independence and reducing greenhouse gas emissions. Consequently, the central government, through the National Development and Reform Commission (NDRC), has created programs to encourage the domestic production of biofuels (ethanol and biodiesel) as a means to diversify sources of energy. The NDRC strictly regulates both the supply of and demand for biofuels, and only state-owned enterprises are involved in production.

The NDRC plans for domestic biofuel production to increase from 1 million mt in 2005 to 12 million mt in 2020.[a] If these policy goals are met, domestic biofuel production would satisfy up to 15 percent of China's transportation energy needs. Ethanol production alone reached 1.35 million mt in 2007.[b] Four large state-owned ethanol plants were constructed in Heilongjiang, Jilin, Henan, and Anhui provinces in 2001, and by 2009, eight plants with a total capacity of 2.2 million mt had been built.[c] In Guangxi Province, China built another ethanol plant based on cassava, which started operation in early 2008.

Biodiesel production in China is still too small to meet a significant portion of the nation's overall energy demands, but the central government is promoting biodiesel using a variety of feedstocks. By the end of 2007, 10 biodiesel plants were operating in China, most utilizing industrial waste oil and waste cooking oil. The total annual production capacity for all of these plants was less than 200,000 mt, but by 2009 China had a total of some 2.1 million mt of biodiesel-producing capacity.[d] Given the domestic supply constraints on the feedstocks needed for these plants, as well as the continued large-scale imports of vegetable oil for human consumption, China would like to develop forestry-based biodiesel, based on sources such as jatropha seeds and Chinese dogwood nuts.[e]

Ethanol producers benefit from a number of financial incentives, including the refund of VAT, exemption from a 5 percent consumption tax, a profit guarantee of RMB 100 ($13) per mt, preferential supplies of grain stocks, and compensation for losses.[f]

Until May 2006, government subsidies were limited to fuel ethanol, at which time the central government outlined the creation of a special fund to encourage the development of renewable energy resources, including ethanol and biodiesel. More recently, in an effort to move away from grain-based ethanol production and into alternative feedstocks, the government subsidy was reduced from RMB 1,883 per mt in 2005 ($29 per barrel (bbl)), to RMB 1,628 per mt ($25.70 per bbl) in 2006 and RMB 1,373 per mt ($22.7 per bbl) in 2007. In 2008, the grain-based ethanol subsidy was eliminated.[g]

USDA estimates that food security objectives may clash with energy independence and environmental concerns in China, eventually limiting the development of biofuels.[h]

Under such a scenario, biofuels would affect China's agricultural production mix while diverting output away from human and animal consumption into use as a fuel and contributing to increased imports of feedstock sources.

This clash of policy objectives may already be occurring. In 2010, China became a net corn importer, buying more than 1 million mt, the most since 1995–1996.[i] Ethanol in China is produced mostly from corn (80 percent of fuel ethanol production in 2005). The NDRC asserts that the targeted biofuel production will not threaten China's grain security, but feedstock sources may be expanded to include sugar, oilseeds, sweet sorghum, wheat, and cassava.[j]

[a] Kwiecinski and von Tongeren, "Quantitative Evaluation of Agricultural Policy Reforms in China," July 8–9, 2007, 208.
[b] Qiu et al., "Bioethanol Development in China and the Potential Impacts on Its Agricultural Economy," 2010, 77.
[c] O'Kray and Kang, "Biofuels in China," second quarter 2010, 22.
[d] Ibid.
[e] USDA, FAS, *China, Bio-fuels*, August 8, 2010, 21.
[f] Ibid., 11.
[g] O'Kray and Kang, "Biofuels in China," second quarter 2010, 22.
[h] Lohmar et al., *China's Ongoing Agricultural Modernization,* April 2009, 19.
[i] Bloomberg.com, "Japan to Boost Farm Output as China Net Corn Buyer," August 31, 2010.
[j] Kwiecinski and von Tongeren, "Quantitative Evaluation of Agricultural Policy Reforms in China," July 8–9, 2007, 208.

In the agricultural sector, China generally encourages FDI that enhances productive capacity or technology aimed at reducing pollution.[147] Restrictions apply to conventional seed development, distribution and retail sales of agricultural production, oilseed processing, biofuels production, and some beverage production. The catalog prohibits FDI in the development and production of genetically modified plants and animals (table 4.7).[148]

Chinese policies encourage FDI in particular agriculture-related areas, and Chinese officials have emphasized the importance of spreading technology through investment in rural areas.[149] For example, modern retailers, middle class consumers, and export markets are increasingly demanding higher quality and food safety, and the result is that food processors are increasingly demanding high-quality products and assurances of food safety from their suppliers (i.e., farmers). Investments in areas such as higher-quality storage, transportation, and cold chain infrastructure reduce waste through spoilage, maintain the quality of fresh fruits and vegetables for a longer period, and increase distribution options.[150] Foreign investment is certainly one source of the necessary capital for these expensive improvements.

The 2007 changes to China's foreign investment catalog included several changes relevant to agriculture. Five agricultural industries face new restrictions. First, new investment in the development, breeding, and production of new varieties of crops and seeds must be through a joint venture under the control of a Chinese partner. New restrictions also apply to FDI in soybean processing and biofuels production. Restrictions on FDI in carbonated beverage production now apply to all brands, rather than just local brands. Foreign investors are now also restricted to a minority stake in retailers, wholesalers, and other distributors with over 30 outlets that sell grain, cotton, vegetable oil, sugar, medicines, tobacco, pesticides, and chemical fertilizer.

Table 4.7. China: FDI policies in agriculture-related industries, 2007

Sector	Encouraged	Restricted	Prohibited
Agriculture, forestry, animal husbandry, fisheries	• Transformation of low-yield farmland • Cultivation, development, and production of woody material, edible oil, spices and industrial raw materials • Serialization, development, and production of vegetables (including edible mushrooms and melon), tea cultivation technology, and pollution-free products • Technology development and production of sugar, fruit trees, and forage crops • Construction and operation of flowers and nursery production base • Rubber, sisal, and coffee cultivation • Medicine cultivation and aquaculture (limited to joint ventures, cooperation) • Comprehensive utilization of crop straw and organic fertilizer production resources • Wood (bamboo) cultivation improvements and development of new varieties of trees • Aquatic breeding (excluding China's unique varieties) • Combating desertification and soil erosion, e.g., by planting trees and protecting native grass protection • Aquatic products breeding, sea-cage aquaculture, industrial aquaculture, and ecological marine species aquaculture	• Development, breeding and production of new varieties of crops and seeds (Chinese partner must hold controlling share) • Processing of precious species of lumber (limited to joint ventures, cooperatives) • Processing of cotton (unginned)	• Breeding and cultivation of China's rare and unique species (including crop cultivation, animal husbandry, aquaculture) • Development or production of transgenic plant seeds, livestock, poultry, and aquatic breeds
Food processing	• Biological feed, straw, fodder, and the development of aquatic feed production • Aquatic and shellfish products and the development of functional seaweed foods • Storage and processing of vegetables, and livestock products	• Soybeans and rapeseed edible oil (Chinese partner must hold controlling share) and corn deep processing • Production of biological liquid fuel (fuel ethanol, biodiesel) (Chinese partner must hold controlling share)	

Table 4.7. (Continued)

Sector	Encouraged	Restricted	Prohibited
Food industry	• Development and production of functional foods for infants and elderly persons • Development, production, and processing of forest food • Production of natural food additives and food ingredients production (limited to joint ventures, cooperation)		
Beverage manufacturing	• Development and production of fruit and vegetable drinks, protein drinks, tea drinks, coffee beverages, and plant beverages	• Rice wine and famous liquor production (Chinese partner must hold controlling share) • Carbonated beverage production	
Tobacco industry	• Processing of cellulose acetate tow (limited to joint ventures, cooperation) • Tobacco flake production (limited to joint ventures, cooperation)	Leaf tobacco processing and production	
Textile industry	• Use of high-tech industries with special textile producers • Luxury fabrics and finishing • Processing to meet environmental protection requirements of special natural fiber (including the addition of other inputs outside wool, flax, bamboo, silk, cotton, etc.)		
Leather, fur, feathers (cashmere) and its products	• Clean leather- and fur-processing technology • Post finishing leather-processing technologies • Processing of luxury leather (sofa leather, car cushion leather)		
Timber processing, wood, bamboo, rattan, and grass products industry	• Comprehensive utilization of new technologies in wood and bamboo product development and production		

Sector	Encouraged	Restricted	Prohibited
Paper and paper products industry	• The integration of a single forestry paper production line having an annual output of 300,000 tons • Above-scale chemical wood pulp with a single production line having an annual output of 100,000 tons • Above-scale chemical mechanical pulp, and the simultaneous construction of high-grade paper and paperboard production facilities (limited to joint ventures, cooperation)		
Special equipment manufacturing	• Agricultural machinery: agricultural facilities and equipment (greenhouse automatic irrigation equipment, nutrition liquid fertilizer automatic configuration and equipment, highly efficient vegetable nursery equipment, soil nutrient analysis apparatus); tractors and ancillary tools with supporting engine power of 120 kilowatts and above, low fuel consumption, low noise, low diesel emissions, and with a large tractor supporting the residual spray machine; high-performance rice transplanter; cotton picking machine; corn combine • Forestry machinery equipment technology • Crop straw and rice husk utilization equipment • Utilization of livestock and poultry agricultural waste		
Wholesale and retail trade		• Wholesale, retail and distribution of grain, cotton, vegetable oil, sugar, medicines, tobacco, motor vehicles, crude oil, pesticides, plastic sheeting, and chemical fertilizer (any chain having over 30 outlets with sales from a number of suppliers of different types and brands of commodities is to be controlled by the Chinese partner)	

Source: USDA, FAS, China Changes Agriculture Investment Restrictions, November 23, 2007.

Note: Information in this table was taken from China's 2007 Industrial Catalog for Foreign Investment, as reported by USDA. The original document is written in Chinese, and any unusual terms reflect the translation. Commission staff are unable to further interpret the meaning of those terms.

In contrast, the production of natural food additives and food ingredients is now encouraged, and will likely be open to incentives. Under the new catalog, FDI is now prohibited in the development and production of genetically modified breeds of domestic animals and aquatic products.[151]

According to USDA's Foreign Agricultural Service (FAS), three restricted investment areas pose particular concerns for U.S. companies: seed production and development, soybean processing, and distribution services. Chinese companies face strong competition from foreign multinationals in all of these areas. Restrictions on FDI in seed development are seen as a method for protecting small domestic firms and technology from foreign competition, and may conflict with China's stated goals of increasing.

According to USDA's Foreign Agricultural Service (FAS), three restricted investment areas pose particular concerns for U.S. companies: seed production and development, soybean processing, and distribution services. Chinese companies face strong competition from foreign multinationals in all of these areas. Restrictions on FDI in seed development are seen as a method for protecting small domestic firms and technology from foreign competition, and may conflict with China's stated goals of increasing agricultural productivity. Restrictions on soybean processing reflect Chinese concerns about consolidation in the industry, and the advantages that multinational companies gain from efficient, modern equipment. Restrictions on distribution services make it more difficult for foreign companies to leverage their ability to engineer solutions to delivery and storage problems, especially as multinationals expand into new markets in China's smaller cities.[152]

Bibliography

Aubert, Claude, and Xiande Li. "'Peasant Burden': Taxes and Levies Imposed on Chinese Farmers." In *Agricultural Policies in China after WTO Accession*, edited by Wilfred Legg, Xiande Li, Michèle Patterson, Sally Taylor, and Alexandra Trzeciak-Duval. Paris: OECD, 2002.

Bezlova, Antoaneta. "Buying Farmland Abroad, Ensuring Food Security," *Interpress Service News Agency,* May 9, 2008.

Bloomberg.com. "Japan to Boost Farm Output As China Net Corn Buyer," August 31, 2010. http://www.businessweek.com/news/2010-08-31/japan-to-boost-farm-output-as-china-net-cornbuyer.html.

British Broadcasting Corporation (BBC). "China to Tackle Rural Problems," February 22, 2006. http://news.bbc.co.uk/2/hi/asia-pacific/4739866.stm.

———. "Chinese Melamine Scandal Widens," October 31, 2008. http://news.bbc.co.uk/2/hi/7701477.stm.

———. "Tainted Milk Scandal Resurfaces in China," January 25, 2010. http://news.bbc.co.uk/2/hi/8478195.stm.

———. "Timeline: China Milk Scandal," January 25, 2010. http://news.bbc.co.uk/2/hi/asia-pacific/7720404.stm.

Borlaug, Norman. "Biotechnology and the Green Revolution." Interview by ActionBioscience.org, November 2002.

Central Intelligence Agency (CIA). *CIA World Factbook: China*, May 20, 2010. https://www.cia.gov/library/publications/the-world-factbook.

———. *CIA World Factbook: Brazil*, August 19, 2010.

———. *CIA World Factbook: India*, August 19, 2010.

———. *CIA World Factbook: United States*, October 28, 2010.

Chen, Luming. "China Issues New Policies to Attract Foreign Investment." Jones Day. http://www.martindale.com/print.aspx (accessed August 24, 2010).

Chen, Xiwen. "A Study on China's County and Township Public Finance and Farmer Income Growth." Taiyuan: Shanxi Economic Press, 2003. in Tao, Ran, and Ping Qin,"How Has Rural Tax Reform Affected Farmers and Local Governance in China?" *China & World Economy* 15, no. 3 (2007).

Dairy Markets. "Beijing Increases Financial Aid for Dairies," October 25, 2010. http://www.agra-net.com.

———. "Rabobank Warns of Tighter Dairy Markets Next Year," November 25, 2010. http://www.agranet.com.

Dickinson, Steve. "China Real Estate Laws, Part I." *China Law Blog*. Harris & Moure, PLLC. http://www.chinalawblog.com/2007/05/china_real_estate_laws (accessed October 4, 2010).

Economist. "Buying Farmland Abroad: Outsourcing's Third Wave," May 21, 2009.

Economist Intelligence Unit. "Country Profile 2008: China," 2008. http://www.eiu.com/index.asp?layout=displayIssue&publication id=140000814.

European Commission (EC). Directorate General for Agriculture and Rural Development (DGA). "China: Out of the Dragon's Den?" Monitoring Agri-trade Policy (MAP) No. 01-08, May 2008.

Fan, C. Cindy. "China's Eleventh Five-Year Plan (2006-2010): From 'Getting Rich First' to 'Common Prosperity.'" *Eurasian Geography and Economics* 47, no. 6, 2006, 708–23.

Farmer's Daily. "Nongye Bu Fabu 2010 Nian Qiang Nong Hui Nong Zhengce Huilan [Ministry of Agriculture Announces 2010 Policies Strengthening Agriculture and Benefiting Farmers]," March 11, 2010. http://farmer.com.cn/zt/qnhn/ (accessed November 24, 2010).

People's Republic of China. Ministry of Finance. *2009 Central Financial Revenue Structure*, n.d.

Food and Agricultural Organization of the United Nations (FAO). FAOSTAT (accessed July 27, 2010, and August 26, 2010)

Food and Agricultural Organization of the United Nations (FAO). *How to Feed the World 2050*, Brochure n.d. http://www.iaahp.net/uploads/media Investment.pdf (accessed October 29, 2010).

Fuller, Frank, John Beghin, and Scott Rozelle. "Consumption of Dairy Products in Urban China: Results from Beijing, Shanghai, and Guangzhou," November 29, 2006. http://iis-db.stanford.edu/pubs/21640/dairydemand 3cities final manuscript.pdf.

Gale, Fred, Bryan Lohmar, and Francis Tuan. "How Tightly Has China Embraced Market Reforms in Agriculture?" *Amber Waves*, June 2009.

Gale, Fred. "China's Future: Trade for Sustainable Food." Paper presented at the USDA Outlook Forum Washington, DC, February 19, 2010.

Gale, Fred. "China's Statistics: Are They Reliable?" In *China's Food and Agriculture: Issues for the 21st Century* edited by Fred Gale. U.S. Department of Agriculture, Economic Research Service. AIB775, April 2002, 50–53. http://ageconsearch.umn.edu/bitstream/33723/1/ai020775.pdf.

Gale, Frederick H. "China's Growing Affluence: How Food Markets are Responding." *Amber Waves*, June 2003. http://www.ers.usda.gov/AmberWaves/June03/Features/ChinasGrowingAffluence.htm.

Gale, Fred, Francis Tuan, and Chao Peng. *China's Increasing Farm Subsidies*. Economic Research Service, USDA, unpublished.

Gale, Fred, Francis Tuan, Xiaohui Wang, and Zhi Cao. *China Is Using More Corn for Industrial Products*. U.S. Department of Agriculture, Economic Research Service, FDS-09K-01, December 2009. http://www.ers.usda.gov/publications/fds/2009/12dec/fds09k01.

Gifford, Rob. "Food Fears Persist in China 2 Years after Milk Scare." NPR, October 26, 2010. http://www.npr.org/templates/story/story.php?storyId=130826464.

Graham-Harrison, Emma. "China's Water Pollution Level Higher Than Estimated in 2007." *The Washington Post,* February 10, 2010. http://www.washingtonpost.com/wp-dyn/content/article/2010/02/09/AR2010020903572.html.

Hoffman, Richard. "China's VAT System." *Beijing Review*, August 3, 2009. http://www.bjreview.com.cn/business/txt/2009-08/03/content 210354.htm.

Huang, Jikun, and Scott Rozelle. "China's Agriculture: Drivers of Changes and Implications to China and the Rest of the World." Paper presented at the 27th International Association of Agricultural Economists Conference, Beijing, China, August 2009. http://ageconsearch.umn.edu/bitstream/53207/2/Jikun%20China.pdf.

Huang, Sophia, and Fred Gale. *China's Rising Fruit and Vegetable Exports Challenge U.S. Industries*. U.S. Department of Agriculture, Economic Research Service, FTS-32001, February 2006. http://www.ers.usda.gov/Publications/FTS/2006/02Feb/FTS32001/fts32001.pdf.

Hudson, Durren, Suwen Pan, Maria Mutuc, Samantha Yates, and Don Ethridge. "Crop Subsidies in Foreign Countries: Different Paths to Common Goals." Cotton Economics Research Institute (CERI). Lubbock, TX: Texas Tech University, April 2009.

ISI Emerging Markets. CEIC database (accessed May 24, 2010).

Ito, Junichi. "Inter-regional Difference of Agricultural Productivity in China: Distinction Between Biochemical and Machinery Technology."*China Economic Review*, March 12, 2010. http://www.sciencedirect.com/science?_ob=MImg&_imagekey=B6W46-4YKGJ5Y-1-C&_cdi=6534&_user=2928083&_pii=S1043951X10000246&_origin=search&_coverDate=09%2F30%2F2010&_sk=999789996&view=c&wchp=dGLbVtz-zSkzV&md5=1b44a9e382af7335bce20047dda8e569&ie=/sdarticle.pdf.

Keidel, Albert. "The Causes and Impact of Chinese Regional Inequalities in Income and Well-Being." Carnegie Endowment for International Peace, December 2007. http://carnegieendowment.org/files/Keidel2.pdf.

Kugelman, Michael, and Susan Levenstein, eds. "Land Grab? The Race for the World's Farmland." Washington, DC: Woodrow Wilson International Center for Scholars, 2009.

Kwiecinski, Andrzej, and Frank von Tongeren. "Quantitative Evaluation of Agricultural Policy Reforms in China: 1993–2005." In China's Agricultural Trade: Issues and Prospects; Papers Presented at the International Agricultural Trade Research Consortium, July 8–9, 2007, Beijing, China, edited by Ian Sheldon. IATRC, 2009.

Lardy, Nicholas. "China's Consumption-Driven Growth Path." In China's Agricultural Trade: Issues and Prospects; Papers Presented at the International Agricultural Trade Research Consortium, July 8– 9, 2007, Beijing, China, edited by Ian Sheldon. IATRC, 2009.

Li, Zongzhang and Xiaomin Liu. "The Effects of Rural Agriculture Infrastructural Development on Agricultural Production Technical Efficiency: Evidence of the Data from Second National Agricultural Census of China." Paper presented at the International Association of Agricultural Economists Conference, Beijing, China, August 2009.

Lieberthal, Kenneth. "China's Governing System and Its Impact on Environmental Policy Implementation." China Environment Series I. Washington, DC: Woodrow Wilson International Center for Scholars, 1997.

Lohmar, Brian, Fred Gale, Francis Tuan, and Jim Hansen. *China's Ongoing Agricultural Modernization: Challenges Remain after 30 Years of Reform.* U.S. Department of Agriculture, Economic Research Service, Economic Information Bulletin no. 51, April 2009. http://www.ers.usda.gov/Publications/EIB51/EIB51.pdf.

MacDonald, Stephen, and Sarah Whitley. *Fiber Use for Textiles and China's Cotton Textile Exports.* Outlook report no. CWS-081-01. U.S. Department of Agriculture, Economic Research Service, March 2009.

National Bureau of Statistics of China. *China Statistical Yearbook,* 2006.

Nath, Prem. *The FAO Field Programme and Agricultural Development in Asia and the Pacific.* Bangkok: FAO, 1999. ftp://ftp.fao.org/docrep/fao/005/ac621e/ac621e00.pdf.

Newsweek. "Where Rivers Run Dry," April 16, 2007. http://www.newsweek.com/2007/04/15/wherechina-s-rivers-run-dry.html.

O'Kray, Caleb, and Wu Kang. "Biofuels in China: Development Dynamics, Policy Imperatives, and Future Growth." *IAEE Energy Forum,* Second Quarter 2010.

Organisation for Economic Co-operation and Development (OECD). *2009 OECD Factbook: Economic, Environmental and Social Statistics.* Paris: OECD, April 2009.

Organisation for Economic Co-operation and Development (OECD). 2010 OECD Producer Support Estimates (PSE) database. http://stats.oecd.org/Index.aspx?DataSetCode=PSE 2010 (accessed October 21, 2010).

———. *Agricultural Policies in Emerging Economies: Highlights.* Paris: OECD, 2009.

———. *Agricultural Policies in Emerging Economies: Monitoring and Evaluation.* Paris: OECD, 2009.

———. *Agricultural Policies in non-OECD Countries.* Paris: OECD, 2007.

Owen, Elise. "Standards in China: Behind the Headlines." *China Business Review,* January – February 2010. http://www.chinabusinessreview.com/public/1001/owen.html.

People's Daily Online. "China Increase Subsidies to Grain Growers," April 12, 2006. http://english.peopledaily.com.cn/200604/12/eng20060412 257763.html.

———. "China to Further Boost Farm Produce Distribution," October 22, 2010. http://english.peopledaily.com.cn/90001/90776/90882/7173800.html.

People's Republic of China (PRC). "About China: Topography." www.GOV.cn (accessed November 4, 2010).

People's Republic of China (PRC). Ministry of Commerce (MOFCOM). "Reference Data on FDI Distribution in Major Sectors." Annual FDI Data. http://www.fdi.gov.cn/pub/FDI (October 28, 2010).

Promar International. *The Chinese Potato Industry in Transition.* Alexandria, VA: Promar, 2007. Qinghui, Pang. "China's GM Gamble: Seeds of Change." *News China,* May 5, 2010.

Qiu, Huanguang, Jikun Huang, Jun Yang, Scott Rozelle, Yuhua Zhang, Yahui Zhang, and Yanli Zhang. "Bioethanol Development in China and the Potential Impacts on Its Agricultural Economy." *Applied Energy* 87, no. 1 (January 2010), 76–83. http://en.ccap.org.cn/uploadfile/2010/0209/20100209055001176.pdf.

Resurreccion, Lyn. "China 'Bt' Rice OK to Boost Supply." *Business Mirror*, March 2, 2010.

Reuters. "China Arrests 96 Over Tainted Milk," January 13, 2011. http://news.yahoo.com/s/nm/canada us china melamine.

Roberts, Ivan, and Neil Anders. *Developments in Chinese Agriculture.* Australia: Abare eReport, July 2005.

Rozelle, Scott. *Vegetables in China: Final Report to the Western Growers Association,* March 2006.

Rozelle, Scott, Daniel A. Sumner, Mechel Paggi, and Jikun Huang. "Rising Demand, Trade Prospects, and the Rise of China's Horticultural Industry," n.d. http://iisdb.stanford.edu/pubs/21634/Rising_Demand_Trade_Prospects_ Rise_of_Hort_NAAMIC_workin g_paper.pdf (accessed January 13, 2011).

Smil, Vaclav. "China's Thirsty Future." *Far Eastern Economic Review*, December 2005.

South East Farm Press. "China on Track for Good Grain Harvest," October 5, 2010. http://southeastfarmpress.com/grains/china-track-good-grain-harvest?cid=nl sefpd.

Spencer, Richard. "China Looks Abroad to Grow Its Own Food." *The Telegraph,* May 9, 2008.

Srivastava, Mehul, and Subramaniam Sharma. "Indian Co. Leases Land in Ethiopia the Size of Rhode Island." *Ethiopian Review,* November 12, 2010.

Stewart & Stewart. Written testimony submitted to the U.S. International Trade Commission in connection with inv. no. 332-518, *China's Agricultural Trade: Competitive Conditions and Effects on U.S. Exports,* June 3, 2010.

The U.S.-China Business Council. "Recent Developments in China's Food Safety Regime," 2010.

Thomson Financial News. "China Considers Buying Land Overseas to Secure Food Supply," May 8, 2008.

U.S. Department of Agriculture (USDA). Economic Research Service (ERS). China Agricultural and Economic Data: National Data (accessed June 11, 2010).

———. *Briefing Rooms: China,* updated March 12, 2009. http://www.ers.usda.gov/Briefing/China

———. State Fact Sheets: United States, updated September 10, 2010. http://www.ers.usda.gov/statefacts/us.htm#FC.

U.S. Department of Agriculture (USDA). Foreign Agricultural Service (FAS). *China, Peoples Republic of: 2010 Agriculture Policy Directive*, by Chanda Beckman. GAIN Report no. CH10004,

February 18, 2010. http://gain.fas.usda.gov/Pages/Default.aspx.

———. *China, Peoples Republic of: Agricultural Situation; Livestock and Products, 2008*, by Zhang Jianping and Chanda Beckman. GAIN Report no. CH0878, December 22, 2008.

———. *China, Peoples Republic of: Agricultural Situation; Oilseed Situation Update*, by Mark Petry and Wu Xinping. GAIN Report no. CH9081, October 23, 2009.

———. *China, Peoples Republic of: Agricultural Situation; 2009 Agricultural Policy Directive, 2009*, by Mark Petry. GAIN Report no. CH9026, April 7, 2009.

———. *China, Peoples Republic of: Biofuels Annual*, by Chanda Beckman and Jiang Junyang. GAIN Report no. CH9059, July 17, 2009.

———. *China, Peoples Republic of: Bio-fuels, an Alternative Future for Agriculture 2006*, by K. Latner, C. O'Kray, and J. Jiang. GAIN Report no. CH6049, August 8, 2006.

———. *China, Peoples Republic of: China Changes Agriculture Investment Restrictions, 2007*, by Mark Petry. GAIN Report no. CH7087, November 23, 2007.

———. *China, Peoples Republic of: China's Corporate Income Tax Exemption for Agricultural Enterprises, 2008*, by Mark Petry and Zhang Lei. GAIN Report no. CH9003, January 13, 2009.

———. *China, Peoples Republic of: Cotton and Products; Annual, 2007*, by Mark Petry and Wu Xinping. GAIN Report no. CH7033, May 1, 2007.

———. *China, Peoples Republic of: Cotton and Products; Annual, 2008*, by Jorge Sanchez and Wu Xinping. GAIN Report no. CH8037, May 23, 2008.

———. *China, Peoples Republic of: Cotton and Products Annual 2009*, by Jorge Sanchez and Wu Xingping. GAIN Report no. CH9035, April 15, 2009.

———. *China, Peoples Republic of: Cotton Update*, by Wu Xinping. GAIN Report no. CH10033, June 29, 2010.

———. *China, Peoples Republic of: Citrus; Annual, 2007*, by Wu Bugang, Jane Li, Eric Trachtenber, and Chanda Beckham. GAIN Report no. CH7090, November 29, 2007.

———. *China, People's Republic of: Citrus Annual, 2008*, by Chanda Beckman, Wu Bugang, Wang Tong, and Seth Peavey. GAIN Report no. CH8110, December 1, 2008.

———. *China, Peoples Republic of: Citrus; Annual, 2009*, by Jorge Sanchez and Jiang Junyang. GAIN Report no. CH9013, March 3, 2009.

———. *China, Peoples Republic of: Citrus; Annual, 2009*, by Joshua Lagos, Wang Tong, and Wu Bugang. GAIN Report no. CH9122, December 1, 2009.

———. *China, Peoples Republic of: Fertilizer*, by Chanda Beckman, Katie Rohm, and Bao Liting. GAIN Report no. CH9082, December 14, 2009.

———. *China, Peoples Republic of: Fresh Deciduous Fruit; Annual, 2006*, by Jorge Sanchez and Wu Bugang. GAIN Report no. CH6078, September 22, 2006.

———. *China, Peoples Republic of: Fresh Deciduous Fruit; Annual, 2007*, by Wu Bugang, Evid Liu,

Eric Trachtenber, and Chanda Beckham. GAIN Report no. CH7085, November 15, 2007.

———. *China, People's Republic of: Fresh Deciduous Fruit; Annual, 2008*, by Chanda Beckman, Wu Bugang and May Liu. GAIN Report no. CH8108, November 15, 2008.

———. *China, People's Republic of: Dairy and Products; Annual Report, 2007*, by Jianpin Zhang, Eric Trachtenberg, and Tanya Franke. GAIN Report no. CH7080, October 12, 2007.

———. *China, Peoples Republic of: Grain and Feed; Annual, 2008*, by Jorge Sanchez and Jiang Junyang. GAIN Report no. CH8012, March 1, 2008.

———. *China, Peoples Republic of: Grain and Feed; Annual, 2009*, by Jorge Sanchez and Jiang Junyang. GAIN Report no. CH8012, March 3, 2009.

———. *China, Peoples Republic of: Grain and Feed; Annual, 2010*, by Chanda Beckman and Jiang Junyang. GAIN Report no. CH10008, March 1, 2010.

———. *China, People's Republic of: Livestock and Products; Semi-Annual Report, 2007*, by Jianpin Zhang and Eric Trachtenberg. GAIN Report no. CH8014, March 1, 2008.

———. *China, Peoples Republic of: Livestock and Products; Semi-Annual Report, 2009*, by Mike Woolsey and Jianping Zhang. GAIN Report no. CH09017, March 9, 2009.

———. *China, Peoples Republic of: Livestock and Products; Annual Report, 2009*, by Mike Woolsey and Jianping Zhang. GAIN Report no. CH9069, September 14, 2009.

———. *China, Peoples Republic of: Livestock and Products; Semi-Annual, 2010,* by Michael Woolsey and Jianping Zhang. GAIN Report no. CH10009, March 2, 2010.

———. *China, People's Republic of: Oilseeds and Products Annual Part 1 of 2—Analysis, 2008*, by Mark Petry and Wu Xinping. GAIN Report no. CH8010, March 1, 2008.

———. *China, People's Republic of: Oilseeds and Products Annual Part 1 of 2—Analysis, 2009*, by Mark Petry and Wu Xinping. GAIN Report no.CH9030, April 15, 2009.

———. *China, Peoples Republic of: Poultry and Products Annual; Annual Report,* by Michael Woolsey and Jianping Zhang. GAIN Report no. CH9070, September 16, 2009.

———. *China, Peoples Republic of Poultry and Products Semi-Annual, 2010*, by Michael Woolsey, Joshua Lagos, and Jianping Zhang. GAIN Report no. CH10010, April 14, 2010.

———. *China, People's Republic of: Sugar; Annual 2008*, by Mark Petry and Jiang Junyang. GAIN Report no. CH8023, April 10, 2008.

———. *China, People's Republic of: Sugar; Semi-Annual*, by Mark Petry and Jiang Junyang. GAIN Report no. CH9084, November 3, 2009.

———. *China, People's Republic of: Trade Policy Monitoring; VAT Protections: The Rest of the Story, 2007,* by Kevin Latner and Jianping Zhang. GAIN Report no. CH7018, March 19, 2007.

U.S. Department of Agriculture (USDA). Production Supply and Distribution (PSD) Online (accessed various dates).

United Nations. *Food Prices Issues in the People's Republic of China.* Occasional Paper. China: United Nations Resident Coordinator, 2008. http://www.un.org.cn/public/resource/782a457602152e7b768a16b5b8488e0e.pdf.

Waldron, Scott A., Colin G. Brown, and John William Longworth. *Rural Development in China: Insights from the Beef Industry*. London: Ashgate Publishing Limited, 2003.

Wang, Hui, Ran Tao, and Juer Tong. "Trading Land Development Rights under a Planned Land Use System: The 'Zhejiang Model' and Its National Implications." *China & World Economy* 17, no. 1 (2009), 1–17. www.ccap.org.cn/uploadfile/2009/0706/20090706102418524.pdf.

Wang, Jinxia, Jikun Huang, Scott Rozelle, Qiuqiong Huang, and Lijuan Zhang. "Understanding the Water Crisis in Northern China: What the Government and Farmers are Doing." *International Journal of Water Resources Development* 25, no. 1 (March 2009): 141–58.

Wee, Sui-Lee, and Sabrina Mao, "China's Incentives for Renewable or Clean Energy." Reuters, August 27, 2010. http://www.reuters.com/article/idUSTRE67Q0YV20100827.

World Bank. Data: China. http://data.worldbank.org (accessed various dates).

———. "Mid-Term Evaluation of China's Eleventh Five-Year Plan." Report no. 46355-CN, December 18, 2008.

Xinhua News Agency. "Investment in Farming Increased to $37 Bln," June 27, 2006. http://www.gov.cn/english/2006-06/27/content 321628.htm.

———. "Wen Hears Farmers' Concerns on 11th Five-Year Plan," March 20, 2006. http://www.china.org.cn/english/2006/Mar/162330.htm.
Zhang, Yongsheng. "To Achieve the Goals of China's 11th Five-Year Plan through Reforms." Beijing: Development Research Center, Beijing, November 2006.
Zhong, Yongling. "Analysis of Wheat Market in 2009 and Outlook for the Furture [in Chinese]." *Agricultural Outlook* 5, no. 11 (2009): 5–10.

5. Competitive Factors Affecting China's Agricultural Sector

Overview

This chapter analyzes competitive factors across the full range of Chinese agricultural production. As a general matter, none of these factors exist in isolation. To the extent possible, any discussion of China's farm sector competitiveness must address the interaction between competitive factors—e.g., manual labor vs. mechanization—the most efficient mix of productive inputs, and the role of government policies.[1] China's rising farm labor costs and rural migration to urban areas will likely result in new capital investment and land consolidations which could lead to farm efficiency gains. But the chances are low that these gains will suffice to satisfy China's increasing food needs, so China will also likely import more agricultural goods, and government policies will heavily influence which foodstuffs are imported.

The first section of this chapter presents an analytical framework describing the factors that influence relative competitiveness in agricultural goods for any country, using China as the focus. The second section describes and analyzes the primary factors influencing competitive conditions in China's farm sector, breaking out the relevant factors into three categories—delivered cost, product differentiation, and reliability of supply. [2] This chapter focuses on the primary factors shaping competitive conditions in China's agriculture sector, but it will also attempt to anecdotally address interactions between factors and allocative efficiency. Table 5.1 compares selected economic indicators for the farm sectors in China and the United States. Relative to the United States, China's labor rates and percentage of irrigated land are favorable, but its availability of arable land and water are unfavorable, as are China's per-hectare cereal yields.

Analytical Framework

Whether in the Chinese market or elsewhere, agricultural competitiveness is measured by comparing delivered cost, product differentiation, and reliability of supply of domestically produced goods against those of imports, both in the domestic market and in third-country markets.[3] Anything that determines or influences delivered cost, product differentiation, and reliability of supply can be considered a competitive factor affecting the agricultural sector. Figure 5.1 identifies several specific competitive factors for agriculture.

Figure 5.1 does not list government policies and foreign direct investment (FDI) as competitive factors because they have the potential to influence all three categories (delivered cost, product differentiation, and reliability of supply). For example, government programs that subsidize fertilizers, seeds, and machinery lower the delivered cost of domestic producers, while government sanitary and phytosanitary (SPS) and labeling regulations on imported products raise the delivered cost for importers. Government food safety regulations, as well as government-mandated grades and standards requirements, provide a mechanism for product differentiation. Government intervention can influence the reliability of supply through publicly funded or subsidized marketing and transportation infrastructure, as well as by imposing supply and export controls on producers. See box 5.1 for further discussion of FDI's impact on delivered cost, product differentiation, and reliability of supply.

Delivered Cost

For many globally traded agricultural products, delivered cost is the most important criterion in making purchasing decisions. For producers to be competitive in their domestic market, they must be able to supply products to purchasers at or below the price offered by importers and other domestic producers. The price-competitiveness of domestic suppliers therefore depends on factors that tend to lower or raise their delivered costs vis-à-vis the delivered costs of imported and other domestic products in their home market.

Table 5.1. China and U.S. agricultural indicators

Indicator	China	United States
Arable land (hectares per capita)	0.1	0.6
Average farm size (hectares)	0.6	169
Wage per day of on-farm labor	\$8-\$13	\$86.56
Cropland per agricultural worker (hectares)	0.4	78.5
Annual per capita renewable water resources (m^3)	2,138	10,231
Percent of harvested land irrigated	47	18
Tractors per 100 square km of arable land	146.7	247.6
Cereals yield (kg per hectare)	5,535	6,624

Sources: USDA, NASS, "2007 Census of Agriculture: United States Summary and State Data," December 2009, 17, 66; Industry officials, interviews by Commission staff, Beijing and Shandong province, China, September 6–15, 2010; USDA, NASS, "Farm Labor," August 19, 2010; The World Bank, Data: World Development Indicators; USDA, ERS, "Briefing Rooms: China: Basic Information."

Notes: (1) Daily wage rates for on-farm labor in the United States are based on an 8-hour workday and the \$10.82 hourly average reported by USDA, NASS, "Farm Labor," August 19, 2010. (2) Cereal yield includes wheat, rice, corn, barley, oats, rye, millet, sorghum, buckwheat, and mixed grains. The Chinese average does not account for double-cropping, so Chinese yields are likely much lower on a per-planting basis than represented in the table.

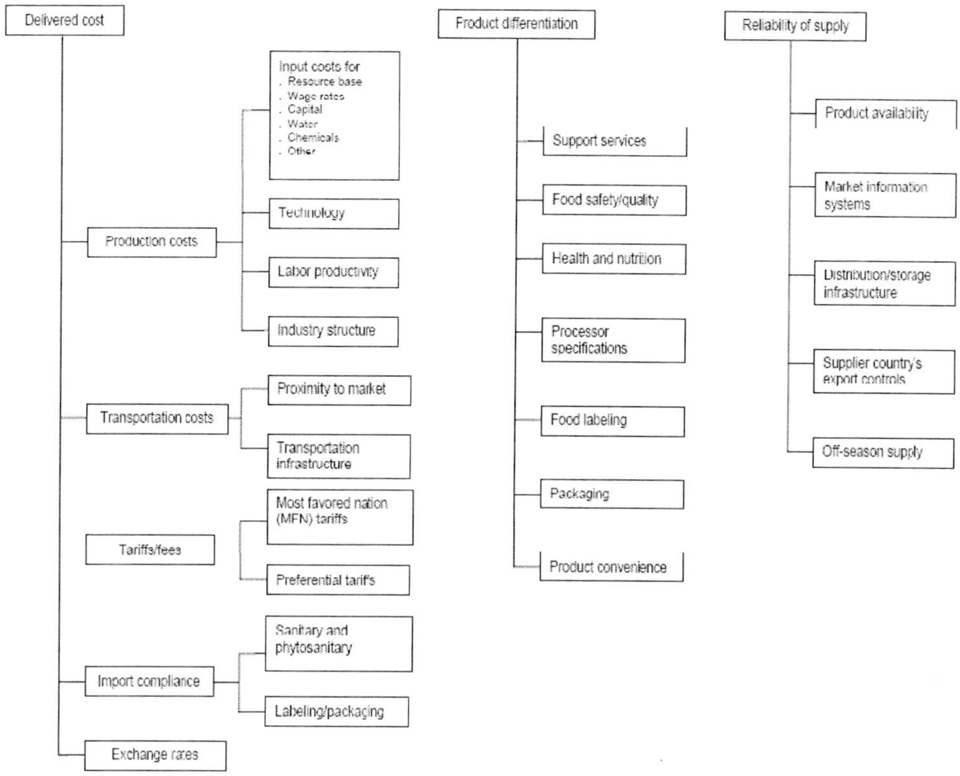

Figure 5.1. A broad range of factors affect competitiveness in any agricultural market.

BOX 5.1. Foreign Direct Investment in China's Agricultural Sector is Limited

Foreign direct investment (FDI) supports the economic development of countries and can spur the advancement of specific industries. In the ITC's analytical framework, FDI cuts across all three factor categories (delivered cost, product differentiation, and reliability of supply). In the agricultural and food processing sectors, FDI can lower costs by improving production efficiency, as foreign producers introduce new growing techniques or manufacturing processes. FDI may also lead to greater product differentiation through improvements in food quality, branding, access to global managerial skills, and better agricultural practices. Lastly, FDI may be able to provide China's agricultural sector, particularly in food processing, with capital for investments in modern production and distribution systems that can more reliably supply high-quality goods in the Chinese market.

FDI in China's agriculture and food processing industries typically represents a small percentage of total FDI in China, accounting for only 1.9 percent of the total value in 2008. For most of 2005–09, FDI in food processing accounted for a greater share of both the reported value and the number of projects than FDI in agriculture.[a] Throughout the period, Hong Kong and the British Virgin Islands were China's largest sources of agricultural FDI; the United States ranked fifth or below as a source of agricultural FDI every year between 2005 and 2009.[b] Country data for investment in the food processing sector are not available.

The industry that received the most FDI between 2004 and 2009 within the agriculture or food processing sectors was the alcoholic beverage industry, but FDI also played an important role in the dairy, meat, and non-alcoholic beverage industries.[c]

China FDI, number of projects and realized FDI value, 2008

	Number of projects	Share of total projects	Realized FDI value (million	Share of total FDI value
Total agriculture	628	2.3	838.8	0.8
Total food processing	386	1.4	1,214.1	1.1
Total food-related FDI	1,014	3.7	2,052.9	1.9
Total FDI	27,537	100.0	108,312.4	100.0

Source: PRC, MOFCOM, "The Survey of Foreign Investment in China's Agriculture Industry."

[a] PRC, MOFCOM, "Reference Data on FDI Distribution in Major Sectors." The pattern of FDI, with significantly higher values in 2004 and 2005 than in the following years, follows the pattern of overall FDI in China during those years.

[b] Data for the United States as a share of the total were not available for 2009 and 2007, but in those years, the United States was not ranked among the top five FDI sources. PRC, MOFCOM, "The Survey of Foreign Investment in China's Agriculture Industry."

[c] Examples abound in China of foreign mergers and acquisitions (M&A) and greenfield investment projects in food processing during 2005–9.

In alcoholic beverages, two examples include Asahi Breweries' (Japan) $668 million investment for a 20 percent stake in Tsingtao in January 2009, and InBev SA's (Belgium) takeover of Fujian for $733.1 million in January 2006. In the dairy sector, two of the biggest M&A investment projects during 2005–009 were Uni-President's (Taiwan) acquisition of Beijing Sanyuan for $111.4 million in July 2006 and Fonterra's (New Zealand) acquisition of Sanlu Group for $106.5 million in December 2005. In addition, Yakult Honsha (Japan) invested $96 million in September 2006 in a greenfield dairy project in Shanghai. For poultry, Tyson's (USA) invested $31.8 million in February 2008 in a greenfield project in Haiman, China. ISI Emerging Markets, CEIC Database.

The delivered cost of domestically produced goods depends on the costs of producing the good and domestic transportation cost from production points to consumption points. Production costs in turn depend on the costs of inputs, such as fertilizer and wages. The application of biological technology, such as high-yielding varieties of seeds, and production technology, such as machinery and irrigation, is also a determinant of delivered cost. Transportation depends on several factors such as fuel costs and the efficiency of the transportation system, such as the quality of roads and ports.

Both Chinese producers and importers incur production costs and transportation costs. However, importers incur additional costs that impact the overall delivered cost. These include the costs of international transportation, exchange rate conversion, trade risk coverage, and Chinese tariffs.[4] The delivered cost of imported products also includes expenditures on import compliance, such as meeting China's sanitary and phytosanitary (SPS) regulations and China-specific labeling and packaging requirements.

China's export competitiveness in third-county markets relies on the same set of factors which influence delivered cost in the domestic market. Products exported from China are competitive if they can be delivered at or below the cost of domestic products in the importing country and competing exporting countries. In this case, cost of production, international transportation costs, tariffs, fees, and compliance costs imposed on Chinese products by the importing country make up the delivered cost of products imported from China.

Product Differentiation

In addition to delivered cost, purchasers compare the level of product differentiation of domestically produced and imported products in making their buying decisions. The more processed and branded the product, the more likely product characteristics and reputation will form the basis of the purchasing decision, thereby making delivered cost less important. Similar products are differentiated from one another through their brand packaging and labeling and their level of convenience, in conjunction with large investments in marketing, promotion, and media advertising. Additionally, in China, especially among middle- and upper-class income groups, product quality (gauged by nutrition, health, brand, and safety characteristics) is increasingly considered in food-purchasing decisions.[5]

Reliability of Supply

Reliability of supply refers to the ability of a supplier to deliver a specified product, of a particular quality and in an agreed-upon volume, to a specified location at a contracted time. It also refers to the ability of overseas suppliers to supply products in the off-season to domestic consumers.[6] Risks inherent in agricultural production (potentially impacting both the quantity and quality of supply) and its political significance in China and many other countries make this criterion particularly important for purchasers to consider. Reliability of supply depends on the efficiency of the supply chain, including storage and transportation infrastructure, as well as market information systems. In agriculture, several factors may disrupt the reliability of supply, such as a government-imposed export control. [7] political unrest, poor transportation infrastructure, [8] and unstable production and exportable surplus.

Primary Competitive Factors Affecting China's Agricultural Sector

Factors Affecting Delivered Cost

Labor Supply and Wage Rates

China has an abundant, inexpensive supply of unskilled labor in rural areas.[9] As a result, when combined with effective labor productivity, China's agricultural sector has a comparative advantage in labor-intensive products. However, rapid industrial development has created competition for labor and is beginning to erode China's cost advantage. Competition for labor has caused many rural laborers to move into urban areas for jobs in other sectors. While the figures vary significantly by source, the number of Chinese rural workers who have migrated to urban areas was estimated to be between 130 million and 250 million in 2009.[10] Because average urban incomes are 3.3 times the average in rural areas, labor migration will continue to reduce the rural labor supply and increase costs until the income gap narrows.[11]

Wages for farm labor in China vary substantially by region and product, but they are generally higher in eastern provinces, where demand for labor from other industries is greater. According to the Chinese National Bureau of Statistics, the average annual income in rural areas was approximately RMB 5,153 ($687) in 2009, up 8.2 percent from 2008.[12]

Furthermore, it was reported that in 2010 rural wages were up almost 19 percent from the prior year.[13] As labor availability has declined, labor costs, both on farms and in food processing facilities, are reportedly increasing by 10 to 20 percent per year. Farms that are large enough (5–10 *mu*) to hire additional workers generally pay laborers RMB 50-60 a day (less than $10). In comparison, the average hourly wage for hired on-farm labor is $10.82 in the United States.[14] In 2010, the average cost of labor for Chinese food processors was RMB 6.5 per hour (approximately $0.87), an increase of 20 percent from 2009.

Employee turnover can increase labor costs for agricultural producers and limit the availability of skilled laborers and management. In China, turnover rates vary significantly by region and product, depending on the extent of competition from other industries. As the textile, machinery, and chemical industries have developed in agricultural provinces like Shandong, annual turnover rates have risen, increasing recruiting and training costs for food processors. In some food-processing industries, annual turnover rates have reached 50 percent.[15] As a result, many firms and other agricultural organizations lack employees with the skills necessary to implement management practices that would increase productivity, such as methods to efficiently allocate agricultural resources through cooperatives.

Despite significant annual increases in Chinese farm wages, low-cost labor is still the primary advantage for the Chinese agricultural sector. As competition for workers from other domestic industries continues to increase, however, a more restricted supply of labor, particularly skilled labor, will likely affect future agricultural productivity gains and result in increased mechanization and capital costs.[16]

Water Cost

Water is a critical input around the world for agricultural production; its abundance or scarcity heavily influences the food countries produce and the final yields from those crops and livestock. China relies more heavily on irrigation than other countries because of low rainfall, particularly in the north and west of the country. While an estimated 80 percent of China's food production comes from irrigated land, the irrigation systems currently in place are inefficient relative to international standards, increasing production costs and reducing the reliability of supply.[17]

Under national law, water is state owned. Local governments have the ability to assess fees to cover the capital costs of delivering the water (i.e., reservoirs, irrigation canals, and pumps), as well as electricity fees for running the pumps. Water costs vary significantly by region, and many farmers do not pay for their water use at all.[18] Even those farmers assessed water fees by their local government reportedly pay approximately 25 percent of the actual cost of the water, thus artificially reducing production costs.[19]

When water fees are collected, farmers generally pay a fixed fee per unit of land.[20] Some localities are reportedly starting to charge fees based on consumption, but this is still not the norm.[21] Water for industrial use, which includes packinghouses and food processing facilities, is typically priced by weight (e.g., per metric ton) when supplied by local governments. Water expenses for most farmers and many processors, however, consist only of the costs of maintaining or drilling a well, the pumps, and the electricity used to pump from the wells. While water costs to farmers and processors have remained stable in many areas during recent years, the true cost of groundwater pumping has been increasing dramatically as groundwater levels continue to decline.[22] Food processors are reportedly budgeting for anticipated annual increases.[23]

While artificially low fees for water currently reduce production costs in China, these fees result in overuse and provide no incentive for adopting more efficient irrigation techniques.[24] Recent water shortages and falling groundwater levels have drawn attention to the inefficiencies in the system. Only 40 percent of China's irrigation networks are considered to be running efficiently, which forces farmers to draw yet more water and further reduce scarce water supplies. [25] In the very near future, raising water prices for all users to cover the true cost of supplying water and encouraging conservation may be the only real choice for Chinese policymakers. To the extent that changes in government policy mean sharp increases in water fees for farmers and processors, the delivered cost for agricultural goods will likely increase, weakening China's competitiveness in many of those goods.

Land Tenure System in China

Land in China is owned by the state, and land use rights are typically provided by local village leaders to households for farming. The fact that farmers are not permitted to individually own or sell their land[26] affects the structure of Chinese agriculture by restricting consolidation and thereby thwarting efficiency gains, while also increasing transaction and marketing costs. The local governments which control the land have the unilateral authority to consolidate land holdings, and in many cases have done so. But decisions by village leaders have the potential to be heavily influenced by political objectives which may or may not encourage economic efficiency in the farm sector.[27]

Another facet of the land tenure system is that farmers are often not charged market rents for the land they use, and sometimes are charged nothing at all.[28] This is clearly beneficial to farmers in lowering production costs relative to farming in other countries. Therefore, the overall impact of China's land tenure system on the competitiveness of agricultural goods produced by Chinese farmers is mixed. Low production costs for the land must be weighed against the long-term economic impact of the system on agricultural productivity. Whether the result of artificially low land rents and inefficiencies within China's farm sector is a net gain or loss to overall competitiveness depends upon the specific goods produced. Over the long term, however, China will likely face difficulties in boosting farmers' incomes, encouraging investment, and increasing agricultural productivity unless reforms are made to the current land tenure system. For more detail on the mechanics of the system, see the following sections on farm industry structure and land cost.

Farm Industry Structure

Land use rights per household in China average 0.6 hectares nationally, but many farms are smaller, ranging between 0.1 and 0.4 hectares.[29] By comparison, the average U.S. farm is 169 hectares (table 5.1). Moreover, Chinese farms usually comprise several unconnected plots (figure 5.2). Due to the size and discontiguous nature of farm holdings, buyers must purchase crops from perhaps thousands of individual farmers.

Most agricultural products in China require multiple layers of middlemen to consolidate large volumes of harvested crops in the supply chain, which invariably increases marketing and distribution costs. Certain product sectors in China have developed similar methods for consolidating their product. In horticulture, for example, approximately 80 to 90 percent of total Chinese production is initially sold to small traders that consolidate product from the farms and transport it to other traders, larger wholesale markets, wet markets, or directly to buyers.[30] Grain and vegetable processors in Shandong province reportedly source their raw inputs from between several dozen and a few hundred traders in order to avoid sourcing from several thousand small household farmers. Each additional consolidation stage increases transaction costs, since prices rise each time the product is re-sold.[31]

Source: Commission staff.

Four small plots, likely belonging to different farmers, are each planted with a distinct crop (e.g. ornamental trees (center), leafy vegetables (foreground), squash on trellises (center right), and another unidentified low growing crop (behind trees)).

Figure 5.2. Crops grow in traditional small plots in Sichuan province, China.

Firms in certain sectors have begun to develop models to overcome the limitations of the current land tenure system. The central government, through institutions such as the Chinese Academy of Agricultural Sciences, is also studying this issue. National law now allows farmers to lease out their land use rights, and companies have begun to lease large pieces of land (200–300 acres) from village leaders. In certain counties in Gansu province, for example, 25 percent of farm land is now leased.[32] While this practice is growing, it is still relatively uncommon, especially in the densely populated regions in eastern provinces where the large number of farmers, as well as the concentration of housing, can prevent the consolidation of land into larger holdings.[33]

Various experiments with farmer cooperatives[34] are taking place as well, particularly in horticulture, in order to consolidate product, streamline product distribution, and reduce transaction costs.[35] Collectively, the land holdings controlled by cooperative members range in size from a few hectares to a few thousand hectares. Cooperatives are commonly formed under village control, like those for vegetables in Shaanxi province,[36] and they seek to rationalize rural land use while trying to minimize disruption to rural communities; [37] the hope is that larger operations will make Chinese agricultural production more efficient. But cooperatives are currently being formed slowly, in part because village leaders often lack the management skills necessary to effectively organize farmers, disseminate resources and revenue, and plan effective product marketing.[38] Whether or not cooperatives become a widespread phenomenon in China, farm labor continues to migrate into urban areas, searching for higher wages. As this trend continues, larger scale operations are likely to become more common, increasing economies of scale and lowering transaction costs.

Land Cost

Land costs for farming vary significantly across China's diverse agricultural regions. In general, however, the cost of agricultural land allocated by village leaders to small household farmers, typically in the form of 30-year leases, is either free or significantly below market rental rates.[39] But consolidating land into larger, more efficient plots in China is both expensive and logistically difficult. Not only do farmers seek market rates from agricultural firms or other farmers consolidating the land into more efficient production plots, but consolidators are forced to negotiate with perhaps hundreds of farmers in separate land lease agreements.[40]

In some areas of China, land lease agreements negotiated between farmers and consolidators may actually be more costly than in the United States, especially in the eastern provinces with high demand for land from real estate developers and industry.[41] Even in western and more rural areas of China, leases for large, contiguous farm plots are attracting big sums of money for farmers that control the land.[42] Rents paid by consolidators vary significantly by region; anecdotal information indicates rents of approximately 300 RMB per mu ($40 per acre) annually in Hebei and southern Inner Mongolia, with rents much higher in Shandong province.[43] By comparison, the national average annual cash rent for cropland in the United States is $102 per acre, though prices range from $28 per acre in Nebraska to $345 per acre in California.[44]

Because all land in China is owned by the state, and ownership rights for farm leases have only existed under the law since 2004, real estate markets are not yet fully open and transparent. This creates business uncertainty for agricultural firms and farmers who seek market prices on which to base leasing negotiations.[45] Even in areas where the rental costs for farmland are low, incentives to invest and improve the land's productivity and the quality of output are limited by the continued uncertainty surrounding the legal status of farmers' leases and their rights to control farmland.[46]

Other Input Costs

Seeds, chemicals, and medicines (for live animal operations) are important inputs in agricultural production and account for a significant portion of farm production costs in China. These costs have reportedly been increasing annually, even as intensive farm management practices and government incentives focus on containing the negative impact that these input costs have on Chinese agricultural production costs.

Typical farm management practices add more to input costs for farmers in China than they do for producers in other countries. Fertilizer, pesticide, herbicide, and fungicide costs are higher in China because of high usage rates; Chinese fertilizer use rates are typically more than double the rates in other major agricultural producing countries (table 5.2).[47] For example, farmers in northern China use about 525 pounds of nitrogen fertilizer per acre annually, approximately six times the U.S. average.[48] The common Chinese practice of double- or triple-planting plots of land without rotating crops is reducing soil nutrients and yields, and driving the overuse of fertilizers. If Chinese farmers switch to global best management practices, including the use of ideal cultivars and the proper timing of fertilizer applications, yields could increase, costs would certainly decrease, and nitrogen fertilizer use could be cut by 20 to 40 percent.[49]

Chinese government programs providing funds to offset fertilizer and seed costs artificially lower the cost of these inputs to farmers. Government funds for seeds and fertilizer totaled approximately $11.2 billion in 2008.[50] Fertilizer subsidies in China are directed to fertilizer producers and indirectly benefit agricultural producers. The payments to fertilizer producers include value-added tax (VAT) holidays; preferential prices for electricity, natural gas, and coal; a preferential transportation price of RMB 80 ($11) per metric ton; duty-free imports of raw materials; and loan subsidies ensuring low rates for producers or distributors who store fertilizers during the winter.[51] In 2007, the comprehensive fertilizer subsidy package averaged about $14.50 per farm household.[52] International fertilizer prices fluctuate significantly from year to year, but Chinese farmers are less sensitive to price fluctuations than farmers in the United States because of these government payments.[53] While the impact of these government payments on retail fertilizer prices is difficult to measure, in early 2008 during a period of high international fertilizer prices, farm prices for urea in China were $0.12–$0.13 per pound compared to $0.28 per pound in the United States.[54] Subsidized seeds available for purchase by farmers include soybeans, rice, wheat, corn, rapeseed, and cotton, as well as hull-less barley and peanuts on a trial basis in 2010.[55] However, many provinces limit payments to purchases of high-quality and high-yield seeds. Seed payments are provided either through direct payments to farmers or to seed suppliers who then pass along the benefit to farmers through discounted seed prices. In provinces that use direct payments, the U.S. Department of Agriculture (USDA) estimates that farmers generally receive RMB 10–15 ($1.33–$2.00) per *mu* to purchase seeds.[56]

Table 5.2. Fertilizer consumption (kilograms per hectare of arable land)

Country	2003	2004	2005	2006	2007
China	3,062	3,069	3,217	3,633	3,311
Brazil	1,666	1,739	1,396	1,451	1,901
USA	1,458	1,594	1,651	1,478	1,712
India	1,051	1,154	1,278	1,364	1,423

Source: World Bank, "Country Data: China" (accessed July 26, 2010).

Farmers also use a wide variety of veterinary drugs to control diseases in livestock. Vaccination costs for live animal operations are reportedly higher in China than in other countries with major livestock production operations, because of widespread overuse and because the majority of these drugs have to be imported.[57]

Mechanization

The use of advanced technology and mechanization is not extensive in Chinese agricultural production compared to other countries, largely because of relatively low labor costs that encourage the substitution of labor for capital (in the form of mechanization and technology) whenever possible.[58] As long as labor costs remain low, the cost advantages of switching to mechanized production and paying for technological upgrades are limited.[59] However, rising labor costs, the migration of Chinese skilled farm labor to urban areas, and government incentives have promoted agricultural mechanization in recent years, raising agricultural productivity and lowering production cost for crops such as corn.[60]

While the use of mechanized harvesters and processing lines has increased, further upgrades are limited by both supply and demand factors. Chinese domestic production of processing and harvesting equipment is expanding, but much of the equipment is not available domestically and must be imported, increasing the cost of modernization.[61] At the same time, the structure of Chinese farming restricts the demand for new equipment and other modernization efforts because small and discontinuous plots reduce the benefits of mechanization. Furthermore, small households and most cooperatives have limited capital reserves to make equipment purchases. However, pressure to increase mechanization in China's farm sector will likely increase as rural workers shift to other industries, wages rise, and the sector consolidates.

Typically, conventional bank financing options are not available for farmers to make capital purchases. As a result, some food processors and large traders that source inputs from farmers on a contract basis use alternative methods to increase on-farm technological upgrades, such as self-financing equipment purchases. For example, a processor may supply farmers with an irrigation system, pesticide sprayers, and tractors, while requiring only a modest down payment. The processor then receives payments from the farmer toward the price of the delivered input for a defined period of time.[62]

Since equipment subsidy programs were implemented in 2005, government support has made purchases of new machinery more affordable, boosting machinery use. When a farmer, business, or cooperative buys new machinery, the local government provides a direct payment to cover between 20 and 40 percent of the cost of new machinery, depending on the type of machinery purchased and good being produced.[63] Purchasers of harvesting and packing equipment reportedly receive the largest payments. [64] As agricultural mechanization continues to become more common in China, cost savings and harvesting productivity will likely continue to increase. In 2007, the total combined mechanization level for plowing, sowing, and harvesting wheat, rice, and corn accounted for only 43 percent of total production compared to the United States, where almost all production is mechanized.

Access to Credit

Credit is an important source of capital for farmers to finance investments in new equipment and other operating expenses. Chinese farmers, however, have minimal access to credit, a situation that restricts investment and productivity gains. Because land is owned by the state, farmers lack financing collateral.[65] As a result, small-scale farmers have difficulty getting formal loans and investing in mechanization, which increases labor costs and lowers overall productivity. Even though the dominance of state-owned banks typically keeps the cost of credit low in China, lending has historically been directed towards industry and state-owned enterprises that show the most job growth potential.[66] Most farmers therefore acquire needed capital through nonfarm income or personal loans from friends.

Both central and local governments are seeking to improve access to credit for farmers through state-owned banks, rural credit cooperatives, and local programs which encourage microlending.[67] Increased access to credit from banks may allow farmers to buy better seeds, fertilizers, and pesticides, and to invest in capital assets such as mechanized harvesters, greenhouses, and efficient irrigation systems. If Chinese farmers do gain more access to credit, costs will decline as investments lead to more efficient production methods.

The food-processing sector has greater access to capital than farmers do because of greater cash flows and collateral. Food processors finance capital costs through either retained earnings or bank loans. If bank loans are available, interest rates vary significantly depending on the investment's projected return and the company's financial status, but typically range from 5 to 10 percent per year for financially strong companies.[68] The Chinese government also supports the development of certain food industries by making cash investments, giving the government minority holdings in preferred companies. To boost cash flow, the government typically refrains from accepting dividends or other returns on its investment. Reportedly, government investments are made with the expectation of repayment, but the payback terms are often not clearly defined.[69]

Transportation Infrastructure and Time to Market

Upgrades to China's rural infrastructure have improved the agricultural sector's ability to get its products to the market, but the agricultural logistics industry is still fragmented: small traders and service providers are spread across the country, creating inefficiencies and increasing costs. Service companies typically provide individual pieces of the supply chain, such as transport or storage. Very few companies can actually provide a completely integrated transportation network throughout the country, unlike in the United States, where such networks are widely available. As a result, firms must generally hire multiple service or transport providers to get their products to market, making transportation relatively inefficient and expensive.[70]

China has improved its rural transportation infrastructure in recent years as a result of massive investments in road construction, but despite these investments the transportation infrastructure has still not kept pace with growth in demand. Investments have created a road network that is the second-longest in the world behind the United States, with the total length of rural roads expected to increase to 3.5 million kilometers in 2010.[71] Greater access to modern roads generally increases farmers' incomes by lowering either their own transport costs or the transportation costs of the traders who buy their goods and ship them to consumer markets. The net effect is that farmers retain a larger portion of the prices paid by consumers. However, high toll fees on the country's highway networks add significant transportation costs and are generally more expensive than in the northeastern United States.[72] In addition, the lack of available freight transport by rail and the high fees imposed on drivers traveling from one province to another increase costs and create uncertainty.[73]

China's fragmented logistics network increases cold storage and transportation costs relative to other markets. A much higher percentage of total costs (18 percent on average) for agricultural producers in China goes toward transportation and cold storage logistics services than the average (9 percent) in Organisation for Economic Co-operation and Development (OECD) countries.[74] For example, it costs at least RMB 6,000 ($800) to transport 20 metric tons of vegetables about 900 miles from Shouguang, a wholesale center in Shandong province, to Heilongjiang province in northeast China.[75] Because of the high cost and time needed to transport goods throughout China, shipping from the United States to southern China is often cheaper than transporting goods from northern China to the major cities in the south. As a result of the high costs, imports often have a competitive advantage in markets where there is no significant domestic production of a particular good in the region.[76] Cold storage capacity is also limited, despite recent increases, and even when refrigerated services are available, the cost is often prohibitive.[77] Moreover, in cases where the cost of cold storage is affordable, the quality of the equipment generally does not meet Western standards, and the risk of spoilage is higher.[78]

The fragmented structure of the logistics industry in China increases costs for the Chinese agricultural sector and forces many larger firms, whether multinationals or state-owned enterprises, to invest heavily in their own transportation networks. Logistics firms specializing in agricultural transport services are beginning to develop advanced cold storage capabilities and expand across China. As China's transportation infrastructure increases in size and capacity, transportation costs and the time to market will decrease, both of which should lower the cost and increase the quality of Chinese agricultural products.

Exchange Rates

The competitiveness of U.S. exports is affected, in part, by the effect of foreign currency fluctuations on the relative price of goods. China started informally pegging its currency to the U.S. dollar in 1998, largely as a stability measure in response to the Asian financial crisis.[79] The peg (8.3 RMB to the dollar) was removed in 2005, as China injected large amounts of cash into the economy in the wake of the severe acute respiratory syndrome (SARS) epidemic and needed to mitigate inflationary pressures.[80] The renminbi exchange rate relative to the U.S. dollar gradually appreciated over the next three years under a managed float until August 2008 (6.8 RMB to the dollar), when China repegged the renminbi to the dollar in response to the global financial crisis.[81] Chinese authorities decided in June 2010 to allow the renminbi to appreciate once again, and through January 27, 2011, the renminbi appreciated 3.7 percent against the U.S. dollar.[82] When a fixed exchange rate causes the renminbi to be less expensive (relative to the U.S. dollar) than it would be under competitive supply and demand conditions, Chinese exports are relatively inexpensive and U.S. exports to China are relatively expensive. As a result, U.S. exports and the production of U.S. goods and services that compete with Chinese imports fall, at least in the short run.[83]

The renminbi is widely viewed as being significantly undervalued, making imports more expensive in China and Chinese exports less expensive in foreign markets. In February 2010, the International Monetary Fund (IMF) stated that the renminbi remains substantially undervalued relative to the U.S. dollar from a medium-term perspective.[84] In April 2010, the Peterson Institute of International Economics estimated that renminbi undervaluation against the U.S. dollar was about 30 percent.[85] In its July 2010 report on China's currency, the U.S. Treasury Department noted that China's continued foreign reserve accumulation, the limited appreciation of China's real effective exchange rate relative to rapid productivity growth in the traded goods sector, and current account surpluses suggest that the renminbi remained undervalued.[86] According to the U.S. Department of the Treasury's subsequent February 2011 *Report to Congress on International Economic and Exchange Rate Policies*, China's real effective exchange rate has appreciated only modestly over the past decade. China's large increases in productivity in export manufacturing, its improvements in transportation and logistics, and its accession to the WTO all suggest that the renminbi should have appreciated more significantly over this period.[87] If the renminbi were permitted to float freely in the market and strengthened against the dollar, as it did during China's last period of currency float, U.S. goods would likely be less expensive for China to import, and demand for these goods would likely increase.[88]

U.S. agricultural exports to China are no exception to this phenomenon. According to the USDA, whenever the U.S. dollar depreciates against the renminbi, U.S. agricultural commodities can become more price-competitive in China, potentially increasing China's demand for these U.S. goods.[89] In 2007, differences in the amount of goods and services that could be purchased in China and the United States with a given amount of U.S. dollars suggested that the exchange rate was not at its equilibrium value—specifically, the renminbi was undervalued.[90] As a result, imported foods in China were much more expensive than domestic products, and consumer demand for U.S. imports was limited to a handful of commodities (soybeans, cotton, and animal hides) and certain high-end niche items. Even assuming an undervalued renminbi, the United States had a net surplus with China in agricultural trade during 2007, which continues today. In fact, U.S. agricultural exports to China have grown sharply, more than doubling from $6.3 billion in 2005 to $13.4 billion in 2009. If the renminbi appreciates against the dollar in the future, more U.S. agricultural products are likely to become price-competitive in China, and U.S. exports of these goods may grow even faster. Appreciation of the Chinese renminbi will also reduce the competitiveness of China's rising exports of labor-intensive agricultural goods, among them vegetables, fruits, and juices.[91] Several U.S. agricultural companies and trade groups submitted information to the ITC for this report, indicating that the current exchange rate between the renminbi and the U.S. dollar impacts their trade with China. For the most part, they indicated that the Chinese currency is undervalued, making U.S. products more expensive in China and Chinese exports cheaper in the United States.[92]

Food Safety Laws and Regulatory Compliance Costs

In the aftermath of consumer illnesses and several deaths in 2008 owing to the presence of melamine in dairy and other food products, awareness and spending on food safety compliance increased in China's agricultural production system. But despite these changes, in general Chinese food processors reportedly spend less on regulatory and food safety compliance than most firms in the United States.[93] Any cost advantage this may confer on Chinese agricultural products vis-à-vis imports must be weighed against negative perceptions Chinese consumers may have about domestic food safety. Food safety compliance costs are therefore likely to increase in order to meet ongoing consumer demands for a safer domestic food supply.

Since the implementation of the Food Safety Law in 2009, provincial agents from the Chinese Administration of Quality Supervision, Inspection, and Quarantine (AQSIQ) and Chinese Inspection and Quarantine (CIQ) have reportedly increased inspections, and compliance costs for food companies and government funding are reportedly rising in a number of ways.[94] Although the financial impact of investments to build food testing laboratories and install equipment to meet stricter food safety controls varies, estimates are that per-unit costs throughout China's food product system may increase by about 1 percent. Furthermore, these efforts require increased management oversight, the cost of which could be the most expensive aspect of compliance.[95]

Chinese food safety laws are complex and, as with most regulations in China, enforcement varies depending on the distribution channel and the final destination point of the product. Products destined for export markets undergo more thorough and stringent inspections to prevent foreign market closures. Agricultural goods for export must be certified by CIQ; local CIQ officials inspect and sample 15 percent of every export shipment, and shipments that fail to meet the proper standards will be suspended. CIQ charges a per-shipment fee (RMB 200–300 [$27–40] per ton), but this cost is generally viewed by Chinese producers as comparable to certification fees in other countries.[96]

Other Government Programs

Valued Added Tax Exemptions for Chinese Farmers

As discussed in chapter 4, the central government provides VAT exemptions for Chinese farmers, giving them a cost advantage in comparison to imports of similar goods. The tax exemptions are legally permitted at several points in the production chain, including VAT on productive inputs and sales VAT, all of which have the effect of lowering the cost of domestic agricultural production and often pricing certain imported products, such as U.S. wheat, out of the Chinese market.[97] The effect of these tax exemptions is to encourage Chinese purchasers to buy domestic agricultural goods rather than imports.[98]

Agricultural Tax Exemptions

Since the early 2000s, the Chinese government has shifted its policies from taxing agriculture to funding programs that promote rural development and shrinking the income gap between rural and urban workers. By 2004, taxes and fees totaling approximately 8 percent of the value of farm output had been eliminated.[99] In addition, other taxes and fees collected from farmers and agricultural processors are frequently waived by local officials. While these practices vary significantly by province and generally depend on a region's economic development priorities,[100] the net result is to lower production costs for Chinese farmers.

Direct Payments and other Support Programs

In order to achieve its goal of self-sufficiency in food grains, China makes direct payments to farmers who produce certain goods, such as grains, soybeans, and hogs. This increases farmers' net income levels while allowing products to be sold at lower prices in the market. For example, direct grain payments increased the net income of farmers producing grains by $169 per hectare between 2005 and 2009.[101] As with other farm support programs, the implementation of direct payments varies by province. Some provinces give farmers direct payments for grains based on planted area, while others do so based on the quantity produced.[102] Payments also vary by product: in 2009, direct payments to farmers for grain production averaged $129 per hectare, while in 2010 payments to farmers for growing high-oil-content soybeans were 10 RMB ($1.33) per mu and the payment for each pig raised by small producers was 50 RMB ($6.66).[103] While direct payments may allow farmers to sell their products at lower costs, these funds distort the competitive position of Chinese goods in the market.

Factors Affecting Product Differentiation

Like consumers in other countries, Chinese citizens are eating more calories and seeking a wider variety of food as their incomes rise.[104] Rising incomes allow consumers to be less price-conscious in their food choices, opting for better quality or more varied foods at higher prices. The desire for higher volumes of food and differentiated products has the potential to spur imports from the United States and other large agricultural producers, because most of these goods are produced to high quality standards and are seen as premium products by Chinese consumers. A number of factors differentiate Chinese products from imports, including the use of cold storage and refrigerated transportation, industry structure, farm management practices, food safety standards, and branding.

Cold Storage

Cold storage supply chains are critical to ensuring the quality of perishable goods, but as already noted, cold storage is less available in China than in the United States. The lack of an efficient, nationwide cold storage transportation network not only leads to high rates of spoilage, but also lowers the average quality of goods that find their way to market.[105] It is estimated that only 15 percent of meat and 5 percent of fruits and vegetables are transported using cold storage in China, compared to more than 90 percent for both in the United States.[106] The development of supply chains with complete cold storage in China is hindered by many factors, including the fragmented structure of the food industries. Small farm plots scattered throughout China result in significant handling outside of existing cold storage supply chains. As farmland consolidation and village cooperatives become more common in China, investment in cold storage facilities is expected to increase. As cold storage is implemented throughout China's supply chain, the quality of domestically produced perishable goods will improve and increase competition with high-quality imports.[107]

When perishable foods, such as fresh fruits and vegetables, are transported outside of a chilled environment, shelf life is quickly reduced as the product ripens. The result is diminished average quality when the goods arrive at market and higher volumes that must be destroyed or converted into animal feed because of spoilage. China's farm sector suffers from both problems. Moreover, even in cases where cold storage facilities exist, the lack of knowledge among workers about proper storage temperatures, chilling procedures, and ideal controlled atmosphere conditions reduce the effectiveness of cold storage throughout China.[108]

Industry structure and consumer preferences also impact the use and availability of cold storage. Particularly in China's horticultural industries, thousands of small-scale household farmers and small traders consolidate products after harvest, and financial resources for investing in cold storage are typically not available. Fresh fruits and vegetables generally do not enter into cold storage until further up the supply chain, when products reach wholesale markets in larger cities after several days in transit.[109] In addition, consumer preferences in China have slowed the expansion of cold storage facilities even for fresh-chilled products because consumers still prefer local wet markets, where perishable products are sold at room temperature.[110] For example, Chinese consumers buy larger cuts of meat from butchers at wet markets because they perceive meat sold at room temperature to be fresher than prepackaged meats chilled and sold at larger retail outlets.[111]

Industry Structure and Land Tenure System

As previously mentioned, the land tenure system and the structure of the Chinese agricultural industry increase production costs. But the industry structure also restricts industry-wide product quality improvements. The use of modern production techniques and good agricultural practices have increased the volume of high-quality products that can compete with imports, but the supply of these products is small because implementing these practices consistently on many small household farms is expensive and difficult. Information distribution in China remains inefficient, and government extension services are typically focused on producing higher output and not on quality improvements.[112] In addition, because farmers have limited land ownership rights, few incentives exist for farmers to invest in efficient and environmentally sustainable agricultural practices. If China's farm sector is able to consolidate land holdings, either through land leasing or village cooperatives, good agricultural practices will likely become more widespread, increasing the ability of Chinese producers to compete with higher-quality imports.

Farm Management Practices

Advanced farm management practices can significantly improve the quality of production. For example, drip irrigation, proper thinning, pruning, weed control, and mulching can improve the size, color, and flavor of fruits while also increasing disease and insect control. In China, the use of advanced farm management practices has been increasing, but these types of practices are still not implemented universally.[113] On many farms, lack of information is restricting the implementation of advanced farm management and thus reducing product quality.

Limited knowledge of modern farming techniques is also causing soil degradation and restricting yield growth in China. In order to boost output from their tiny plots of land, farmers use intensive agricultural practices, including heavy fertilizer applications and double- or triple-cropping.[114] Poor on-farm management of soil organic matter, water applications, and fertilizer use reduces soil fertility and nutrient levels needed to maintain high yields and cost-effective crop production. Inadequate awareness of the proper balance and mix of different fertilizers to use, as well as of appropriate application practices, result in inefficient releases of soil nutrients, suboptimal growth rates, and water and soil pollution.[115] The widespread sale of counterfeit and poor-quality farm chemicals, feeds, and veterinary drugs also undermines output.[116]

Food Safety

China's 2009 food safety law has begun to increase compliance costs for some domestic agricultural processors, but the law is also beginning to result in higher-quality and safer products. Despite some improvements in food quality and safety, however, most Chinese consumers still typically view processed foods as less safe and lacking the nutrients that fresh products provide.[117] Chinese food processors have begun marketing campaigns in order to change the negative perception of processed foods.[118]

Chinese consumers also typically view imports, both processed and fresh, as higher in quality than domestically produced food and safer to eat. As consumer awareness of food safety has increased and import volumes of higher-quality products have risen, Chinese agricultural producers have been forced to improve their product quality to compete with imports. For example, pressure from imported apples has resulted in the wider availability of higher-quality, domestically produced fresh apples.[119] With increased competition from high-quality imported products, Chinese producers are expected to continue to improve their product quality.[120]

Branding

As per capita income continues to grow in China, consumers are increasingly looking to buy products that are differentiated by a strong brand reputation.[121] Internationally recognizable food brands, often produced locally in China, are some of the most recognized brands to Chinese consumers.[122] Overall, Chinese food processors have not developed strong well-known national brands because they normally produce items based on customer specifications, often under other companies' independent labels. This industry structure lowers processors' profit margins. Additionally, not having recognizable brands generally restricts processors' ability to develop new markets, especially at the global level.[123] However, some Chinese companies, particularly in the beverage and noodle sectors, have developed brands that are gaining customer recognition, and are thus strengthening their market position.[124]

Factors Affecting Reliability of Supply

Reliability of supply is an important determinant of a supplier's competitive position in the Chinese market. The availability and supply of agricultural products are first shaped by weather conditions, but a number of other factors affect a supplier's reliability, including storage and transportation infrastructure, marketing information systems, and water and land availability.

Storage and Transportation Infrastructure

For agricultural products produced in China, a reliable supply is generally less assured than it is for products coming from other suppliers because of poor or insufficient storage and transportation infrastructure. Despite China's heavy investments in infrastructure in recent years, the growth in road networks has not kept up with demand growth. And, as noted earlier, the lack of availability of quality storage, including cold storage throughout the supply chain, restricts the ability of Chinese producers to reliably supply their customers.[125]

The limited use and availability of cold storage and refrigerated transport in China, which is critical to reliably supplying perishable products, results in significant product losses during delivery and increases uncertainty for buyers. As a result, according to the Chinese National Development and Reform Commission, on average 15 to 20 percent of agricultural products are lost during transportation. For some products in certain regions, losses could run as high as 30 percent, compared with losses of less than 2 percent in the United States.[126]

While grains are not highly perishable, lack of quality storage impacts the reliability of the Chinese grain supply. Although China has ample grain storage facilities, including the capacity to store 200 million metric tons of wheat and paddy rice, losses of grain in storage to mold and insects reach 8 to 10 percent annually,[127] a much higher percentage than for most large developed grain-producing countries. However, China's use of mechanized handling and preservation techniques reportedly extends the shelf life of grain much more effectively than some other large grain-producing developing countries, such as India.[128]

The reliability of on-time deliveries of agricultural products in China is compromised by inefficient transportation networks. An inefficient freight rail system and traffic jams that slow down product movement along the supply chain increases uncertainty for buyers of domestic products.[129] Because of the time needed to transport goods throughout China, imports often have a competitive advantage in markets where there is not significant domestic production in the region.[130]

Marketing Information Systems

The lack of integrated automated information systems also creates uncertainty in the timing of product deliveries.[131] Chinese producers have not yet modernized their modes of commerce to provide more efficient and reliable product marketing and distribution. Most small enterprises and traders use cell phones to talk to clients and suppliers, but the majority do not use integrated information systems and do not have automated storage, transportation, or billing systems. For distribution and sales, it is difficult for Chinese marketers to use e-commerce because the law does not recognize e-signatures.

Water Resource Depletion and Degradation

As noted above, China's farm sector does not have an abundant supply of water, and resources are under strain from overuse and degradation due to pollution. This situation may limit the ability of Chinese farmers to reliably supply sufficient quality and quantities of agricultural products, thus creating increased demand for imports. China already has a restricted supply of water, with resources per capita that are only about 30 percent of the world average.[132] In addition, water resources in China are unevenly distributed by region, with south China (the Yangtze River basin and to the south) accounting for 80 percent of the nation's naturally available water resources, but only 53 percent of the population and 35 percent of its arable land.[133] Demand growth from the expanding industrial and urban sectors has increased competition for these limited water resources, while overuse by the agricultural sector is driving a drop in groundwater tables and causing rivers to run dry.[134]

The water that is left is also being degraded by industrial, urban, and agricultural pollution, which is already harming agricultural output. One-fourth of the water sampled along China's two largest rivers—the Yangtze and Yellow—was found to be too polluted even for farm irrigation.[135] In 2007 it was estimated that 70 percent of China's rivers were severely polluted.[136] However, this estimate may have underestimated water pollution by more than half because it did not take into account agricultural pollution.[137] China does not yet have effective water management polices and practices to deal with these challenges.[138]

Land

The Chinese government has set goals to try to maintain a minimum level of land devoted to agriculture, but agricultural land is being threatened by ongoing urbanization, industrialization, and desertification.[139] As economic development continues and demand for land increases, future agricultural output may become uncertain because of reduced land availability.

The total amount of agricultural land has not yet fallen as a result of the increased competition for land, but the land in use has shifted into areas with marginal productivity. Land that was previously left fallow by farmers has been pushed back into production.[140] As production shifts onto less productive land, future increases in yield and output to meet growing demand may be difficult to achieve, in effect limiting the reliability of Chinese supply. This may result in increased Chinese demand for land-intensive agricultural crops from international suppliers.

Bibliography

Batson, Andrew. "Farm Wages Trip Beijing's March Against Inflation." *The Wall Street Journal*, December 23, 2010. http://online.wsj.com/article/SB10001424052748704447604576007240671670976.html.

British Broadcasting Corporation (BBC). "Timeline: China Milk Scandal," January 25, 2010. http://news.bbc.co.uk/2/hi/asia-pacific/7720404.stm.

Bolton, Jamie, and Wenbo Liu. "Creating an Effective China 'Cold Supply Chain': Current Status, Challenges and Implementation Considerations." *Accenture*, 2006. http://www.accenture.com/Global/Research_and_Insights/By_Subject/Supply_Chain_Mgmt/Crea tingChinaColdSupplyChain.htm.

Brown, Lester."Aquifer Depletion." *Encyclopedia of Earth,* January 23, 2010. http://www.eoearth.org/article/Aquifer depletion.

Bureau van Dijk. Zephyr Mergers and Acquisitions Database. https://zephyr.bvdep.com (accessed various dates).

Charron, Martin. "Agriculture, Food and Beverages Sector Profile: Beijing, China." *Agriculture and Agri-Food Canada*, December 2008. http://www.ats.agr.gc.ca/asi/4625-eng.htm.

China Daily. "China's Urban, Rural Income Gap Widens," January 22, 2010. http://www.chinadaily.com.cn/bizchina/2010-01/22/content 9361049.htm.

———. "Poor Irrigation Facilities Threaten Grain Production," March 12, 2009. http://www.chinadaily.com.cn/china/2009npc/2009-03/12/content 7571569.htm.

———. "Chinese Farmers Lease Their Farmland, Take Advantage of Govt Policy," June 16, 2010. http://www.chinadaily.com.cn/business/2010-06/19/content 9992788.htm.

"China's Farmers Could Use Land as Collateral." *Wall Street Journal*, July 29, 2010. http://online.wsj.com/article/SB10001424052748703940904575396343463631042.html.

European Commission (EC). Directorate General for Agriculture and Rural Development (DGA). "China: Out of the Dragon's Den?" Monitoring Agri-trade Policy (MAP) no. 01-08, May 2008.

Fei Ping and Xu Li-qing. "Dynamic Relationship between Agricultural Trade and FDI in China." *Asian Agricultural Research* 1, no. 4 (2009): 27–31.

Financial Times. fDiMarkets Investment Database. http://www.fdimarkets.com (accessed various dates).

Food and Agricultural Organization of the United Nations (FAO). "Synthesis of Asia and the Pacific Region's Perspective on Nutrient Management and Soil Productivity," May 14, 2007. http://www.fao.org/docrep/010/ag120e/AG120E04.htm.

———. "A Competitive Analysis and Strategy." Chapter 6 in *Global Agriculture Marketing Management.* Rome: FAO, 1992.

Ford, Peter. "China's Land Reform Aims to Revolutionize 750 Million Live: Beijing Hopes the Policy Will Improve Farming and Free Peasants to Seek a Better Livelihood." *The Christian Science Monitor*, October 27, 2008. http://www.csmonitor.com/World/Asia

Gale, Fred, and Francis Tuan. *China Currency Appreciation Could Boost U.S. Agricultural Exports.* Report no. WRS-0703. U.S. Department of Agriculture, Economic Research Service, August 2007.

Gale, Fred, Brian Lohmar, and Francis Tuan. *China's Ongoing Agricultural Modernization: Challenges Remain after 30 Years of Reform.* Economic Research Bulletin no. 51. U.S. Department of Agriculture, Economic Research Service, April 2009.

Gifford, Rob. "Food Fears Persist in China 2 Years after Milk Scare." NPR, October 26, 2010. http://www.npr.org/templates/story/story.php?storyId=130826464.

Gonzalez, Julio, Jose Luis Guasch, and Tomas Serebrisky. "Latin America: High Logistics Costs and Poor Infrastructure for Merchandise Transportation." *Consulta de San Jose Solution Paper,* 2007.

Guang-xin, Zhang, and Deng Wei."Groundwater Crisis and Sustainable Agriculture in Northern China." *Water Engineering & Management* 149, no. 4 (April 2002). Republished in *Water and Wastes Digest*, 2010. http://wwdmag.com/Groundwater-Crisis-and-Sustainable-Agriculture-in-NorthernChina-article3037.

International Monetary Fund (IMF), Meetings of G-20 Deputies, Seoul, Korea, *Global Economic Prospects and Policy Challenges*, February 27, 2010. http://www.imf.org/external/np/g20/pdf/090309.pdf.

ISI Emerging Markets. CEIC Database. http://www.ceicdata.com (accessed May 21, 2010).

Karhl, Fredrich, Li Yunju, David Roland-Holst, Xu Jianchu, and David Zilberman. "Toward Sustainable Use of Nitrogen Fertilizers in China." *Agricultural and Resource Economics Update, University of California*, 14 no. 2, November/December 2010.

Law, Violet. "China: Chairman Mao's Dream Water Project Beijing of Farmers." One Water, Knight Center for International Media, University of Miami. http://www.onewater.org/stories/story/china3 (accessed November 10, 2010).

Lee, Ju Young. "Is Water Too Cheap in China?" Water Matters (blog), The Earth Institute, Columbia University, May 22, 2009. http://blogs

Liao, Hui. "The Evolution of Fertilizer Subsidies in China." Presented at the International Fertilizer Association Crossroads Asia-Pacific 2008 Conference. Melbourne, Australia, December 16–18, 2008.

Li, Bing. "Shandong Agricultural Product Exports: Growth, Problems, and Countermeasures." *International Journal of Business Management* 4, no.3 (March 2009), 62–65.

Liu, Coco. "More Cold Chain Gives More Choice in China." *Poultry International*, September 2009.

Liu, Zhixiong. "Researches on the Effect of Foreign Direct Investment in the Food Processing Industry in China." Business School of China University of Political Science and Law, n.d. http://www.seiofbluemountain.com/upload/product/201003/2010cygchy01a22.pdf.

Morrison, Wayne M., and Marc Labonte. "China's Currency Peg: A Summary of the Economic Issues." Report no. RS 21625. Congressional Research Service, April 25, 2005. http://assets 20050425.pdf.

Mullan, Katrine, Pauline Grosjean, and Andreas Kontoleon. "Land Tenure Arrangements and Rural-Urban Migration in China." *World Development* 39, no. 1 (January 2011), 123–33.

National Cotton Council. Written submission to the U.S. International Trade Commission in connection with inv. no. 332-518, *China's Agricultural Trade: Competitive Conditions and Effects on U.S. Exports*, September 15, 2010.

Newsweek. "Where Rivers Run Dry," April 16, 2007. http://www.newsweek.com/2007/04/15/wherechina-s-rivers-run-dry.html.

Organization for Economic Co-operation and Development (OECD). "Rural Policy Reviews: China 2009," May 22, 2009.

Pearce, David W., ed. *The MIT Dictionary of Modern Economics*. 4th ed. Cambridge, Massachusetts: The MIT Press, 1997.

People's Daily Online. "China's Rural Roads to Cover 3.5 Million Km by 2011," October 28, 2010. http://english.peopledaily.com.cn/90001/90776/7180264.html.

People's Republic of China (PRC). Ministry of Agriculture (MOA). "General Surveys: Agricultural Mechanization," June 25, 2009.

People's Republic of China (PRC). Ministry of Commerce (MOFCOM). "Reference Data on FDI Distribution in Major Sectors." Annual FDI Data, 2004–09. http://www.fdi.gov.cn /pub/FDI_EN/Statistics/AnnualStatisticsData/default.jsp (October 28, 2010).

———. "The Survey of Foreign Investment in China's Agriculture Industry." Annual FDI Data, 2004–09. http://www.fdi.gov.cn/pub/FDI_EN/Statistics/Annual StatisticsData /default.jsp (October 28, 2010).

Poultex News. "Half of China's Antibiotics Are Fed to Animals," December 1, 2010. Powell, Simon. "Falling Up! Water Prices Must Meet the True Costs." *CLSA,* May 2010.

Qian, Wang. "Water Prices Need to Shoot Up." *China Daily*, March 10, 2009. http://www.chinadaily.com.cn/china/2009-03/10/content 7557856.htm.

Reuters. "China No Longer Pegging Yuan to Dollar," July 21, 2005. http://www.foxnews.com/story/0,2933,163183,00.html.

Roberts, Dexter. "China Faces a Water Crisis." *Bloomberg Businessweek*, April 15, 2009. http://www.businessweek.com/globalbiz/content/apr2009/gb20090415 032220.htm.

Scheinson, Andrew. *China's Internal Migrants.* Council on Foreign Relations, May 14, 2009. http://www.cfr.org/publication/12943/chinas internal migrants.html.

Schwartz, Mark. "Balancing Act in Global Fertilizer Use: Science Report." Stanford University Program on Food Security and the Environment, June 29, 2009. http://foodsecurity.stanford.edu/news/balancing_act_in_global_fertilizer_use__science_ report_20 090629.

Southeast Farm Press. "China, India Waking Up to Fertilizer with Huge Subsidies," n.d. http://westernfarmpress.com/management (accessed November 2, 2010).

Stewart and Stewart. Written submission to the U.S. International Trade Commission in connection with inv. no. 332-518, *China's Agricultural Trade: Competitive Conditions and Effects on U.S. Exports*, June 29, 2010.

Subramanian, Arvind. "New PPP-Based Estimates of Renminbi Undervaluation and Policy Implications." *Peterson Institute of International Economics*. Policy Brief Number PB10-8, April 2010, http://www.piie.com/publications/pb/pb10-08.pdf.

Tada, Minoru, Dinghuan Hu, and Ruangrai Tokrisna. "The Role of Foreign Direct Investment and Food Processing Industry toward High Value Agriculture in Asia: Cross Country Statistical Analysis and Farm Level Evidence." *JARQ* 43, no. 4 (2009): 317–322.

Taige, Li. "Drought without End?" *China Dialogue*, March 3, 2009. http://www.chinadialogue.net/article/show/single/en/2803.

Terreri, April. "Go West with Caution." *World Trade,* November 1, 2010. http://www. worldtrademag.com/Articles/CoverStory/BNPGUID9-5-2006A100000000 00000931673.

Tollroad News. "Central Government Struggling to Control Tollroads in China: Market Forces Dominant," February 29, 2008. http://www.tollroadsnews.com/node/3417.

Ulics, Rozi, and David Mead. "Current Price Topics: China's Exchange Rate Policy Reflected in U.S. Import Prices." *Focus on Prices and Spending: Import and Export Prices*. 1, no. 5 (August 2010).

U.S. Department of Agriculture (USDA). Economic Research Service (ERS). "Briefing Rooms: China; Basic Information," December 19, 2009. http://www.ers.usda.gov/Briefing/China

U.S. Department of Agriculture (USDA). Foreign Agricultural Service (FAS). *China, Peoples Republic of: Grain and Feed; Annual, 2010*, by Chanda Beckman and Jiang Junyang. GAIN Report no. CH10008, March 1, 2010, 16.

———. *China, Peoples Republic of: Grain and Feed; Annual, 2008*, by Jorge Sanchez and Jiang Junyang. GAIN Report no. CH8012, March 1, 2008.

———. *China, Peoples Republic of: Fertilizer,* by Chanda Beckman, Katie Rohm, and Bao Liting. GAIN Report no. CH9082, December 14, 2009.

———. *China, Peoples Republic of: Trade Policy Monitoring; VAT Protections: The Rest of the Story 2007*, by Kevin Latner and Jianping Zhang. GAIN Report no. CH7018, March 19, 2007.

U.S. Department of Agriculture (USDA). National Agricultural Statistics Service (NASS). "2007 Census of Agriculture: United States Summary and State Data," December 2009.

———. "Farm Labor," August 19, 2010.

———. "Land Values and Cash Rents Summary," August 2010.

U.S. Department of Treasury, Office of International Affairs. "Report to Congress on International Economic and Exchange Rate Policies." July 8, 2010. http://www.treasury.gov/resourcecenter/international/exchange-rate-policies/Pages /index.aspx.

———. *Report to Congress on International Economic and Exchange Rate Policies,* February 2011.

U.S. Wheat Associates. Written submission to the U.S. International Trade Commission in connection with inv. no. 332-518, *China's Agricultural Trade: Competitive Conditions and Effects on U.S. Exports*, September 16, 2010.

Wang, Hongling, Xiaoxia Dong, Scott Rozelle, Jikun Huang, and Thomas Reardon. "Producing and Procuring Horticultural Crops with Chinese Characteristics: The Case of Northern China." *World Development* 37, no. 11 (November 2009): 1791–1801.

Wolverson, Roya. "Confronting the China-US Economic Imbalance." Council on Foreign Relations Backgrounder, October 19, 2010. http://www.cfr.org/publication/20758/confronting the chinaus economic imbalance.html.

The World Bank. Data: World Development Indicators, http://data.worldbank.org/data-catalog/worlddevelopment-indicators (accessed November 9, 2010).

Xiang, Li. "PE Firms Eye Rich Harvest in Agricultural Logistics." *China Daily*, June 11, 2010. http://www.chinadaily.com.cn/bizchina/2010-06/11/content 9965536.htm.

Xie, Jian. *Addressing China's Water Scarcity: Recommendations for Selected Water Resource Management Issues.* Washington, DC: World Bank, 2009. http://www-wds.worldbank.org/external/default/main?page PK=64193027&piPK=64187937& theSitePK=523679&menuPK=64187510&searchMenuPK=64187283&theSitePK=52367 9&entityID=00033303 7_20090114011126&searchMenuPK=64187283&theSitePK=523679.

Zhang, F.S. "Improving Fertilizer Use Efficiency through Management Practices: China Experience." China Agriculture University, n.d. http://www.faidelhi.org/FAI F%20Zhang.pdf.

Zhang, Jialin. "China's Slow-motion Land Reform." *Policy Review*, February/March 2010. http://findarticles.com/p/articles/mi qa3647/is 201002/ai n52374979.

6. Competitive Factor Case Studies

Overview

This chapter describes competitive conditions in selected Chinese agricultural sectors: fresh apples, pork, processed foods, and wheat. These sectors were chosen to illustrate China's competitive factors, and to explain which factors have been used to capitalize on rapid changes to China's food production system. Salient competitive factors affecting China's agricultural sector as a whole may not include some factors that are key to U.S. exporters entering the Chinese market (box 6.1).

China is a major horticultural producer because many horticulture products require significant labor resources to plant, tend, and harvest. The Chinese fresh apple industry illustrates how low labor costs benefit China's farm sector, especially in terms of lower-quality exports. The U.S. apple industry, also one of the world's largest producers and exporters, competes with Chinese apples in the high-quality segment of export markets and has been increasing exports to China to take advantage of China's growing wealth and consumption.

Pork is the most widely consumed meat in China and accounts for approximately 65 percent of animal protein consumption there. Chinese government policies have encouraged domestic production of pork and insulated many small-scale, low-quality producers from global market conditions. However, rising labor and feed costs, as well as increasing demand among Chinese consumers for higher-quality pork, may lessen Chinese pork competitiveness in the future, creating opportunities for global exporters from the United States and Europe.

China is one of the world's largest processed food exporters, behind only the European Union (EU-27) and the United States. Rising disposable income and urbanization are driving demand and consumption growth in China. Processed food production in China is shifting from a reliance on low-cost labor to more capital-intensive and heavily mechanized production processes that are similar to the advanced methods used in other countries that are significant producers of processed foods. Many U.S. food processors compete with Chinese products both in China and in third-country export markets.

Wheat is a staple food that China has identified as important to China's food security. Despite limited land and water resources for wheat production, the central government has set domestic and trade policies to maintain domestic production and control import supplies. In the aggregate, government policies have boosted China's domestic wheat production and insulated producer and consumer prices from volatile global wheat markets.

Fresh Apples

The Chinese apple industry is competitive in both domestic and export markets primarily because of its abundant supply of low-cost labor. Nonetheless, other factors, such as low quality and industry structure, restrict its ability to compete in certain market segments. Chinese growers primarily supply market segments that focus on cost, while the U.S. industry provides high-quality apples to higher-income market segments through established retail outlets. Chinese producers, however, have been improving their quality; they are entering higher-quality segments of the market, and they have started to capture larger shares of import growth than U.S. apples in certain export markets, such as India. If the Chinese industry continues to improve the quality of its production by upgrading its orchard management techniques and post-harvest treatment, Chinese apples may continue to expand their market share in some countries. On the other hand, production costs in China are rising quickly, particularly for labor and certain inputs, and domestic volumes are heavily concentrated the Fuji variety.[1] If demand shifts away from the Fuji variety, if production costs rise, and if the improved farm management methods cease to be cost-effective, China's low-cost advantage in producing apples may diminish.

BOX 6.1. Certain Chinese Import Restrictions Negatively Affect the Competitiveness of U.S. Exporters More Than They Benefit That of Chinese Producers

For many U.S. exports to China, the shipped volumes are extremely small relative to Chinese production and consumption. Thus, even if a given Chinese import restriction on U.S. exports were removed (such as a tariff or nontariff measure) and resulted in a significant percentage change in U.S exports, the change would still be small relative to China's domestic consumption.

Further, many U.S. exports to China do not directly compete with the majority of Chinese domestic production because of their high quality, high cost, or other differences.

China's markets for apples and for pork can be used to illustrate the interaction between imports and domestic production in China. China produces nearly half of global production of both pork and apples. In 2009, China produced approximately 32 million metric tons (mt) of apples and imported 61,000 mt. In that year, China produced 48.9 million mt of pork and imported 600,000 mt.[a] For both apples and pork, U.S. exports account for a large share of China's imports, but are small relative to consumption.[b] U.S. apple exports to China are restricted by the lack of pest risk assessments on most varieties from most U.S. states. U.S. pork exports to China are restricted by differences in sanitary requirements, and in 2009 were further restricted by bans related to the presence of A1H1 influenza in the United States.

Because of the small volume of imports relative to Chinese production, changes in China's import policies for apples or pork that would have significant effects on U.S. export volumes would be expected to have little effect on the competitive conditions facing China's producers.

Furthermore, for both apples and pork, imports largely do not compete directly with the bulk of Chinese production. Traders report that imported apples are sold through different channels of distribution, and that changes in the volume in one segment have little impact on prices in the other.[c] Most Chinese consumers prefer pork sold fresh through wet markets, a channel into which frozen imports do not enter.[d]

[a] Sum of reported global exports of fresh or chilled pork, frozen pork, edible pork offal, and prepared pork products from all sources; GTIS, Global Trade Atlas database.

[b] In 2009, U.S. exports accounted for 40 percent of global apple exports to China on a quantity basis, but much less than 1 percent of China's consumption. According to official statistics, U.S. pork exports to China in 2009 accounted for 6 percent of global pork exports to China but were equivalent to much less than 1 percent of consumption.

[c] Industry official, interview by Commission staff, Beijing, China, September 6, 2010.

[d] Industry official, interview by Commission staff, Washington, DC, May 6, 2010.

Industry Overview

Global Position

China is the world's largest apple grower, accounting for approximately 50 percent of global production in 2009/10 (table 6.1). Rising per capita income levels in China have been driving increased consumption of fresh horticultural products, including apples. In response, Chinese apples growers have expanded their acreage and production in order to meet this growing demand, a trend that is expected to continue in the future. Recent growth in acreage has occurred primarily in the Shaanxi, Shanxi, and Gansu provinces because of ideal growing conditions and a wider availability of land for expansion.[2]

Table 6.1. Fresh apples: Production, consumption, and trade, selected countries, marketing year 2009/10 (1,000 metric tons)

Country	Total production[a]	Fresh consumption	Imports	Exports
China	31,680	24,941	61	1,201
EU-27	12,210	8,298	594	1,216
United States	4,403	2,319	182	768
Rest of World	14,365	13,744	4,133	2,048
Total	62,658	49,202	4,970	5,233

Source: USDA, FAS, PSD Online (accessed January 13, 2011).

Note: Consumption does not include apples that are further processed into products such as juice and slices. [a]Total production includes apples for the fresh market as well as for processing.

In China, the Fuji apple variety accounts for almost 70 percent of total production.[3] Over time, consumer preferences among apple varieties have remained relatively stable, so the Fuji variety is expected to continue to account for a high percentage of production. Instead of shifting new plantings into different varieties to match consumer preferences which are beginning to widen, the industry is focused on upgrading the quality of production of Fuji apples through improved rootstocks [4] and orchard management techniques.[5]

Because of its large supply of low-cost labor, China is increasingly competitive in international markets and is the world's primary low-cost apple supplier.[6] China is also the world's largest single-country apple exporter by volume, even though exports accounted for only 4 percent of domestic production in 2009.[7] While the quality of production has improved over time, the Chinese apple industry typically competes in the lower-quality segment of the global market, where competition among suppliers is based primarily on price. In addition to export markets, another major outlet for Chinese apple production is the juicing sector. Typically between 4 and 5 million metric tons (mt) of apples are processed each year, primarily into juice, depending on the relative prices for the fresh market versus for processing. Chinese imports of fresh apples, primarily from the United States and Chile, more than doubled in value between 2005 and 2009 in response to strong consumer demand for high-quality imported apples.

Industry Structure

The structure of the Chinese apple industry consists of millions of small household farms that grow apples. This differs from the structure in the United States, which is characterized by orchards on plots of several hundred hectares. Farms that grow apples in China generally range in area from 5–10 *mu* (about 0.33–0.66 hectares), but many are even smaller. The typical household farm produces only about 20 mt of apples per year.[8] The small scale of apple farming in China results in inefficient supply chains in comparison to other major producers.[9]

Domestic Marketing Outlets

Product distribution within the Chinese market occurs through four different types of outlets: traditional wet markets and fruit stalls, larger urban markets, supermarkets, and large wholesale export markets. Traditional local wet markets are the most common marketing outlet for the Chinese apple industry. Most of the fruit is first collected by fruit traders, who consolidate the product from a number of farms and then sell it to other traders or vendors.[10] Apples sold through local wet markets are almost never placed in cold storage. Sales of apples through supermarkets have increased greatly in recent years, but this outlet still only accounts for a small share of consumption.

The Chinese fresh apple market consists of two distinct channels. The first channel serves the wealthier, mainly urban segment of the population, in which high-quality apples are sold through supermarkets, restaurants, hotels, and smaller urban fruit stalls. Apples sold through this channel consist mostly of imports, but also include higher-quality domestically produced apples. The second channel serves most other Chinese consumers, with the majority of sales through traditional local wet markets. Apples sold in this channel are domestically produced and of mixed quality. Because of these distinct channels, there is little correlation between supplies and prices of domestically produced and imported apples. The price premiums of imported over domestically produced apples vary, but generally reflect the volume of apple imports rather than the price of domestic supply.[11]

Factors Affecting Competitiveness

Labor Costs

Low wage rates provide Chinese apple producers primary competitive advantage because labor accounts for the largest portion of production costs.[12] Apple production is a labor-intensive process, since pruning, thinning, and harvesting must be done manually. Because Chinese producers have access to an abundant supply of low-cost labor, growers are able to use significantly more labor per hectare than in other major producing countries while remaining one of the world's low-cost suppliers.

Despite the cost advantage currently provided by low wage rates, competition for labor from other industries has reduced labor availability in many apple-producing provinces and increased wages. Many small household farmers harvest their own apples and do not need to hire outside labor, but larger apple producers (5–10 *mu*) that hire laborers typically pay on average between 50 RMB ($6.66) and 60 RMB ($8.00) per day.[13] However, labor costs vary by province, and they have dramatically increased in the major apple-producing provinces of northeastern China in 2009. For example, in Shaanxi province, labor costs increased by 20 percent in 2009 to an average of $8.80 per day per worker; in Shandong, one of the largest apple-producing provinces, daily labor costs almost doubled in 2009 to $13.25 as a result of continued economic development in the province.[14] In comparison, the average wage in the United States for field labor in 2010 was $10.12 per hour.[15]

The lack of mechanization in Chinese apple production increases the use of labor, as growers often pollinate by hand, cover individual apples with bags, apply pesticides and fertilizers from hand-pumped backpacks, and transport apples in bags through the orchard on foot. These practices increase the average number of hours worked per acre in Chinese apple orchards to over 1,300 hours per year, compared with an average of 200 hours in Washington state.[16] Nonetheless, low wages still reduce China's total cost relative to other producers.

Industry Structure

As discussed in chapter 5, China's land tenure system results in a farm structure comprising many small farms, many fragmented into still smaller plots. For the apple industry, small-scale production creates an inefficient supply chain involving high transaction and distribution costs compared to other major global apple-producing countries that are able to reap the benefits of economies of scale. This structure weakens China's competitiveness in global apple markets.

Because apples are grown on millions of small household farms, getting the product to the market requires several links in the supply chain that consolidate the product, each incurring additional cost. Approximately 80 to 90 percent of apples are first sold to small traders who buy directly from farmers in the field.[17] Even at the second point of sale, the product is generally sold to other small traders before being sold to packinghouses, supermarkets, processors, or specialized supply firms. Despite their increasing prevalence in urban areas, large supermarkets are still almost completely disconnected from production and buy the majority of their produce at wholesale markets after the produce has been compiled by smaller traders.[18] The price paid by the average Chinese consumer is significantly higher than the cost of production because the product changes hands so many times before getting to the consumer.

While the industry structure results in high transaction costs in the marketing and distribution system, it also gives producers less reason to invest in improved farm management techniques and lowers the quality of apples produced. Incentives for farmers to invest and improve their land's productivity and quality are limited by the continued legal uncertainty about their land use rights.[19] Further, because there are so many small producers, the distribution system lacks consistent quality oversight and standardization.[20]

Farmers' cooperatives are being established in the Chinese apple industry to combat the inefficiencies associated with the current farm structure. With cooperatives, farmers operate individual plots of land but consolidate the product and centralize administrative control through a village leader or "dragonhead." As a result, farmers may be able to control more of the value chain, standardize farm management practices to improve quality, centralize resource distribution, finance investments in new machinery, and reduce transaction costs. However, the effectiveness of cooperatives is currently restricted by a lack of skilled village managers,[21] and while the number of cooperatives is increasing they still only account for a small share of total Chinese apple production.[22]

Product Quality

Product quality is a key factor affecting Chinese competitiveness in both the domestic and overseas apple markets. The flavor, appearance, size, shelf life, and overall condition of Chinese apples is restricted by traditional methods of irrigation, fertilization, and farm management.[23] For example, in Shandong province production practices include flood irrigation and the application of manure and top-dressed fertilizer, two production practices that are considered inefficient. Growers also rarely use improved rootstocks. Product quality is also affected by the lack of financial resources for small-scale farmers to invest in on-farm cold storage. As a result, apples generally do not enter into cold storage until they reach wholesale markets in larger cities, a delay that lowers product quality. Reportedly, only 5 percent of total apple production is transported using cold storage, and most of that is destined for export markets.[24] In comparison, imports into China use a complete cold storage supply chain, which increases costs but maintains the quality of the product.

Although imports and domestic production typically use different marketing channels, increased import volumes have forced the Chinese industry to make changes.[25] Growers have begun implementing methods to improve production and quality. For example, some Shandong apple farmers are applying fertilizers more prudently, and fertilizer application rates are slowly falling as packinghouses and other traders increase their oversight.[26] In addition, in 2005 the government launched a subsidy program to cover individual apples with bags; this improves apples' quality by protecting them from weather and pests.[27] The practice of bagging apples is becoming increasingly common.[28]

As producers continue to implement better, but more costly, orchard management practices, Chinese quality has been improving to the extent that some domestic production now competes with U.S. imports in terms of color, shape, and general appearance.[29] The prices of the highest-quality Chinese apples have risen to levels that are almost comparable to imported apple prices.[30]

Pork

In the past, Chinese pork products were competitive with imports owing to the low cost of labor. As China's pork industry shifts to a production model using more capital (through mechanization) and fewer laborers, the advantage of low labor costs has declined. Feed costs now account for a greater share of the delivered cost than labor. As feed costs rise because of greater demand, Chinese producers become less competitive relative to other global pork-producing countries because China's domestic grain prices are relatively high. Currently, government payments and trade policies—e.g., tariffs, non-tariff measures (NTMs), and value-added tax (VAT) policies—have kept the delivered cost of Chinese pork competitive with imports. Without these government measures, the delivered cost of domestic pork would be far higher, and imports would likely capture a greater share of China's market.

Industry Overview

Global Position

China is the world's largest producer and consumer of pork, accounting for nearly one-half of global production and consumption in 2009 (table 6.2).[31] At the end of 2009, there were an estimated 469 million swine in China and 49 million productive sows.[32] China is the world's third-largest pork-importing country, with 5 percent of global imports in 2009, and with 4 percent of global exports, it is the fourth-largest pork exporter behind the United States, [33] the EU-27, and Brazil.

Industry Structure

In China, production of the primary input into pork production, live swine, is of three main types: backyard producers, "specialized households" producing from 30 to several hundred hogs per year, and commercial operations producing 500 or more hogs per year (figure 6.1). In 2003, it was estimated that backyard producers supplied 80 percent of Chinese swine production, specialized households 15 percent, and commercial operations 5 percent.[34] Since then, commercial operations have increased rapidly, while the share of production accounted for by backyard producers has declined. By 2008, China's Ministry of Agriculture estimated that 56 percent of hogs slaughtered in China were produced by farms with 50 or more hogs.[35]

Table 6.2. Pork: Production, consumption, and trade, selected countries, 2009 (1,000 metric tons)

Country	Total production	Consumption	Imports	Exports
China	48,905	48,823	[a]270	232
EU-27	22,159	20,782	38	1,415
United States	10,442	9,013	378	1,857
Brazil	3,130	2,423	0	707
Russia	2,205	3,049	845	1
Rest of World	13,632	16,251	3,980	1,429
Total	100,473	100,341	5,511	5,641

Source: USDA, PSD, accessed November 2, 2010.

[a] Data is reported in carcass weight equivalent. Chinese pork imports are largely edible offal. Imports on a product weight basis in 2009 were 579,000 metric tons. GTIS, Global Trade Atlas database.

Domestic Marketing Outlets

In China, the major marketing outlets for pork include wet markets, supermarkets, and food consumed away from home. A significant share of China's meat is consumed away from home; recent estimates vary, but are as high as 50 percent.[36] Consumption away from home includes roadside stands, as well as quick-service and other restaurants. These outlets may source pork from wet markets, supermarkets, directly from pork processors, or through imports.

The vast majority of meat consumed in China, including pork, is purchased fresh, rather than chilled or frozen. Larger cuts and whole carcasses are displayed in the wet market and may be cut to order. In general, Chinese consumers prefer that food, including meats, be as fresh as possible. However, purchases of chilled and frozen pork from supermarkets are increasing with improvements in the cold chain[37] and are now considered by some Chinese consumers to be safer than purchases from wet markets.[38] Imports, which are all either chilled or frozen, currently account for a small share of China's pork consumption. The majority of imported pork is limited to frozen edible offal (such as organ meat), further reducing the direct competition between imports and the majority of pork produced in China.

Factors Affecting Competitiveness

Government Support

Much of China's recent support for pork production was a response to shortages of pork in 2007/08 following the widespread outbreak of blue ear disease, later identified as porcine reproductive and respiratory syndrome (PRRS). In 2007 and 2008, China lost as many as 10 million swine to PRRS.[39] In an effort to rebuild the herd, the Chinese government expanded the level of subsidies available to swine producers, especially commercial operations. In 2009, government subsidies for infrastructure improvements to commercial hog farms were estimated at $366 million. Other subsidies available to swine producers in 2009 included a per-head subsidy for producing sows, insurance, and subsidies for genetic improvements.[40]

Additionally, the Chinese government implemented a market intervention program that purchases pork for its central reserve system whenever the price of pork falls below six times the price of grain.[41] In April 2010, the hog-to-grain price ratio fell below that level, and some provincial governments began purchasing pork for reserves. The price of pork subsequently rebounded.[42]

Source: Commission staff.

At this specialized household, swine are housed in barns which are open to the elements. Workers live next to the barns in close proximity to the pigs. This is distinct from commercial operations, where pigs are isolated from the elements and are separated from human dwellings.

Figure 6.1. Swine production at a specialized household in Sichuan province, China.

Labor and Feed

China's swine producers, particularly its smaller producers, use more labor and less feed grain than producers in the United States. China's smaller producers substitute some crop residues and food scraps for feed grains, essentially substituting labor (used to collect and haul the residues and waste to the farm) for feed costs. Typically, as the scale of operation increases, the share of labor in total costs falls, while that of feed rises. In 2007, it was estimated that feed accounted for 43 percent of the total cost of production for backyard swine producers in China, while labor accounted for 20 percent. For specialized households, feed accounted for 55 percent of total swine production costs, while labor accounted for 7 percent.[43] By comparison, feed costs accounted for 41 percent of total costs for U.S. "feeder-to-finish"[44] swine producers in 2007, and labor accounted for 5 percent.[45]

Two trends have eroded the production cost advantage formerly enjoyed by China's domestic swine industry. First, labor has grown more expensive, as more off-farm employment opportunities have become available. As the opportunity cost of labor has increased, China has moved towards larger, more modern swine production facilities. The second factor eroding China's cost advantage is that feed accounts for a larger share of total cost in modern swine production facilities, and the price of feed has increased substantially in recent years, both globally and in China.

Feed has the largest effect on delivered cost because it accounts for the majority of "farrow-to-finish"[46] production costs in both China and the United States.[47] Swine feed is often formulated from corn and soybean meal. China's feed costs are thus heavily influenced by corn and soybean meal prices, which in turn are affected by a variety of factors including weather, demand levels, and government policies in China, as well as international prices. Feed costs fluctuate throughout the year, but are reportedly higher in China than in the United States because corn and soybean meal are costlier in China (table 6.3).[48]

Feed costs are also affected by feed conversion ratios (FCR), or the amount of feed needed to generate a unit increase in animal weight. The lower the FCR, the lower the cost of swine delivered to market. China's backyard swine producers have traditionally used breeds of swine that are less efficient at converting feed into weight gain, thus requiring more feed per pound of weight gain.

As the price of China's feed grains has increased, the production cost advantage of Chinese swine producers has declined. Before 2003, comparisons of the delivered cost of swine in China and the United States found that this cost was lower in China. More recent estimates have increasingly found that swine production costs in China now generally exceed those in the United States (table 6.4).

Table 6.3. Comparison of corn and soybean meal prices in China and the United States (dollars per metric ton)

Items	2007	2008
Corn		
China[a]	231	242
United States[b]	147	201
Soybean meal		
China[a]	408	555
United States[c]	226	370

Sources: USDA, FAS, China: Grain and Feed; Annual, 2009, March 3, 2009, 22; USDA, ERS, Feed Yearbook, Table 12; USDA, FAS, China: Oilseeds and Products; Annual, 2008, March 1, 2008, 39; USDA, FAS, China Oilseeds and Products; Annual, 2009, April 15, 2009, 38; USDA, ERS, Oil Crops Yearbook, Table 4.

[a] Average wholesale prices in both a producing region and a consuming region for the calendar year. [b]Central Illinois cash price for no. 2 yellow corn for marketing year beginning the previous September 1. [c]Decatur, Illinois, price for 48% protein soybean meal for marketing year beginning the previous October 1.

As shown in table 6.2, China is largely self-sufficient in pork production. However, China increasingly imports inputs for swine feed in the form of soybeans,[49] imports of which have more than doubled in volume over the past 5 years.[50] Chinese government import policies encourage the import of this primary input into pork production, while discouraging the import of pork; China's tariffs on soybeans are low, at less than 3 percent, compared to China's 12 percent tariffs on most pork imports. In addition, a VAT is not assessed on soybeans or corn imported by state trading enterprises in China, further lowering the cost of soybeans relative to imported pork. Other NTMs discussed in greater detail in chapter 9 further limit China's pork imports. A reduction in government support for swine producers, a reduction in pork tariffs, or elimination of NTMs affecting pork would make Chinese pork less competitive compared to imports.

Table 6.4. Swine production: Comparison of delivered cost between China and the United States

China	United States	Data date	Source
$0.852/kg (average)	$0.866/kg	1998	Wang, "China: Pork Powerhouse of the World," 2006.
$0.845/kg (commercial)	$0.902/kg	2002	Fabiosa, Hu, and Fang, "A Case Study," 2005.
$1.270/kg	1.082/kg	2003	Hartog, "Global Perspective on Integrated Pork Production," 2005.
$1.654–$1.764/kg	$1.058/kg	2010	Industry representative, interview by Commission staff, 2010.

Source: Calculations by Commission staff, based on referenced publications.

Note: Average annual exchange rate data are from International Monetary Fund (IMF) database (accessed December 8, 2010).

Quality of Supply

Backyard swine farmers labor under several disadvantages compared to large commercial Chinese pork producers. Hogs from traditional backyard producers vary widely in composition and overall size, hampering mechanized processing. Backyard producers are less able to control disease. They are also less likely to contract with slaughterhouses to ensure a consistent supply.[51] As each backyard producer produces only a few head per year, processors that depend on consistent deliveries in order to maximize their capacity utilization must either contract for delivery from many small producers or add an additional link (and cost) into the procurement chain in the form of a consolidator. Further, traditional backyard producers do not have large capital investments in production facilities. This makes it easier for them to exit the industry when returns are low, contributing to the volatility of swine supplies in China.[52]

Consumer Preferences

Many consumers in China display a relatively greater preference for cuts considered less desirable in the West, such as offal and cuts from the shoulder as compared to loin cuts, which are typically preferred in the United States and Europe. Many Chinese consumers also prefer cuts with a higher fat content.[53] This may confer some advantage on Chinese producers raising traditional breeds of swine that have a higher ratio of fat to lean, and may offset some of the disadvantage of lower feed efficiency.

Processed Foods

For the Chinese food processing sector to meet the growing demand—both domestically and in export markets—for high-quality, low-cost processed foods, it will have to do two things: improve the quality of its raw materials through better farm management practices, and find cost-effective ways to invest in capital-intensive production lines to tamp down rising labor costs. Foreign direct investment (FDI) and government assistance have provided the technology and capital needed for many food processing industries in China to mechanize. However, the small scale of Chinese farming, rising farm labor costs, the lack of advanced farm management practices, and poor transportation infrastructure reduce the quality of raw materials and increase transaction costs. In addition, while food safety regulations and regulatory compliance have generally improved, a widespread consumer perception still persists that Chinese processed foods lack food safety controls. This perception may restrict global demand for Chinese products in the future.

Industry Overview

Global Position

China is among the world's largest producers and exporters of processed foods[54] and is the leading exporter of several products, including apple juice, canned peaches and pears, instant noodles, and dehydrated garlic. Chinese processed food production has been growing at about 20 percent annually in recent years, largely in response to rising domestic disposable income, urbanization, and the trend toward higher quality and more convenience in food purchases.[55] Throughout the world, processed food production is mostly a capital-intensive, highly mechanized activity. But in China food processors typically employ relatively more labor and less mechanization. Despite the greater use of labor, China is able to compete in global processed foods markets because of labor's availability at a low cost.

Industry Structure

The Chinese processed food sector consists of about 40,000 firms.[56] Operations vary dramatically in their use of labor and capital in production, depending on the nature of the end product. Manufacturers source raw agricultural products from the farming sector as inputs into processed food production. As discussed in chapter 5, the small size of the average farm and the fragmented geographic location of the farm sector in China results in a variety of models for sourcing raw materials. Small farm sizes typically mean that processors are required to source their materials from a large number of farmers, each supplying a small volume of their input needs. Food processors typically source from a variety of traders or village leaders who procure and consolidate the product for them.[57] Some processors have also begun using a farmer cooperative structure in order to streamline procurement, limit transaction costs, and improve quality.[58] Other processors contract directly with growers, while still others operate farms that supply some of the raw material in-house.

Domestic Marketing Outlets

Chinese food processors market their product through a number of different channels, primarily through supermarkets, restaurants, and smaller stand-alone retail outlets (mostly in urban areas) as well as further processors. A much lower share of total food consumption is processed in China than in Western countries, but urban Chinese consumers are increasingly looking for the convenience offered by processed foods, which has driven domestic sales growth.[59] The importance of fast food restaurants as an outlet for certain processed foods, such as frozen french fries (FFF) and processed poultry, is growing. For example, KFC is opening almost one new outlet per day in China, while McDonald's is opening 150–175 new outlets each year.[60]

Many food processors in China are primarily export-oriented and have not developed marketing channels within China. For processors that also serve consumers in China, the marketing channels they serve—hotels, restaurants, and high-end grocery chains, to name a few—can affect how they organize procurement, how far they mechanize production, and how well they implement food safety precautions. Less price-sensitive channels that focus on quality and food safety may force processors to use a different mix of capital and labor to meet those standards. Higher quality and food safety precautions often require more mechanization and less labor to consistently meet stringent specifications.

Factors Affecting Competitiveness

Procurement of Quality Raw Materials

The quality and cost of raw agricultural products used as inputs in food processing are crucial to manufacturers. There are two aspects of quality that are important to them. The first is the quality of the raw product itself—for example, whether it has the required taste and freshness. The second is the product's suitability for processing, particularly when processing involves mechanization. Processors typically need products that are of consistent quality and that meet the specifications needed to process the product efficiently. For example, the FFF processing industry requires a long potato that has the appropriate moisture and sugar levels and will show minimal browning, to ensure high slicing yields and a high-quality final product.[61] Mechanized poultry processing facilities require birds of a consistent size for the production line to operate efficiently.

Chinese food processors encounter problems in procuring quality raw materials. Because processors often have to source product from many small traders, they find it difficult to oversee quality and control product consistency on the farm. It is also difficult to move farmers away from traditional farming methods to more modern approaches that allow better control of production. For poultry, the lack of advanced growing operations in China means that birds are not of uniform size, which makes mechanical cutting inefficient and results in Chinese poultry slaughtering and processing facilities that rely much more heavily on manual labor.[62] In crop farming, many Chinese farmers are not aware of the varieties they are planting, do not use advanced or appropriate fertilizers, lack disease prevention practices, and often use older seeds saved from previous plantings that lower yields.[63] These problems are compounded by the long distances between the processors and the farm, coupled with poor transportation infrastructure in some parts of the country.[64] The longer products are kept in storage before processing, the more the quality is reduced. In addition, limited cold storage capacity during transport and at the plants reduces the volume of the raw materials that can be processed at a later time.

Faced with all of these constraints, food processors in China use various approaches to improve the quality of the raw materials. Increasingly, processors are trying to consolidate production by reducing the number of growers or livestock producers that they must deal with. They are also implementing a number of methods that make oversight manageable, as well as providing financing options to improve quality. Processors have also invested in educating farmers to implement practices that, while higher-cost, will also produce higher-quality inputs that are fit for processing.[65]

Some of the methods food processors are using in China to improve quality are illustrated in the FFF and poultry industries. As mentioned, fresh potatoes for FFF production must be the right shape and contain the right amount of moisture and sugar to ensure high slicing yields and a superior final product.[66] To achieve this level of quality, individual FFF processors have reduced the number of Chinese farmers from whom they source from thousands to a few dozen or less, by shifting production to more remote areas where large-scale leases are possible and by sourcing on a contract basis once growers adopt more efficient production methods.[67] While some growers in the sector still farm small plots of land, this strategy has resulted in some potato farms that are 121 hectares or more in size.[68] Once the number of source farms has been reduced, processors typically employ field managers that oversee the production work of a few growers each.[69] FFF processors that source product on a contract basis also provide financing to improve access to technology. For example, one FFF processor supplied farmers with irrigation systems, sprayers, and tractors while requiring only a modest down payment before taking annual payments over a certain repayment period.[70] Despite these efforts, potato quality still keeps processing yields for Chinese FFF processors lower than in other major producing countries.[71]

In the Chinese poultry industry, the use of advanced technology in grow-out houses, such as automated feeders and water systems, good ventilation, and climate control, is limited by a lack of access to credit and results in inconsistent bird sizes.[72] To counter this, larger poultry firms give farmers loans to build modern houses and to buy advanced equipment. Other processors would like to increase the percentage of broilers raised in-house to ensure a reliable supply of high-quality birds. Expanding grow-out operations is problematic, however, because of difficulties in leasing land created by the land tenure system, the length of negotiations (often up to six months), and the cost of land rents in some provinces.[73]

Cost of Procurement of Raw Materials

In addition to quality, another major concern for processors is the cost of raw inputs delivered to the plant. The cost of the raw input to the processor consists of the cost of the input itself plus the cost of procurement and transportation to the processing facility. While the small scale of farming in China restricts quality improvements, it also increases the cost of procuring the product.

Because of the small size of farms, processors typically are forced to source the product from traders or village leaders who consolidate the product. For example, one grain processor in Shandong province reportedly obtains its raw inputs from between several dozen and a few hundred traders in order to avoid sourcing from several thousand small household farmers.[74] In the dehydrated garlic industry, where a typical grower's farm is between 2 and 3 *mu*, larger processors source from as many as 300 village leaders who represent about 100,000 farmers.[75] This additional consolidation increases transaction costs because prices increase each time the product is resold.[76] Small and fragmented farms in China also increase transportation costs, because processors have to source product from a wider geographical area.

In addition to raising the cost of procurement, the fragmented farm structure also affects the reliability of input supply for processors in China. This is because, unlike in other countries, farmers typically do not grow their product exclusively for processing.[77] For example, in China's dehydrated garlic industry, processors generally contract with growers. However, because the contracts provide the farmer with the right of refusal if the fresh market is offering a higher price, processors may need to offer additional services in order to entice farmers to sell to them.[78]

Firms in certain sectors have begun to develop models to overcome the limitations of the current land tenure system and to lower procurement costs. Some processors, primarily in the canned mandarin, peach, and pear sectors, have begun using farmer cooperatives in order to streamline procurement, to avoid traders, and to limit transaction costs, while also improving quality. Large-scale leasing of land, as in the FFF industry, is also becoming more common as companies look to consolidate the number of farmers and reduce transaction costs. Although growing, the use of cooperatives and large-scale leasing of land are still not widespread among food processors in China.[79]

Mechanization

Chinese food processors vary widely in the level of mechanization and technology they use in the production process. Processing facilities range from labor-intensive production lines where employees cut and sort by hand to modern, capital-intensive, mechanized production lines that use minimal labor. Variation exists not only between industries, but also within specific industries themselves. For example, in the processed fruit sector, some canned fruit manufacturers employ a cutting process that is fully mechanized, similar to processing lines in the United States, while others still cut the fruit by hand. Also, fruit canning operations are generally more labor-intensive than other types of fruit processors, such as apple juicers.

As a whole, the Chinese food processing sector is slowly moving toward becoming more capital-intensive, as labor costs rise and mechanization improves.[80] For example, Chinese poultry-slaughtering and processing facilities rely heavily on manual labor.[81] Historically, processing facilities have not mechanized because of low labor costs, inconsistent bird sizes, and difficulties in obtaining equipment. As labor costs continue to rise, mechanization is becoming more attractive, and processors are beginning to invest in processing technology. It is true that acquiring modern processing technology requires significant capital and access to credit, which present barriers to many small and medium-sized food manufacturing firms. However, capital for improved mechanization is available from a number of sources, most notably from FDI and government assistance. As a result, Chinese food processing is becoming less reliant on labor and more capital intensive.

FDI in food processing has enabled greater access to technology and accelerated the mechanization of the Chinese sector. Because the machinery needed for efficient food processing is costly, capital from foreign firms has allowed certain sectors to expand quickly and supply larger volumes of product both domestically and in export markets. Both the FFF and the poultry industries offer examples of FDI's impact in the food processing sector. The FFF industry has developed and mechanized as a result of FDI from three firms that responded to expansion by Western fast food restaurants into the Chinese market. As a result of the investments, domestic FFF production now supplies a large percentage of the market, originally supplied only by imports, primarily from the United States. In addition, as FFF consumption continues to grow and these food outlets become more accepted in second-tier urban markets, domestic production is expected to supply most of that growth.[82] In the poultry sector, investment from international players, such as Tyson, Keystone Foods, OSI, and Cargill, has been driving the industry's increasing use of modern processing lines and grow-out barns as well as the adoption of stronger food safety precautions.

Chinese government policies can also influence the level of mechanization in the food processing sector. The government provides assistance through cash investments for holdings in food processing firms, viewed as important in providing jobs and development in certain regions.[83] For example, the government has invested in the canned peach industry to boost various firms' cash flow and allow investments in machinery.[84] The government typically refrains from receiving dividends or other returns on its investment. Reportedly, government investments are made with the expectation of repayment, but the payback terms are often uncertain and not clearly defined.[85] In addition, local governments provide tax incentives for companies to invest in advanced production methods and machinery.[86]

However, other government policies hinder mechanization. For example, foreign-owned firms with Chinese operations often have access to used equipment from abroad, but the large number of certifications and inspections, fees, taxes, and tariffs required can make importing such equipment uneconomical. Reportedly, in the poultry industry companies are also required to use the imported equipment in a specific plant for a certain, often extended, length of time.[87] This effectively eliminates a resale market and hinders firms' ability to increase their level of mechanization.

Food Safety

In recent years, consumer concerns about the safety of Chinese processed foods has reduced the competitiveness of Chinese products in both domestic and export markets.[88] Stronger government oversight since the introduction of a new food safety law in 2009, increasing consumer awareness of food safety issues, and the greater involvement of internationally branded companies are all encouraging Chinese food processors to improve their product safety. Despite improvements, a general lack of confidence in the safety of Chinese processed goods could continue to depress global demand for Chinese products in the future.[89]

Safety inspections of Chinese food processors have increased in recent years, and while Chinese processors spend significantly less capital installing production facilities that ensure food safety than producers in the United States, compliance costs are rising.[90] The enforcement of Chinese food safety laws differs depending on the final destination of the product. Products destined for export markets typically undergo more thorough and stringent inspections to prevent market closures abroad. For example, local Chinese Inspection and Quarantine agents sample, inspect, and certify 15 percent of every export shipment of a certain garlic processor in Shandong, while processors in other provinces are rarely tested if they do not export.[91]

FDI in the Chinese food processing sector has also improved Chinese food safety standards[92] because investors typically install stringent screening and testing methods that often exceed Chinese standards. For example, international dehydrated garlic producers have increased levels of product sorting, high-tech screening, metallic testing, and employee sanitation in their Chinese plants. They have also compartmentalized plants so that an individual employee does not have contact with multiple points on the production line, which decreases the chances of introducing foreign materials. [93] International companies that source inputs for their products also require their suppliers to comply with strict food safety precautions. For example, FFF producers that supply McDonald's must implement McDonald's food safety measures.

In the Chinese meat sectors, residues from veterinary medicines given to animals are found in meat at levels that harm product quality and threaten consumer health in varying ways.[94] Largely because of overmedication, certain diseases have become resistant to drugs, which causes growers to medicate the animals even more. As a result, there have been periods when both Japan and the EU-27 have banned imports of Chinese poultry because of excessive antibiotic residue.[95] Foreign firms operating in China, however, have implemented precautions and stricter oversight to prevent overmedication of animals by growers.

Wheat

China is not naturally competitive in domestic and global wheat markets, compared with major global wheat-producing and -exporting countries. Although China has an advantage with respect to labor costs, it is at a disadvantage regarding land. Wheat is a land-intensive agricultural product, and foreign competitors such as Australia, the United States, and Russia hold a substantial competitive advantage in that production factor vis-à-vis China. China's limited and unbalanced water resources also weaken China's relative competitiveness. Furthermore, the small scale and diffuse structure of China's wheat sector limits efficiencies, raises delivered costs, and compromises the reliability of supplies.

China has taken measures to address its competitive disadvantages in the wheat sector and to implement a transition to a large-scale, efficient industry structure. Recent government emphasis on rural development and concerns about food security have precipitated initiatives such as the National Plan for Expansion of Grain Production Capacity (National Grain Plan) and recent No. 1 Documents.[96] These initiatives, coupled with a trade policy that maintains limited market access for wheat imports, will benefit China's wheat producers and enhance and maintain their competitiveness in the domestic market. Other developments, such as changes in domestic market preferences for wheat which could result in pressure from internal actors, such as private millers and traders, as well as from trading partners, could lead to increased market access for high-gluten wheat imports in the future. The impact of such developments on the competitiveness of China's wheat is less clear, as rural income and self-sufficiency in grains likely will remain long-term government concerns.

Industry Overview

Global Position

China is the leading global consumer of wheat, accounting for approximately 21 percent of the total quantity in 2009/10 (table 6.5). It is also the second leading global wheat producer, trailing only the EU-27,[97] and is a minor wheat trader. Imports account for a small share of consumption and exports for a small share of production (each less than 1 percent). These levels reflect China's policy of self-sufficiency in wheat.

Industry Structure

Wheat is produced in every Chinese province except Hainan, but production is concentrated in the northeastern provinces.[98] Winter wheat accounts for the bulk of Chinese wheat acreage—93 percent in marketing year 2009/10.[99] Wheat milling is less concentrated, with the leading provinces accounting for about 60 percent of total mill use of wheat compared with three-quarters of wheat production.[100]

Wheat production in China has risen in recent years, from about 99.6 million mt in 2000/01 to 115.0 million mt in 2009/10.[101] The increase resulted from rising yields, as the amount of land used to grow wheat has been relatively stable. Data are not available on the number of wheat farms or farmers in China. However, the number likely is substantial, as individual farms are small compared with those of competing wheat producers. The average per capita area of cultivated land in the major wheat-producing provinces ranged between 1.14 and 1.94 *mu* in 2009.[102] Assuming an average household size of 4 people, the average farm size is about 7 *mu*, or one-half hectare.

Table 6.5. Wheat: Production, consumption, and trade, selected countries, marketing year 2009/10 (1,000 metric tons)

Country	Production	Consumption[a]	Imports	Exports
		1,000 metric tons		
EU-27	138,051	124,500	5,480	22,117
China	115,120	107,000	1,394	892
India	80,680	78,201	([b])	([b])
Russia	61,700	42,000	([b])	18,556
United States	60,366	30,932	3,228	23,977
Rest of World	226,685	267,066	123,127	70,104
Total	682,602	649,699	133,684	135,746

Source: USDA, PSD Online (accessed January 14, 2011).
[a] Wheat held in stocks is not accounted for in consumption. [b]Less than 500 metric tons.

The Chinese wheat processing (milling) sector comprises a large number of firms, most of which are relatively small in scale. In the mid-2000s, there were approximately 9,800 wheat millers with a capacity of over 50 mt per day, of which 500 had a capacity of 200– 400 mt per day and 80 had a capacity of at least 400 mt per day. [103] The Chinese wheat milling sector has been undergoing major restructuring in recent years. After a long-term market liberalization in the grain sector, the number of mills has declined while their average size has increased. The largest miller now accounts for approximately 3 percent of the wheat used for processing; the top three millers account for about 6 percent.[104] Concentration among the three leading millers is expected to reach 25 percent within 5– 10 years. The leading miller, the Wudeli Group, is a privately held Chinese firm; the second is the state trading enterprise (STE) COFCO; and the third is a major foreign agribusiness group, Wilmar Industries. Competition is expected to intensify among the leading millers in the future.

Domestic Marketing Outlets

There are two domestic marketing outlets for Chinese wheat farmers—the milling sector, as described above, and the government. The milling sector comprises private companies as well as STEs. The government procures and stores a share of wheat output under a program that guarantees a minimum price to farmers, then auctions the stocks to millers. Its role is to manage supplies, prices, and strategic reserves. The State Administrator of Grain of the National Resources Development and Reform Commission is responsible for overall grain management, including storage and distribution.[105] China Grain Reserves Corporation (Sinograin) is an STE that has primary responsibility for grain storage and transportation.[106] COFCO, China's largest food products STE, is the second-largest wheat processor and holds 90 percent of China's wheat import tariff-rate quota (TRQ).[107] COFCO has grain storage capacity of 3 million metric tons.[108] Imports are marketed to the private milling sector or are used by COFCO.

Factors Affecting Competitiveness

Land

Land issues in China define the competitiveness of the wheat sector. Wheat is a land-intensive crop and requires relatively little labor. China's land resources suitable for wheat production have been fully utilized and are under increasing pressure from other agricultural commodities that are higher in value and profitability for farmers, as well as from increasing industrialization and urbanization.[109] Rising land rental rates, which for wheat production increased by two-thirds during 2003–08, [110] have contributed to increased delivered costs for wheat farmers and diminished their competitiveness vis-à-vis major foreign producers such as the United States and Australia. [111] Another land issue affecting the competitiveness of China's wheat sector is the land tenure system, which results in small individual household plots. This system undermines China's competitiveness, as it limits economies of scale and increases production, quality control, and marketing costs. The Chinese government recognizes these issues and has recently proposed measures in the National Grain Plan to improve the sector's competitiveness to maintain food security. Measures in the plan affecting land issues include establishing core production zones, shifting a larger share of production to the northeastern provinces, and reclaiming and consolidating land to expand farmland area (figure 6.2).[112]

Water and other Inputs

Water is another issue threatening the competitiveness of the Chinese wheat sector. Increasing competition and pressure on water resources from agricultural, industrial, and urban activities; declining water quality, owing mainly to heavy use of fertilizers and pesticides as well as to industrial pollution; supply volatility caused by weather, mainly droughts and flooding; and uneven water distribution, as the bulk of suitable land for wheat production is located away from water surplus areas, all have increased water scarcity and costs for wheat farmers. Irrigation fees rose by 25 percent during 2003–08, while a water fee was established in 2004.[113] The Chinese government has identified water as a priority issue in the National Grain Plan and plans to construct and upgrade irrigation and drainage infrastructure.[114] Thus, increases in water costs on input prices could decrease the competitiveness of the Chinese wheat industry regarding delivered cost.

Costs have risen substantially for other wheat farm inputs as well. Labor costs rose by one-third during 2003–08, while fertilizer costs rose by 103 percent and seed costs by 53 percent.[115] However, wheat is not a labor-intensive crop, and the response to rising labor costs appears to have been a reduction in the utilization of this production factor, as labor use declined by one-third during 2003–08; [116] thus, the impact of rising farm labor costs on the competitiveness of the Chinese wheat sector has been minimal. Farmers also reduced their use of seed, which declined by 10 percent during the period, but they compensated for this reduction by a 21 percent rise in fertilizer use.[117]

Technology

The use of biological and production technology in the Chinese wheat sector is relatively high compared with the country's other agricultural sectors and has enhanced competitiveness by lowering delivered costs. A relatively high level of mechanization,[118] extensive use of irrigation, and the use of higher-yielding, disease-resistant strains have helped to increased wheat yields in recent years. However, China's wheat yields are inconsistent and lag those of major competitors. The Chinese government has indicated it will increase support to improve technology in the sector.[119] Any increase in the application and quality of technology will improve China's competitive stature in the domestic and global wheat markets by lowering delivered costs and reinforcing the reliability of supply.

Market Preference

The Chinese wheat market has been changing in recent years. Declining per-capita consumption, owing to a shift from carbohydrates to animal protein, and changing consumer preferences for convenience foods, which use higher-gluten wheat, have limited the competitiveness of Chinese wheat producers.[120] The Chinese wheat sector focuses on the lower-gluten wheat used in traditional products, such as steamed and boiled flour products.[121] Chinese wheat processors import high-gluten wheat to produce convenience foods, such as instant noodles and baked goods.[122] The Chinese government has recognized this shift and has introduced measures to develop new uses for wheat, improve wheat quality, and increase the share of high-gluten wheat in total production.[123] These measures likely will enhance Chinese wheat producers' future competitiveness in the domestic market by improving their product differentiation and ability to supply high-gluten wheat. In addition, export competitiveness likely will be enhanced, as there is a substantial regional market for high-gluten wheat.[124]

Government Policy

Government policy has had a substantial positive effect on the competitiveness of the Chinese wheat sector. Recent domestic policy has focused resources on the grain sector in order to maintain self-sufficiency. Trade policy generally has applied restrictive TRQs and high over-quota tariffs, as well as excluded wheat from free trade agreements.[125] These policies generally have kept domestic wheat prices at levels high enough to motivate farmers to stay in wheat production. In addition, government direct payments, input subsidies, technology assistance, and infrastructure development have lowered Chinese wheat producers' delivered cost and improved their reliability as suppliers in the domestic market.

Source: Commission staff.

Unlike the structure of traditional farms (figure 5.2), these plots have been consolidated to grow one crop. The plots likely belong to multiple farmers but, under the direction of village leaders, were consolidated to improve efficiency and likely increase the mechanization of planting and harvesting.

Figure 6.2. Farm plots consolidated to grow one crop, Shandong province, China.

Issued in November 2009, the National Grain Plan calls for a 50 million metric ton increase in grain production capacity during 2009–20.[126] Objectives outlined in the plan include improving technology (irrigation, farming methods, seed varieties, and extension services); modernizing sector structure (land preservation and consolidation, and grain production, storage, and distribution infrastructure); and increasing market regulation and monitoring (minimum purchase price, grain reserve controls, and market information). In addition, Document No.1, which addresses rural issues such as support for grain farmers, specified that future direct subsidies will be focused on large farmers and specialized farmers' cooperatives.[127] The U.S. Department of Agriculture estimates that the aggregate subsidy effect of direct payments and input subsidies accounted for 30 percent of grain farmers' net profit margins in marketing year 2009/10.[128] In addition, the government's procurement of wheat at the minimum floor price accounted for an average of 37 percent of total production annually during the past three marketing years.[129] Government policy also promotes large-scale wheat processors and distributors. The National Grain Plan specifies the development of grain storage, logistics, and processing capacity as a major objective and further directs the promotion of large-scale processing as well.[130] China's import policy reserves 90 percent of the wheat TRQ for COFCO, further enhancing the government's control of wheat supplies and market prices.

Bibliography

Baecker, Geoffrey. *Food and Agricultural Imports from China.* Congressional Research Service Reports for Congress. Washington, DC: September 26, 2008.

Barboza, David. "China's Contaminated Food Scandal Widens." *The New York Times,* October 31, 2008.

British Broadcasting Corporation (BBC). "Chinese Melamine Scandal Widens," October 31, 2008. http://news.bbc.co.uk/2/hi/7701477.stm.

———. "Timeline: China Milk Scandal," January 25, 2010. http://news.bbc.co.uk/2/hi/asia-pacific/7720404.stm.

Calvin, Linda, Fred Gale, Dinghuan Hu, and Bryan Lomar. "Food Safety Improvements Underway in China." *Amber Waves*, November 2006. http://www.ers.usda.gov/AmberWaves/November06/Features/FoodSafety.htm.

China Grain Reserves Corporation (Sinograin). "Company Overview." http://www.sinograin.com.cn/eng/guanyu_Eng_sub2.jsp (accessed July 28, 2010).

CNNMoney.com. "China: The New Fast Food Nation," July 13, 2010. http://money.cnn.com/2010/07/13/news/companies/Yum Brands/index.htm.

COFCO. "Agricultural Primary Products Storage and Logistics." http://www.cofco.com/en/industry storage.aspx?con id=3290 (accessed November 10, 2010).

———. "Wheat Division." http://www.cofco.com/en/about_cofco/bu_all.aspx?con_id=3314 (accessed November 10, 2010).

Den Hartog, Leo. "Global Perspective on Integrated Pork Production." Paper presented at the London Swine Conference, "Production at the Leading Edge," London, Ontario, April 6–7, 2005.

Fabiosa, Jacinto F., Dinghuan Hu, and Cheng Fang. "A Case Study of China's Commercial Pork Value Chain." MATRIC Research Paper 05-MRP 11, August 2005.

Food and Agriculture Organization of the United Nations (FAO). FAOSTAT database. http://faostat.fao.org/site/567/default.aspx#ancor (accessed November 2, 2010).

———. *Livestock's Long Shadow: Environmental Issues and Options.* Rome: FAO, 2006. Financial Times. FDI Markets database (various dates).

FruitGrowersNews.com. "Editor Studies the Chinese Apple Industry with IFTA," n.d. http://fruitgrowersnews.com/index.php/magazine/article/Editor-Studies-the-Chinese-AppleIndustry-With-IFTA (accessed November 9, 2010).

Gale, Fred, and Kuo Huang. *Demand for Food Quantity and Quality in China.* Economic Research Report no. 32, U.S. Department of Agriculture. Economic Research Service, January 2007.

Gallardo, Karina, Mykel Taylor, and Herb Hinman. "2009 Cost Estimates of Establishing and Producing Gala Apples in Washington." Washington State University Extension, Fact Sheet FSOOSE, n.d.

Gao, Lili. "Regulating Trade with a Systems Approach: The Case of Chinese Fresh Apples," 2008.

Gao, Lili, and Suzanne Thornsbury. "Increasing Food-Safety Protection: Fresh Apple Markets in China." *Journal of Food Distribution Research* 39 (1): March 2008.

Gifford, Rob. "Food Fears Persist in China 2 Years after Milk Scare." NPR, October 26, 2010. http://www.npr.org/templates/story/story.php?storyId=130826464.

Global Trade Information Service, Inc. (GTIS). Global Trade Atlas database (accessed November 10, 2010).

Honglin Wang, Xiaoxia Dong, Jikun Huang, Scott Rozelle, and Thomas Reardon. "Producing and Procuring Horticultural Crops with Chinese Characteristics: Why Small Farmers Are Thriving and Supermarkets Are Absent in Rural China." *World Development* 37, no. 11 (November 2009): 1791–1801.

Hughes, Mark, and Yang Ning. "Farmers Slowly Cultivate a New Image." *China Daily*, May 31, 2010. http://www.chinadaily.com.cn/business/2010-05/31/content 9910739.htm.

International Monetary Fund. International Financial Statistics. http://imfstatistics.org (accessed December 8, 2010).

Latner, Kevin. "Who Is Feeding China?" Presentation, USDA Agricultural Outlook Forum, February 19, 2010.

Lomar, Bryan, and Fred Gale. "Who Will China Feed?" *Amber Waves,* June 2008. http://www.ers.usda.gov/AmberWaves/June08/Features/ChinaFeed.htm

Lohmar, Brian, Fred Gale, Francis Tuan, and Jim Hansen. *China's Ongoing Agricultural Modernization: Challenges Remain after 30 Years of Reform.* Economic Information Bulletin no. EIB-51. U.S. Department of Agriculture, Economic Research Service, April 2009. http://www.ers.usda.gov/Publications/EIB51/EIB51.pdf.

Liu, Coco. "China Still Wooing Foreign Investors." *Poultry International,* March 2010.

———. "More Cold Chain Gives More Choice in China." *Poultry International*, September 2009.

Liu, Zhixiong. "Researches on the Effect of Foreign Direct Investment in the Food Processing Industry in China," n.d. http://www.seiofbluemountain.com/upload/product/201003/2010cygchy01a22.pdf (accessed November 2, 2010).

National Chicken Council. "U.S. Broiler Performance," current as of October 22, 2009. http://www.nationalchickencouncil.com/statistics detail.cfm?id=2.

MacDonald, James M. *The Economic Organization of U.S. Broiler Production.* Economic Information Bulletin no. EIB-38. U.S. Department of Agriculture, Economic Research Service, June 2008. http://www.ers.usda.gov/Publications/EIB38.

Martinez, Steve. *Vertical Coordination of Marketing Systems: Lessons from the Poultry, Egg, and Pork Industries.* Agricultural Economic Report no. 807. U.S. Department of Agriculture, Economic Research Service, May 2002. http://www.ers.usda.gov/Publications/AER807.

Meat Trade News Daily. "China: McDonalds Leading the Way." News release, August 10 2010.

McKee, David. "Changes in Chinese Flour Milling." *World Grain*, September 1, 2010. http://am1.sosland.com/Olive/ODE/WorldGrain/LandingPage/LandingPage.aspx?href=V0cvMjAxMC8wOS8wMQ..&pageno=NTI.&entity=QXIwNTIwMA..&view=ZW50aXR5

National Bureau of Statistics of China (NBSC). *China Statistical Yearbook 2009*, September 2009. http://www.stats.gov.cn/tjsj/ndsj/2009/indexee.htm.

New Zealand-China Trade Association. "Processed Food for Thought: China's Food Processing Market," n.d. http://www.nzcta.co.nz/chinanow-general/1000/processed-food- (accessed November 10, 2010).

Ortega, David L., H. Holly Wang, and Laping Wu. "Food Safety and Demand: Consumer Preferences for Imported Pork in Urban China." *Journal of Food Distribution and Research* 40, no. 3 (November 2009).

Peng,Y, Y. Peng, P. Shiqi, T. Youguo, S. Ivanova, and H. Magen. "Fertigation Management in Young Apple Trees in Shandong, China." *International Potash Institute*, December 18, 2008. http://www.ipipotash.org/eifc/2008/18/2.

People's Republic of China (PRC). Ministry of Agriculture (MOA). "General Surveys: Agricultural Mechanization." *China Agriculture Yearbook 2008*, June 25, 2009. http://english.agri.gov.cn/sa/sd/yearbook/200906/t20090625 1176.htm

———. National Development and Reform Commission (NDRC). *National Agricultural Product Cost and Profit Materials Compilation*. Beijing: China Statistics Press, 2003–08.

———. National Development and Reform Commission (NDRC). State Administration of Grain (SAG). "Message from the Administrator of SAG," 2002. http://www.chinagrain.gov.cn/english/General%20Situation.html.

Poultex News. "Half of China's Antibiotics Are Fed to Animals," December 1, 2010.

Promar International. "The Chinese Potato Industry in Transition." U.S. Potato Board, July 2007.

Somwaru, Agapi, Zhang Xiaohui, and Francis Tuan. "China's Hog Production Structure and Efficiency." Presentation at the American Agricultural Economics Association Annual Meeting, Montreal, Canada, July 27–30, 2003.

U.S. Department of Agriculture (USDA). Economic Research Service (ERS). Chinese Provincial Data. July 1, 2009. http://www.ers.usda.gov/Data/China

———. "U.S. Hog Production Costs and Returns per Hundredweight Gain." 2004–09 database. http://www.ers.usda.gov/Data/CostsAndReturns/testpick.htm(accessed October 28, 2010).

———. *Feed Yearbook*. http://www.ers.usda.gov/Data/FeedGrains/Yearbook/FG Yearbook Table12-Full.xls (accessed November 1, 2010). ———. *Oil Crops Yearbook*. http://usda.mannlib.cornell.edu/MannUsda/viewDocumentInfo.do?- documentID=1290 (accessed December 6, 2010).

U.S. Department of Agriculture (USDA). Foreign Agricultural Service (FAS). *Canada: Potatoes and Potato Products; Annual, 2010*, by Mihai Lupescu. GAIN Report no. CA0038, October 15, 2010.

———. *China, Peoples Republic of: 2010 Agriculture Policy Directive*, by Chandra Beckman. GAIN Report no. CH10004, February 18, 2010. http://gain.fas.usda.gov/Pages/Default.aspx

———. *China, Peoples Republic of: Fresh Deciduous Fruit; Annual, 2006*, by Jorge Sanchez and Wu Bugang. GAIN Report no. CH6028, September 22, 2006.

———. *China, Peoples Republic of: Fresh Deciduous Fruit; Annual,* by Joshua Emmanuel Lagos, May Liu, and Wu Bugang. GAIN Report no. CH9090, November 17, 2009.

———. *China, Peoples Republic of: Fresh Deciduous Fruit; Annual, 2010*, by Ryan R. Scott, May Liu, and Wu Bugang. GAIN Report no. CH10061, November 5, 2010.

———. *China, Peoples Republic of: Grain and Feed; Annual, 2009*, by Jorge Sanchez and Jiang Junyang. GAIN Report no. CH9013, March 3, 2009.

———. *China, Peoples Republic of: Grain and Feed; Annual*, by Chandra Beckman and Jiang Junyang. GAIN Report No. CH10008, March 1, 2010.

———. *China, Peoples Republic of: Livestock and Products; Annual Report, 2007*, by Jianping Zhang and Eric Trachtenberg. GAIN Report no. CH7076, September 25, 2007.

———. *China, Peoples Republic of: Livestock and Products; Annual*, by Michael Woolsey and Jianping Zhang. GAIN Report no. CH9069, September 14, 2009.

———. *China, People's Republic of: Livestock and Products; Semi-Annual*, by Michael Woolsey and Jianping Zhang. GAIN Report no. CH10009, March 2, 2010.

———. *China, Peoples Republic of: National Plan for Expansion of Grain Production Capacity*, by Chandra Beckman and Jiang Junyang. GAIN Report No. CH10005, February 18, 2010.

———. *China, Peoples Republic of: Oilseeds and Products; Annual, 2008: Part 1 of 2—Analysis*, by Mark Petry and Wu Xinping. GAIN Report no. CH8010, March 1, 2008.

———. *China, Peoples Republic of: Oilseeds and Products; Annual, 2009: Part 2 of 2—Supporting Tables*, by Mark Petry and Wu Xinping. GAIN Report no. CH9030, April 15, 2009.

———. *China, Peoples Republic of: Potato and Potato Products; Annual, 2010*, by Chanda Beckman, Zhang Li, and Zhang Fang. GAIN Report no. CH10051, August 16, 2010.

———. *China, Peoples Republic of: Poultry and Products; Annual, Poultry Report—Final, 2004*, by Casey Bean and Jianping Zhang. GAIN Report no. CH4045, September 24, 2004.

———. *China, Peoples Republic of: Poultry and Products; Annual, Report, 2008,* by Jianping Zhang and Chanda Beckman. GAIN Report no. CH8073, September 1, 2008.

———. *China, Peoples Republic of: Poultry and Products; Semi-Annual; 2009*, by Jianping Zhang. GAIN Report no. CH9014, March 5, 2009.

———. *China, Peoples Republic of: Poultry and Products; Semi-Annual, 2009*, by Michael Woolsey and Jianping Zhang. GAIN Report no. CH9070, September 16, 2009.

———. *China, Peoples Republic of: Poultry and Products; Semi-Annual 2010*, by Michael Woosley, Joshua Lagos, and Jianping Zhang. GAIN Report no. CH10010, April 4, 2010.

———. *China, Peoples Republic of: Poultry and Products; Annual, 2010*, by Michael Woolsey and Jianping Zhang. GAIN Report no. CH10056, September 30, 2010.

U.S. Department of Agriculture (USDA). National Agricultural Statistics Services (NASS). "Farm Labor," August 19, 2010.

U.S. Department of Agriculture (USDA). Production Supply and Distribution (PSD) Online. http://www.fas.usda.gov/psdonline/psdQuery.aspx (accessed various dates).

Wang, Jimin, and Hongbo Xiao. "Development of the Hog Industry and Its Integration in China." In *Pork Production in China: A Survey and Analysis of the Industry at a Lewis Turning Point*, edited by Jimin Wang and Mariko Watanabe. ASEDP No. 77. Chiba, Japan: Institute of Developing Economies, 2007.

Wang, Ruojun. "China: Pork Powerhouse of the World." *Advances in Pork Production* 17 (2006): 33. WATTAgNet.com. "Top Companies for China" (accessed November 2, 2010).

Zhang, A. "Chinese Wheat: Current Situation and Prospects." In *Future of Flour: A Compendium of Flour Improvement*, edited by Lutz Popper, W. Schäfer, and W. Freund. Clenze, Germany: Verlag Agrimedia, 2007. http://www.muehlenchemie.de/downloads-future-of-flour/FoF Kap 08.pdf.

Zeitner, Randy. "China's Apple and Pear Industry: Challenges and Opportunities." Economic Consulting Group, January 2006.

7. Chinese Agricultural Tariff Measures

Overview

China imposes tariffs on imports of agricultural goods in the form of simple tariffs and tariff-rate quotas (TRQs). Other measures, including antidumping and countervailing duties (AD and CVD), may also affect U.S. exports of agricultural products.[1] China substantially reduced its tariffs and replaced absolute quotas with TRQs for various agricultural products in preparation for its World Trade Organization (WTO) accession on December 11, 2001.[2] Most of the tariff concessions for agricultural products were implemented as of January 1, 2004. China's bound and applied tariffs generally are the same, and applied tariffs typically remain constant.

China's agricultural tariffs generally are higher than those for other products, and TRQs exist for agricultural products that are considered vital for food security, such as wheat, corn, rice, and sugar. Most TRQs are held by state trading enterprises (STEs) and have low fill rates. China has entered into several free trade agreements (FTAs) under which it provides reduced tariff rates to qualifying goods from countries that are part of the agreements or arrangements.[3]

Model simulation results prepared by Commission staff, presented in ranges to account for the statistical uncertainty in key economic parameters, suggest that China's food and agricultural tariffs and TRQs reduced U.S. food and agricultural exports to China in 2009 by between $1.3 billion and $2.1 billion. The tariff simulation captures the effects of the removal of Chinese tariffs and TRQs on agricultural imports from all sources. Among U.S. products most affected by China's agricultural tariffs were wheat (U.S. exports to China are estimated to have been reduced by between $489 million and $1.2 billion), poultry ($358–$363 million), pork offal ($51–$84 million), cotton ($28–$71 million), and alcoholic beverages ($32–$43 million). Absent tariffs, in the span of a few years, U.S. exports could expand more rapidly than modeling simulations indicate because of the possible additional effects of economic growth in China and of market development by U.S. exporters, two factors not included in the simulation.

Tariffs

As noted, China significantly reduced its tariffs before its WTO accession. China's simple average tariff[4] for agricultural products fell from 42.2 percent ad valorem in 1992 to 23.6 percent ad valorem in 1998.[5] As a condition of its WTO accession, China reduced its agricultural tariffs to a simple average of 15 percent ad valorem. Most tariff reductions occurred by January 1, 2004, with the remainder completed no later than January 1, 2010.[6] Only a few agricultural products ended their staged reductions in 2010, including fresh strawberries (under subheading 0810.10 in the worldwide Harmonized System [HS]), certain provisionally preserved fruits and nuts (HS0812.90), and certain other fermented beverages and mixtures (HS2206.00).[7] By 2007, China's trade-weighted average tariff[8] had fallen to 12 percent ad valorem for agricultural products.[9]

The government of China announces tariff changes in an annual Tariff Execution Plan. The plan for 2010 made several changes affecting agricultural commodities. Specifically, the plan:[10]

- Finalized most-favored-nation (MFN) tariff reductions on fresh strawberries and provisionally preserved fruit and nuts in accordance with its WTO Accession Agreement;
- Lowered MFN duties on rice wine, other fermented beverages, and wet blue (chrome-tanned) leather;
- Provided details about the administration of TRQs on wheat and cotton;
- Indicated specific and compound rates for several HS subheadings, including frozen chickens; and
- Referenced tariff commitments made under FTAs and PTAs.

Average Tariffs

China's average tariff rates vary significantly by HS chapter. Figure 7.1 shows the ranges and simple averages for Chinese applied tariff rates by HS chapters for agricultural products. Average tariffs are the highest and the ranges the greatest for products in HS chapters 10 (grains), 11 (milled grain products), 17 (sugar), 22 (beverages), and 24 (tobacco). Many products in these chapters are the subject of domestic food security concerns (wheat, corn, rice)[11] or are highly regulated and subject to supplementary consumption taxes (alcohol, tobacco).[12] Table 7.1 shows China's average bound WTO tariffs and applied MFN tariffs (2008), by major agricultural product groups. A relatively small share of China's agricultural imports benefit from duty-free MFN treatment.

Tariff Peaks

China maintains high tariffs on the general product categories of sugars and confectionery, cereals (grains) and cereal preparations, cotton, and beverages and tobacco. Tariff peaks generally exist for agricultural products that China considers to be important to its food security or important for other reasons. With respect to individual agricultural products, China's applied tariffs are highest for corn, wheat, and rice (65 percent ad valorem for over-quota imports); tobacco (57 percent); and raw cane and refined sugar (50 percent for over-quota imports) (table 7.2). Other products with high tariffs include cotton (40 percent for over-quota imports), certain fermented beverages (40 percent), beverage bases (35 percent), and a variety of nuts (30 percent).[13] Some of these peak tariff rates are applied to over-quota quantities related to TRQs, as discussed below.

Tariff Distribution

Most of China's bound and applied agricultural tariffs range between 5 percent and 25 percent ad valorem and generally are higher than those for non-agricultural products (table 7.3). About three-quarters of the value of China's agricultural imports from all sources in 2007 were dutiable at 10 percent ad valorem or less, while about 95 percent of nonagricultural products were dutiable in this range (table 7.3). A significant share (about 16 percent) of China's agricultural imports in 2007 was dutiable between 25 percent ad valorem and 50 percent ad valorem.

Table 7.1. China's final bound WTO tariffs, MFN applied tariffs, and share of imports entering duty free, by agricultural product group

	Final bound tariffs (percent ad valorem)			MFN applied tariffs (percent ad valorem)		Share of imports
Product group	Average	Share that is duty free[a] (percent)	Maximum	Average	Share that is duty free[a] percent)	entering duty free (percent)
Animal products	14.8	9.4	25	14.7	10.1	4.1
Dairy products	12.2	0	20	12.0	0	0
Fruit, vegetables, plants	15.0	4.8	30	14.8	5.9	3.6
Coffee, tea	14.9	0	32	14.7	0	0
Cereals and preparations	23.7	2.6	65	23.9	3.4	0
Oilseeds, fats, and oils	11.6	6.2	30	10.6	5.4	0.1
Sugars and confectionery	27.4	0	50	27.4	0	0
Beverages and tobacco	23.9	2.4	65	22.9	2.2	1.8
Cotton	22.0	0	40	22.0	0	0
Other agricultural products	11.9	10.2	38	11.5	9.4	2.3

Source: WTO and ITC, World Tariff Profiles 2009, 60.

[a] Share of total HS 6-digit subheadings that are duty free.

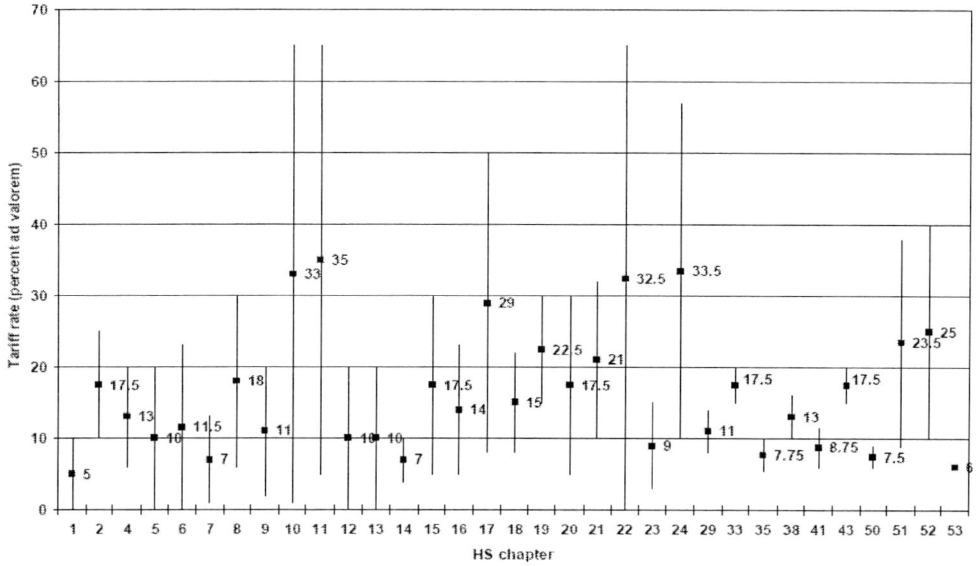

Source: WTO, "Schedule CLII: People's Republic of China."

Note: The vertical line represents the range between the high and low applied tariffs in each HS chapter. The simple average tariff for each HS chapter is indicated by the data point at the mid point of each vertical line. The absence of a vertical line indicates a single tariff for the HS chapter.

Figure 7.1. China's applied agricultural tariffs vary significantly within and among HS chapters.

Table 7.2. China's peak bound and applied tariffs, by HS subheading

		Tariff (percent ad valorem)	
HS subheading	Description	Bound	Applied (2008)
1001.10	Wheat	65	65
1001.90	Wheat	65	65
1005.90	Corn	65	65
1006.10	Rice	65	65
1006.20	Rice	65	65
1006.30	Rice	65	65
1006.40	Rice	65	65
1101.10	Wheat flour	65	65
1103.11	Wheat meal	65	65
1103.13	Corn meal	65	65
1104.23	Corn, otherwise worked	65	65
2205.10	Vermouth and flavored wines	65	65
2205.90	Vermouth and flavored wines	65	65
2403.10	Smoking tobacco	65	57
2403.91	Reconstituted tobacco	65	57
2403.99	Other manufactured tobacco	65	57
1701.11	Raw cane sugar	50	50
1701.12	Raw beet sugar	50	50
1701.91	Refined sugar	50	50
1701.99	Refined sugar	50	50

Source: WTO, "Schedule CLII: People's Republic of China."

Note: For TRQ items, data represent over-quota tariffs.

Table 7.3. Frequency distribution of China's final bound and applied tariffs for agricultural and non-agricultural products

Tariff and product type				Tariff rate (percent ad valorem)				
	Free	0<=5	5<=10	10<=15	15<=25	25<=50	50<=100	>100
Final bound			Share (percent) of total tariff lines[a]					
Agricultural products	5.8	8.1	25.0	25.3	26.3	7.0	2.5	0
Non-agricultural products	6.8	18.3	46.9	14.7	12.1	1.4	0	0
Applied (MFN, 2008)								
Agricultural products	5.9	8.5	26.3	24.5	24.9	7.4	2.6	0
Non-agricultural products	7.6	19.9	46.6	14.3	10.6	1.1	0	0
Imports (2007)			Share (percent) of total imports[a]					
Agricultural products	0.8	41.5	31.4	4.7	4.9	15.9	0.7	0
Non-agricultural products	51.6	19.2	24.0	2.7	2.0	0.4	0.0	0.0

Source: WTO, "China tariff profile," 2009.
[a] HS 6-digit basis.

There is little difference between China's applied and bound tariffs (tables 7.1–7.3), and applied tariffs generally do not fluctuate. One notable exception is cotton, which is subject to a TRQ, as discussed below. China occasionally reduces its over-quota tariff on cotton imports to supplement short domestic supplies.[14]

Tariff escalation is not a major issue regarding China's agricultural tariffs. Average applied tariffs, by HS chapter, show a very slight trend towards tariff escalation, with peaks in chapters containing products with food security concerns (figure 7.1). In some cases, tariff de-escalation exists. For example, the Chinese duty for imports of inshell almonds is 24 percent ad valorem, while that for shelled almonds is 10 percent ad valorem. The United States requested the lower tariff rate for shelled almonds during China's WTO accession negotiations. However, Chinese consumers prefer inshell almonds, which has limited the effectiveness of the tariff reduction.[15]

Tariffs on Priority U.S. Export Products

China substantially reduced tariffs on certain agricultural items that were identified as a priority by the United States during WTO accession negotiations,[16] as shown in figure 7.2. Despite these reductions, tariffs generally remain relatively high and, in some cases, prohibitive for certain major U.S. agricultural export products. Most prominent is the high over-quota tariff on wheat (65 percent ad valorem). This high tariff, coupled with the way the wheat TRQ is administered (described below and in greater detail in chapter 9), limits U.S. wheat exports.[17] U.S. cotton exports are also limited by the relatively high over-quota tariff (40 percent ad valorem) and TRQ administration issues.[18]

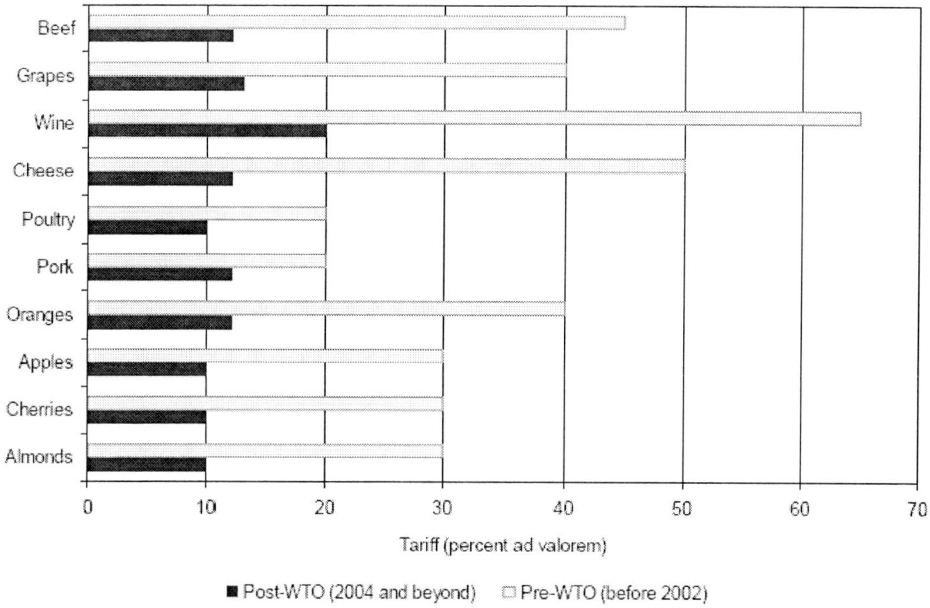

Source: USDA, FAS, "The U.S.-China WTO Accession Deal: A Strong Deal in the Best Interests of Agriculture."

Figure 7.2. Chinese agricultural tariffs were reduced significantly upon joining the WTO.

Tariff-Rate Quotas

China converted absolute quotas to TRQs as a condition of its WTO accession.[19] The main regulation guiding the administration of TRQs is the Interim Measures for the Administration of Import Tariff Quotas of Agricultural Products.[20] TRQ amounts, conditions, and requirements are announced annually,[21] and China notifies the WTO annually regarding its TRQ administration.[22]

Products

Current TRQs for agricultural products are shown in table 7.4. In addition, China maintained TRQs on imports of soybean, palm, and rapeseed oil before 2006; they were eliminated on December 7, 2005, as a result of commitments related to China's WTO accession.[23] However, China reserved the right to establish state trading in these products under the accession protocol.[24] China's TRQs for agricultural products represent about 9 percent of domestic consumption of wheat, 5 percent of corn, 4 percent of rice, 13 percent of sugar, and 2 percent of cotton.[25] These relatively small shares indicate the importance of these products to achieving China's goals in food security through self-sufficiency.

TRQ Fill Rates

China's agricultural product TRQ fill rates vary by product and year (figure 7.3). The TRQ for imports of cotton was consistently and substantially higher than those for other products during 2004–09 and exceeded 100 percent each year.[26] The wool TRQ exhibited the second highest fill rate during the period, and it generally increased during the period. The relatively high fill rates for cotton and wool reflect China's need for these products for further processing. The TRQ fill rate for sugar was relatively stable during the period. The wheat TRQ fill rate declined substantially during 2004–09. TRQ fill rates for corn and short-grain rice typically are small, largely reflecting China's emphasis on food security through self-sufficiency in certain staple food products.

Table 7.4. China's agricultural TRQs

Product	TRQ volume	In-quota tariff	Over-quota tariff (MFN)	State-trading enterprise share of TRQ
	Thousand mt		Percent ad valorem	Percent
Wheat	9,636	1	65	90
Corn	7,200	1	50	60
Rice (short/medium grain)	2,660	1	50	50
Rice (long grain)	2,660	1	50	50
Cotton	894	1	Sliding duty (currently 6–40)	33
Sugar	1,945	15	50	70
Wool	287	1	38	([a])
Wool tops	80	3	38	([a])

Source: USDA, ERS, "China: Policy, International Trade Policies."
Note: Over-quota tariffs have been adjusted to reflect current bound tariffs. [a]Not applicable.

TRQ Administration

The administration of China's TRQs is carried out by the Ministry of Commerce (MOFCOM) and the National Development and Reform Commission (NDRC) under the Regulations on Administration of Import and Export of Goods and the Interim Measures on the Administration of Tariff Rate Quota for Importation of Agricultural Products.[27] MOFCOM administers the TRQs on sugar and wool, while the NDRC administers the TRQs on wheat, corn, rice, and cotton. TRQs are announced annually, although TRQ quantity levels have remained the same over time.[28]

TRQ allocations are based on the number of applicants, historical import levels and shares, domestic production capacity, and other relevant information.[29] Several conditions and requirements are imposed on TRQ holders. TRQ allocations for grain, cotton, and sugar are valid for one calendar year, with an extension available through February of the following year, while the TRQ for wool and wool tops is valid for six months within a calendar year, with an extension available through the following February. TRQ holders for all products are required to return any unfilled quota allocations for reallocation.[30] Any unreturned quotas will be subtracted from the TRQ holder's next allocation. TRQ applications are required to be filed with MOFCOM (for sugar, wool, and wool tops) and NDRC (for grain and cotton).

The TRQ for wool and wool tops is allocated on a first-come, first-served basis. For sugar, a quota holder must process at least 600 metric tons of raw sugar, must have registered capital of at least RMB 1 billion ($0.1 billion), and must have annual sales of at least RMB 3.5 billion ($0.5 billion).[31] For wool and wool tops, a quota holder must have a processing capacity of 5,000 metric tons (mt),[32] while wheat STE quota holders must have a daily processing capacity of 400 mt.[33] Corn quota holders must have a need for 50,000 mt of corn for feed manufacturing or 100,000 mt of corn for other processing. Rice TRQ quota holders must have annual food sales of RMB 100 million ($13 million) and food imports and exports totaling at least $25 million. China administers the cotton TRQ using a "sliding duty" for over-quota quantities. In 2009, this duty was RMB 0.57 ($0.08) per kilogram for cotton valued above RMB 11.397 ($1.52) per kilogram.[34] For cotton valued below that amount, the duty was determined by a formula ensuring that it would go no higher than the bound rate of 40 percent ad valorem.

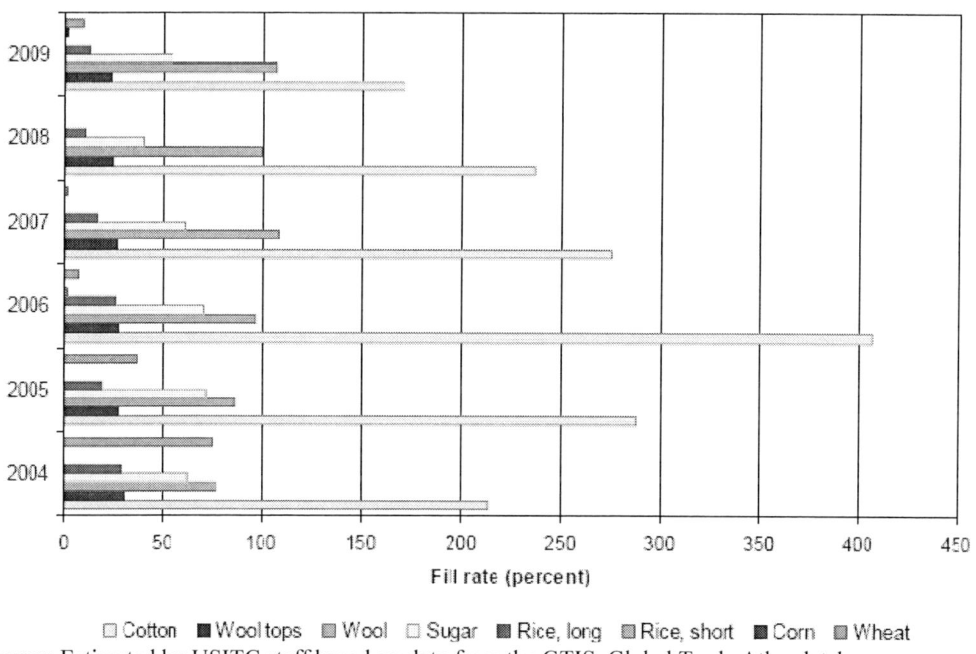

Source: Estimated by USITC staff based on data from the GTIS, Global Trade Atlas database.

Figure 7.3. China's TRQ fill rates vary significantly by product and year.

Most agricultural product TRQs are administered by an STE (table 7.4). Over-quota tariff rates generally are prohibitive. In addition, TRQ utilization rates for most grains are low (figure 7.3). A recent study concluded the principal reasons for the low fill rates were the role of STEs in administering the grain TRQs, increased government support and domestic grain output, and improved domestic grain quality.[35]

China agreed to lower the shares of agricultural TRQs administered by STEs in its WTO accession protocol.[36] However, most of these shares remain relatively large (table 7.4). The share is highest for wheat (90 percent) and lowest for cotton (33 percent). In addition, a lack of transparency in the administration of TRQs is an issue for U.S. agricultural exporters,[37] and transparency remains a general issue for China, as cited by the WTO.[38] Issues regarding the administration of TRQs are discussed in greater detail in chapter 9.

Antidumping and Countervailing Duties

In August and September 2010, China announced final countervailing duty (CVD) and antidumping (AD) determinations, respectively, on imports of certain U.S. poultry products. Actions related to these determinations have adversely affected U.S. poultry exports to China. CVD and AD duties are not traditional tariffs; rather, they are duties to offset foreign subsidies and dumping margins. These determinations resulted in the imposition of CVD rates of between 5.1 percent and 30.3 percent, and AD rates of between 50.3 percent and 105.4 percent, on imports of the subject U.S. poultry into China. [39] The U.S. Trade Representative stated that the United States and U.S. respondents were still reviewing China's final determinations in these investigations, and said that the investigations "had been issued under troubling circumstances, [and that] they are finding, once again, procedures, methodologies and decisions that generate concern, given WTO rules."[40] According to the U.S. Department of Agriculture, China's imports of U.S. broiler meat declined by 80 percent during January–July 2010 as compared with the same period of the previous year.[41] For much of January–July 2010, the U.S. poultry at issue was subject to payment of a bond on entry into China. The bond was required after China announced affirmative preliminary antidumping determinations in February 2010 and affirmative preliminary countervailing duty determinations in April 2010.

Simulated Effects of China's Applied Agricultural Tariffs on U.S. Agricultural Exports

In order to estimate the effects of China's applied agricultural tariffs on U.S. agricultural exports, the Commission conducted an economic model simulation[42] in which China's MFN tariffs and tariff-rate quotas (TRQs)[43] on imports from all countries (i.e., the United States and the rest of the world) were removed. The resulting simulated trade flows were then compared to actual 2009 trade flows. The simulation results, represented by the difference between the two, are characterized as the effects of the removal of China's tariffs and TRQs on its imports, both from the United States and from all other countries. Results were obtained for 131 commodity sectors,[44] a subset of which is presented in table 7.5. This subset includes: (i) products with large actual 2009 U.S. exports to China (e.g., soybeans and animal hides), (ii) products for which the simulation showed large effects on U.S. exports to China (e.g., wheat and poultry), and (iii) products subject to TRQs in China (rice and sugar). For several of the selected products, the volume of imports is small compared with the Chinese domestic market (table 7.6). For example, during 2007–09, imports accounted for less than 1 percent of Chinese consumption for wheat, rice, corn, pork, and apples. Imports also made up a small share of Chinese consumption for poultry, sugar, and several types of fresh fruit. Soybeans and tree nuts are the only imports that account for a large share of Chinese consumption. Because of the low rates of import penetration for many commodities, the effects of tariff elimination on Chinese consumption likely would be muted even when the effects on trade are significant.

Table 7.5 shows actual 2009 Chinese trade-weighted tariff rates for the United States and the world, actual 2009 U.S. exports to China and total Chinese imports, and the Commission's simulated effects for U.S. exports and total Chinese imports of the selected agricultural products. In 2009, total U.S. agricultural exports to China were $10.9 billion (table 7.5). The trade-weighted average import tariff levied by China on agricultural imports from the United States was about 6 percent.[45] The results of the model simulation suggest that China's agricultural tariffs reduced U.S. food and agricultural exports to China in 2009 by $1.3–$2.1 billion; in the absence of China's tariffs, U.S. agricultural exports to China would have been 11–19 percent higher.[46] The effects are presented in ranges to account for the statistical uncertainty in key economic parameters.[47]

In general, the simulated U.S. export effects are positive and are driven by the magnitude of the Chinese tariffs and the degree of sensitivity of China's consumers to prices. According to the simulation, among U.S. products most affected by China's agricultural tariffs were wheat (U.S. exports to China would have been $489 million to $1.2 billion higher without tariffs), poultry ($358–$363 million), pork offal ($51–$84 million), cotton ($28–$71 million), and alcoholic beverages ($32–$43 million). Absent agricultural tariffs, the simulation results also suggest that China's food and agricultural imports from all countries would have expanded by $5.5–$9.8 billion in 2009.

The model simulation shows that total Chinese imports of wheat, poultry, pork, and soybean oil would expand significantly, in dollar terms, in the absence of China's tariffs and TRQs.[48] According to the simulation, the elimination of Chinese tariffs and TRQs would generate more Chinese imports of these products as foreign suppliers take advantage of the cost savings. For example, the model simulation shows an increase in Chinese total imports of poultry of $400–$416 million, with $358–$363 million of this increase accounted for by increased exports of poultry to China from the United States. Likewise, of the simulated increase in total Chinese imports of wheat of $1.2– $3.3 billion, $489 million to $1.2 billion are accounted for by the United States. The increased competition in the Chinese market from lower-cost imports of wheat, poultry, pork, and soybean oil would cause a significant reduction in the supply of those products produced domestically in China.

Changes in the Chinese market for wheat, poultry, pork, and soybean oil would then have knock-on effects in other sectors. These effects are captured in the simulation through the general equilibrium effects, in which factors of production move between sectors in response to policy shocks.[49] In response to the changes, resources (such as land, capital, and labor) would leave the wheat and meat sectors in China and move into sugar crops and wool. At the same time, U.S. producers would respond to increased demand for wheat and meat products in China by drawing resources from other sectors to expand their production.

The model simulation also shows that U.S. exports of soybeans would have been lower, by as much as $7 million, in the absence of the Chinese soybean tariff. This seemingly anomalous result can be traced to the general equilibrium model effects as well. The expansion of meat production in the United States, in response to increased Chinese demand, would increase demand in the United States for animal feed, including soybeans. At the same time, increased demand for wheat in China would cause U.S. wheat production to expand, drawing resources from the soybean sector. Fewer U.S. soybeans would then be available for export, and U.S. exports of soybeans to China would fall slightly or remain constant.

Table 7.5. China: China's tariffs, U.S. and Chinese trade statistics, and simulated effects on U.S. exports to China and Chinese imports from the world in the absence of Chinese tariffs for selected agricultural products, 2009

Product	2009 trade-weighted AVE on U.S. exports to China	Actual 2009 U.S. exports to China	Range of simulated change in U.S. exports to China absent Chinese tariffs		2009 trade-weighted AVE on Chinese imports from World	Actual 2009 Chinese importsfrom the World	Range of simulated change in Chinese global imports absent Chinese tariffs	
	Percent	Million $	Million $	Percent	Percent	Million $	Million $	Percent
Animal products								
Poultry	13	796	358–363	45–46	13	985	400–416	41–42
Pork offal	13	52	51–84	98–162	14	391	385–636	98–163
Bovine hides, other	7	646	24–26	4–4	8	1,398	100–118	7–8
Frozen pork	14	23	12–12	52–52	14	137	72–73	53–54
Whey	6	75	4–5	6–7	7	284	16–20	6–7
Grains								
Wheat	68	84	489–1,192	580–1,415	68	205	1,233–3,269	601–1,595
Corn	46	4	1–4	33–100	46	20	6–16	28–79
Rice	68	0	0–0	0–0	68	5	6–20	119–378
Oilseeds and products								
Soybean oil	42	30	3–3	10–11	42	1,842	1,277–1,627	69–88
Soybeans	2	6,993	(7)–0	0–0	2	18,787	(70)–(63)	0–0
Horticultural products								
Pistachios, walnuts	14	107	18–22	17–21	14	305	53–64	17–21
Grapes	18	59	13–16	22–28	18	189	46–57	24–30
Coconuts and other nuts	22	27	10–13	38–48	18	147	37–47	25–32

Product	2009 trade-weighted AVE on U.S. exports to China	Actual 2009 U.S. exports to China	Range of simulated change in U.S. exports to China absent Chinese tariffs		2009 trade-weighted AVE on Chinese imports from World	Actual 2009 Chinese importsfrom the World	Range of simulated change in Chinese global imports absent Chinese tariffs	
	Percent	Million $	Million $	Percent	Percent	Million $	Million $	Percent
Almonds	15	87	9–11	10–12	15	111	11–14	10–12
Oranges	17	34	8–9	22–27	17	48	11–13	23–28
Apples	14	19	3–4	17–21	14	54	10–12	18–22
Processed foods								
Other prepared foods	18	94	25–26	27–27	21	469	181–189	39–40
Foods prepared from fruits and nuts	18	63	20–21	32–34	18	128	44–45	34–36
Foods prepared from vegetables	16	39	10–11	26–27	18	62	20–21	33–34
Processed rices	68	0	0–0	0–0	68	196	179–871	91–444
Other								
Alcoholic beverages	29	137	32–43	24–31	28	737	153–202	21–27
Cotton	5	803	28–71	3–9	5	2,114	70–185	3–9
Tobacco	16	104	20–22	19–22	16	742	147–166	20–22
Pet food and other feeds	8	100	14–14	14–14	7	241	27–27	11–11
Seeds for planting and other plant parts	6	61	3–3	5–5	5	254	3–3	1–1
Sugar and molasses	0	0	0–0	0–0	0	378	(2)–(1)	(1)–0

Table 7.5. (Continued)

Product	2009 trade-weighted AVE on U.S. exports to China	Actual 2009 U.S. exports to China	Range of simulated change in U.S. exports to China absent Chinese tariffs		2009 trade-weighted AVE on Chinese imports from World	Actual 2009 Chinese importsfrom the World	Range of simulated change in Chinese global imports absent Chinese tariffs	
	Percent	Million $	Million $	Percent	Percent	Million $	Million $	Percent
All other	14	504	101–112	20–22	9	17,757	1,054–1,734	6–10
Total	6	10,942	1,251–2,090	11–19	9	47,986	5,466–9,781	11–20

Table 7.6. China: Imports, domestic consumption, and import share for certain agricultural products, average 2007–09

Commodity	Imports	Consumption	Ratio of imports to consumption
	Thousand metric tons		Percent
Poultry	461	11,899	3.9
Pork	387	46,075	0.8
Wheat	641	106,167	0.6
Rice	311	131,590	0.2
Corn	461	153,333	0.3
Soybeans	43,084	53,561	80.4
Soybean oil	2,245	9,871	22.7
Almonds	8	10	88.8
Apples, fresh	50	21,625	0.2
Grapes, fresh	68	5,087	1.3
Oranges, fresh	69	5,649	1.2
Pistachios	28	23	121.4
Cotton	9,810	46,750	21.0
Sugar	818	14,083	5.8

Source: USDA, FAS, PSD Online (accessed January 10, 2011).

The simulated tariff effects are the marginal effects of China's applied tariffs and TRQs and do not incorporate any other policy changes in China or elsewhere. In the absence of China's tariffs, in a few years U.S. exports would likely expand by more than is indicated in the model simulation results due to economic growth in China and market development by U.S. exporters. This tariff simulation does not consider other policies that reduce demand for U.S. products in China, such as nontariff measures (NTMs). The Commission's economic model simulation of the effects of China's NTMs on U.S. exports is presented in chapter 9.

Bibliography

Blue Diamond Growers. Written submission to the U.S. International Trade Commission in connection with inv. no. 332-518, *China's Agricultural Trade: Competitive Conditions and Effects on U.S. Exports*, June 3, 2010.

Global Trade Information Services, Inc. (GTIS). Global Trade Atlas database (accessed November 2, 2010).

Hertel, T., D. Hummels, M. Ivanic, and R. Keeney. "How Confident Can We Be of CGE-Based Assessments of Free Trade Agreements?" *Economic Modelling* 24, no. 4 (July 2007): 611–635.

National Cotton Council. Written submission to the U.S. International Trade Commission in connection with inv. no. 332-518, *China's Agricultural Trade: Competitive Conditions and Effects on U.S. Exports*, September 15, 2010.

People's Republic of China (PRC). Customs Tariff Commission of the State Council. *Tariff Execution Plan 2010*, December 8, 2009. http://english.mofcom.gov.cn/aarticle/subject/cv/updates/201003/20100306818756.html.

———. "Decree of the State Council of the People's Republic of China (No. 539): Interim Regulations of the People's Republic of China on Consumption Tax." *State Council Gazette*, Issue No. 33, Serial no. 1284, November 30, 2008. http://www.fdi.gov.cn/pub/FDI ns/P020081121397303127411.pdf.

People's Republic of China (PRC). Ministry of Commerce (MOFCOM). "Announcement No. 93, 2005, of the Ministry of Commerce of the People's Republic of China," December 7, 2005. http://english.mofcom.gov.cn/aarticle/policyrelease/domesticpolicy/200602/2006 0201456168.ht ml.

———. "China FTA Network." http://fta.mofcom.gov.cn/topic/enpacific.shtml (accessed July 21, 2010).

People's Republic of China (PRC). Ministry of Commerce (MOFCOM). Department of Western Asian and African Affairs (DWAAA). "Give the Second Batch of African LDCs Exports to China Began to Implement the Policy of Zero-Tariff Treatment [in Chinese]," July 5, 2007. http://xyf.mofcom.gov.cn/aarticle/ghlt/ckzcfg/200707/20070704859086.html (accessed October 27, 2010).

People's Republic of China (PRC). Ministry of Commerce (MOFCOM). *Interim Measures for the Administration of Import Tariff Quotas of Agricultural Products*, September 27, 2003. http://www.fdi.gov.cn/pub/FDIEN/Laws/ImportExport/ImportAdministration/P02006062 03829 11873615.pdf.

———. "Notice No. 51 of 2010," September 26, 2010. http://gpj.mofcom.gov.cn/aarticle/c/201009/20100907158115.html (accessed October 22, 2010).

———. "Notice No. 52 of 2010," August 30, 2010. http://gpj.mofcom.gov.cn/accessory/201008/1283134881743.pdf.

———. "Notice No. 64 of 2010," September 30, 2010. http://www.mofcom.gov.cn/aarticle/b/c/201009/20100907168336.html?1647038752=664 79511 (accessed October 1, 2010).

———. "Notice No. 65 of 2010," September 30, 2010. http://www.mofcom.gov.cn/aarticle/b/c/201009/20100907168332.html?489410848=6647 9511 (accessed October 1, 2010).

People's Republic of China (PRC). Ministry of Commerce (MOFCOM). Investment Promotion Agency (IPA). *Decree of the State Council of the People's Republic of China No. 538*, November 10, 2008. http://www.fdi.gov.cn/pub/FDI EN/Laws/General LawsandRegulations/AdministrativeRegulatio ns/P020081121386712812518.pdf.

People's Republic of China (PRC). Ministry of Commerce (MOFCOM). "US Chicken Products Hit with New Tariffs," April 30, 2010. http://english.mofcom.gov.cn/aarticle/counselorsreport/westernasiaandafricareport/20100 4/20100 406892473.html.

People's Republic of China (PRC). Ministry of Commerce (MOFCOM). Ministry of Finance and State Administration of Taxation (SAT). *Circular, Caishu Notice No. 198*, June 1, 1999.

People's Republic of China (PRC). National Development and Reform Commission (NDRC). "Notice No. 27 of 2010," September 25, 2010. http://www.ndrc.gov.cn/zcfb/zcfbgg/2010gg/t20100929 373772.htm.

People's Republic of China (PRC). State Administrator of Grain (SAG). "Message from the Administrator of SAG," 2002. http://www.chinagrain.gov.cn/english/General%20Situation.html.

People's Republic of China (PRC). State Administration of Taxation (SAT). "Notes on the Taxable Scope of Some Goods for VAT," April 14, 2009. http://www.chinatax.gov.cn/n6669073/n6669088/8991866.html.

Rosen, Daniel H., Scott Rozelle, and Jikun Huang. "China and the WTO Agriculture Agenda." In *Roots of Competitiveness: China's Evolving Agriculture Interests*, 5–17. Washington, DC: Institute for International Economics Development, 2004.

Rumbaugh, Thomas, and Nicolas Blancher. "China: International Trade and WTO Accession." International Monetary Fund. IMF Working Paper WP/04/36, March 2004.

The White House. Office of Public Liason. "Summary of U.S.-China Bilateral WTO Agreement," November 17, 1999. http://www.uschina.org/public/991115a.html.

United Nations Economic and Social Commission for Asia and the Pacific (UNESCAP). "Six Asian Economies Announce Tariff Reductions and Broader Cooperation." Press release no. G/104/2009, December 15, 2009. http://www.unescap.org/unis/press/2009/dec/g104.asp.

U.S. Department of Agriculture (USDA). Foreign Agricultural Service (FAS). *China, Peoples Republic of: Cotton and Products; Annual, 2007*, by Mark Petry and Wu Xinping. GAIN Report no. CH7033, May 1, 2007. http://www.fas.usda.gov/gainfiles/200704/146280975.pdf.

———. *China, Peoples Republic of: Poultry and Products; Annual, 2010*, by Michael Woolsey and Jianping Zhang. GAIN Report no. CH10056, September 30, 2010. http://gain.fas.usda.gov/Recent%20GAIN%20Publications/Poultry%20and%20Products %20Ann ual Beijing China%20-%20Peoples%20Republic%20of 9-30-2010.pdf.

———. Production, Supply, and Distribution (PSD) Online. http://www.fas.usda.gov/psdonline/psdQuery.aspx (accessed June 9, 2010).

———. "The U.S.-China WTO Accession Deal: A Strong Deal in the Best Interests of Agriculture," February 9, 2000. http://www.fas.usda.gov/info/factsheets/China/deal.html.

U.S. Trade Representative (USTR). *2009 Report to Congress On China's WTO Compliance*. Washington, DC: USTR, December 2009. http://www.ustr.gov/webfm_send/1572.

———. *2010 Report to Congress on China's WTO Compliance*. Washington, DC: USTR, December 2010. http://www.ustr.gov/about-us/press-office/reports-and-publications /2010-0.

U.S. Wheat Associates. Written submission to the U.S. International Trade Commission in connection with inv. no. 332-518, *China's Agricultural Trade: Competitive Conditions and Effects on U.S. Exports*, September 16, 2010.

World Trade Organization (WTO). Working Party on the Accession of China. "Accession of the People's Republic of China." WT/L/432, November 23, 2001. http://www.wto.org/english/thewto e/acc e/completeacc e.htm.

———. Working Party on the Accession of China. "Schedule CLII: People's Republic of China." WT/ACC/CHN/49/Add.1, October 1, 2001. http://www.wto.org/english/thewto e/acc e/completeacc e.htm.
———. "China and the WTO." http://www.wto.org/english/thewto_e/countries_e/china_e.htm (accessed December 20, 2010).
———. "China Tariff Profile." http://stat.wto.org/TariffProfile/WSDBTariffPFView.aspx?Language=E&Country=CN (accessed June 4, 2010).
———. "Members and Observers." http://www.wto.org/English/thewto_e/whatis_e/tif_e/org6_e.htm (accessed June 8, 2010).
———. *Responses to Questionnaire on Import Licensing Procedures.* G/LIC/N/3/CHN/3, September 30, 2004.
———. *Responses to Questionnaire on Import Licensing Procedures.* G/LIC/N/3/CHN/4, September 30, 2005.
———. *Responses to Questionnaire on Import Licensing Procedures.* G/LIC/N/3/CHN/8, October 20, 2009.
———. *Trade Policy Review: China.* WT/TPR/S/230, April 26, 2010. http://www.wto.org/english/tratop e/tpr e/tp330 e.htm.
World Trade Organization (WTO) and International Trade Centre (ITC). *World Tariff Profiles 2009.* Switzerland: WTO Secretariat, 2009. http://www.wto.org/english/res e/booksp e/tariff profiles09 e.pdf.
World Trade Organization (WTO). "WTO Successfully Concludes Negotiations on China's Entry." WTO Press release 243, September 17, 2001. http://www.wto.org/english/news e/pres01 e/pr243 e.htm.
Zhou, Zhang-Yue, and Xia Kang. "Uncover the Causes of China's Low Utilisation of Grain TRQ." *Australasian Agribusiness Review,* 17 (2009): 22–46. http://www.agrifood.info/review/2009/Zhou-Kang.pdf.

8. China's Trade Agreements

Overview

In recent years China has negotiated free trade agreements (FTAs)[1] with its immediate neighbors, several Southeast Asian countries, and a number of countries around the Pacific rim. China is now negotiating agreements with several countries in southern Africa and northern Europe. This effort appears to have been driven, in many cases, by China's goal of integrating its economy into the larger Asian economy and/or deepening existing trading relationships.[2] The FTAs that China has implemented to date are mainly with regional trading partners or with less economically developed countries. These agreements enable China to diversify its trade and reduce economic dependence on developed countries. They also contribute to closer political relations with neighbors in its region.

China's FTAs first developed with nearby partners and have spread outward. The first FTAs (effective January 2004) were the Closer Economic Partnership Agreements (CEPAs) with two special administrative regions[3] (SARs) of China—Hong Kong and Macau. These agreements were followed by a set of FTAs with the Association of Southeast Asian Nations (ASEAN) (July 2005) and then FTAs with Chile (October 2006), Pakistan (July 2007), New Zealand (October 2008), Singapore (January 2009), Peru (March 2010), and Costa Rica (April 2010). Another FTA (the China-Asia Pacific Trade Agreement), implemented in December 2009 with five countries—Bangladesh, India, Korea, Laos, and Sri Lanka—essentially solidified and extended a prior agreement.

China's FTAs vary in scope. In general, tariff concessions on agricultural products exclude sensitive products associated with food security concerns, mainly grains. China has given the most extensive agricultural product concessions to partners with relatively small agricultural sectors, such as Hong Kong, Macau, and Singapore. Concessions to other partners generally correspond to products that are complementary or offer limited competition to domestic Chinese agricultural producers.[4]

China's agricultural trade with its FTA partners has been increasing in recent years.[5] The value of its agricultural imports from ASEAN countries increased by 131 percent during 2005–09, reaching $8.6 billion in the latter year. The value of its agricultural imports from New Zealand doubled during the period to $1.2 billion in 2009. China's agricultural exports to ASEAN countries increased by 121 percent in value during 2005–09, reaching $4.6 billion the latter year.

Model simulation results prepared by Commission staff, presented in ranges to account for the statistical uncertainty in key economic parameters, suggest that the effects of China's FTAs on U.S. food and agricultural exports to China in 2009 may have ranged from a contraction of up to $21 million to an expansion of up to $48 million. The simulated effects range from negative (a contraction of U.S. exports) to positive (an expansion of U.S. exports) because they are small; as a result, the range for the aggregate, as well as for some individual products, straddles zero. This simulation assumes the full implementation of tariff and market access provisions for manufactured and agricultural goods in China's ratified FTAs; U.S. tariffs remain unchanged while China and its FTA partners experience tariff elimination or reductions.

Among U.S. agricultural exports shown to be negatively affected by China's FTAs were wheat (U.S. exports to China were reduced by as much as $37 million), whey (a reduction of between $9 million and $12 million), grapes (a reduction of $9–$11 million), and apples (a reduction of $2–$3 million). Among U.S. agricultural exports positively affected by China's FTAs were poultry (an expansion of $63–$68 million), cotton (an expansion of $18–$24 million), and soybeans (an expansion of $15 million).

Free Trade Agreements

Currently, China has completed and implemented 10 FTAs (both bilateral and multilateral) all centered in the Asia-Pacific region (table 8.1).[6] One additional FTA is pending,[7] five are under negotiation, and four are under consideration. These potential FTAs expand China's trade liberalization interest to Europe, Africa, and the Gulf region.

China's FTAs generally provide tariff concessions for products that China does not produce or for those of which it produces too little to meet domestic demand. Concessions have also been given for products not produced in partner countries. China's FTAs generally maintain high tariffs on food security-sensitive products (such as wheat, corn, rice, and sugar) where there is a potential for competition. However, no products are excluded from the tariff reductions under China's FTAs with Hong Kong, Macau, ASEAN, and Singapore (table 8.2).[8]

There are considerable variations in the scope of tariff reductions among China's FTAs. The least restrictive FTAs in terms of tariff reductions are those China has with its SARs, Hong Kong and Macau, and with Singapore, which impose few import tariffs. The most restrictive FTA is with Pakistan. In addition, the Cross-Strait ECFA is limited in scope to concessions by China on a small number of horticultural products.[9] The China-Asia Pacific Trade Agreement involves relatively minor tariff reductions by most of China's partners on a limited number of agricultural products, while more extensive reductions by China generally are from relatively low base tariffs.

Mainland and Hong Kong Closer Economic Partnership Arrangement

Background and Description

The Mainland and Hong Kong Closer Economic Partnership Arrangement (MHKCEPA) was signed on June 29, 2003. It was China's first FTA. Hong Kong is a SAR of the People's Republic of China. As a SAR, Hong Kong maintains its own customs territory. The MHKCEPA provides duty-free treatment for substantially all trade between Hong Kong and China, subject to rules of origin and other requirements. The benefits of the MHKCEPA accrue mostly to Hong Kong, as China had already received duty-free treatment for its exports to Hong Kong.

Table 8.1. Overview of China's trade agreements

	Participants	Effective date	Comments
Agreements in force			
Mainland and Hong Kong CEPA	China, Hong Kong SAR	January 2004	Comprehensive coverage of goods and services, agreement to facilitate investment
Mainland and Macau CEPA	China, Macau SAR	January 2004	Comprehensive coverage of goods and services, agreement to facilitate investment
China-ASEAN Free Trade Agreement (ACFTA)	China, Brunei, Burma, Cambodia, Indonesia, Laos, Malaysia, Philippines, Singapore, Thailand, Vietnam	July 2005 (goods)	Services and investment agreements signed later
China-Chile FTA (CCFTA)	China, Chile	October 2006	Services agreement signed later; agreement to promote investment
China-Pakistan FTA	China, Pakistan	July 2007 (goods)	Services agreement signed later
China-New Zealand FTA	China, New Zealand	October 2008	Comprehensive
	Participants	Effective date	Comments

Table 8.1. (Continued)

	Participants	Effective date	Comments
China-Singapore FTA	China, Singapore	January 2009	Accelerates liberalization of goods under ASEAN FTA; broadens services trade liberalization under ASEAN FTA
China-Asia Pacific Trade Agreement	China, Bangladesh, India, Korea, Laos, Sri Lanka	December 2009	Amended the Bangkok Agreement (1975)
China-Peru FTA	China, Peru	March 2010	Comprehensive
Cross Strait Economic Cooperation Framework Agreement (ECFA)	China, Taiwan	(a)	Signed June 29, 2010; comprehensive
Agreements pending China-Costa Rica	China, Costa Rica	(a)	Signed April 2010; comprehensive
Agreements under negotiation			
China-Southern Africa Customs Union	China, Botswana, Lesotho, Namibia, South Africa, Swaziland	(a)	Launched June 2004; comprehensive
China-Gulf Cooperation Council	China, Bahrain, Kuwait, Oman, Qatar, Saudi Arabia, United Arab Emirates	(a)	Launched July 2004; comprehensive
China-Australia	China, Australia	(a)	Launched April 2005; comprehensive
China-Iceland	China, Iceland	(a)	Launched April 2007; comprehensive
China-Norway	China, Norway	(a)	Launched September 2008; comprehensive
Agreements under consideration			
China-India Regional Trade Arrangement	China, India	(a)	Joint Study Group established in 2003; feasibility study completed in October 2007
China-Japan-Korea	China, Japan, Korea	(a)	First meeting held May 6–7, 2010. Two subsequent meetings.
China-Korea	China, Korea	(a)	Unofficial feasibility study launched November 2004; official study launched November 2006, completed June 2010
China-Switzerland	China, Switzerland	(a)	Joint feasibility study completed August 9, 2010.

Source: Compiled by Commission staff from PRC, MOFCOM, China FTA Network, various schedules of tariff concessions. [a] Not applicable.

Table 8.2. Agricultural products excluded from China's trade agreements

Agreement name	Partners' exclusions	China's exclusions
Mainland and Hong Kong CEPA	None	None
Mainland and Macau CEPA	None	None
China-ASEAN FTA (ACFTA)	None	None
China-Chile FTA	Lactose, rice, sugar, and wheat	Grain and vegetable oils
China-Costa Rica	Pork, poultry, onions, rice, lard, glucose, sugar syrup, chocolate bars, prepared cereals, jams and jellies, fruit juice, animal feed preparations	Coffee, grains (corn, rice, wheat) vegetable oils (soybean, peanut, palm, sunflower, cotton seed, rapeseed), sugar, orange juice, tobacco, wool, and cotton
China-New Zealand FTA	None	Cotton, grains (corn, rice, wheat), vegetable oils (palm, soybean, sunflower seed), wool
China-Pakistan FTA	Egg products, meat, milk products, and nuts (477 products at Harmonized System 6-digit level)	Coffee, cotton, fruits, grains, meat, nuts, and wool
China-Peru FTA	None	Coffee, grains (corn, rice, wheat), vegetable oils (canola, palm, peanut, soybean, sunflower), and tobacco and tobacco products
Agreement name	Partners' exclusions	China's exclusions
China-Singapore FTA	None	None

Source: Compiled by Commission staff from PRC, MOFCOM, China FTA Network, various schedules of tariff concessions.

Several of China's FTAs cover large volumes of trade, including those with Hong Kong, Macau, and the ASEAN countries. These are described in detail below.

Tariff Concessions

Under the MHKCEPA, China's goods continue to receive duty-free treatment in Hong Kong.[10] China agreed to provide duty-free treatment on its imports from Hong Kong in stages, beginning January 1, 2004. The only agricultural product receiving immediate tariff elimination was ice cream. [11] China agreed to eliminate duties on remaining products by January 1, 2006, subject to implementation procedures.[12] These procedures involve application, verification, confirmation, consultation, promulgation, and implementation of rules regarding the Hong Kong origin of goods. An annual list of products verified as having Hong Kong origin, and thus, subject to duty-free treatment, is published. The most recent list includes a limited number of agricultural products.[13] Notably absent are live animals; fresh, chilled, or frozen meat; most horticultural products, including nuts; and grains and oilseeds and their products. Most approved products are those that can be processed in Hong Kong from imported inputs.

Rules of Origin

Rules of origin (ROOs) are an important element of the MHKCEPA. The MHKCEPA sets forth ROO requirements as a condition for duty-free treatment for directly imported goods.[14] Other ROO criteria may be mutually developed and agreed to by both parties.

The general ROO principles are that goods must either be wholly obtained from the relevant region or be substantially transformed there. Wholly obtained agricultural goods are those harvested or collected, born and raised, hunted, or produced or processed from such products in the relevant region.[15] Substantial transformation involves processes or treatment more extensive than transportation, storage, packaging, or display. Substantial transformation criteria include manufacturing or processing operations, a change in tariff heading, specified value-added content, other criteria, and mixed criteria. [16] Manufacturing or processing operations must confer essential characteristics on finished goods. A change in tariff heading requires that there be a shift in 4-digit Harmonized System (HS) headings between non-originating inputs and finished products, and that processing occur in the relevant party's territory. The value-added content rule is that the total value of raw materials, component parts, labor costs, and product development costs be equal to or greater than 30 percent of the free on board value of the exported good, and that the final manufacturing or processing be completed in the relevant party's territory. "Mixed criteria" involves the use of two or more specified ROO criteria in the MHKCEPA. The *Schedule on Rules of Origin for Hong Kong Goods Benefiting from Tariff Preference for Trade in Goods*, which is updated annually, specifies products for which origin has been conferred and, thus, for which duty-free treatment applies to Hong Kong goods.[17]

Sanitary and Phytosanitary Requirements

The MHKCEPA specifically addresses animal and plant health, as well as food safety. The agreement requires that each party will strengthen cooperation and existing mechanisms regarding animal and plant inspection and quarantine, health and quarantine monitoring, and conformity assessment (including certification, accreditation, and standardization).[18]

Investment

The MHKCEPA provides for investment cooperation and promotion under a Joint Steering Committee. [19] The agreement specifies that each party will strengthen cooperation in sharing information regarding investment policies and regulations; hold consultations regarding common investment problems; promotion of mutual and overseas investment; and conduct exchanges on other issues of mutual concern.[20]

Mainland and Macau Closer Economic Partnership Arrangement

The Mainland and Macau Closer Economic Partnership Arrangement (MMCEPA) was signed on October 17, 2003, shortly after the MHKCEPA.[21] The MMCEPA is similar to the arrangement with Hong Kong. Identical ROOs apply, with some differences in the list of Macau products certified as eligible for duty-free treatment by China.[22] Agricultural products granted immediate duty-free treatment by China under the MMCEPA included prepared or preserved fruit and nuts, sugar confectionery, pasta, ice cream, and certain alcoholic and non-alcoholic beverages. The arrangement is virtually identical to the MHKCEPA with respect to sanitary and phytosanitary (SPS) requirements and investment.

China-ASEAN Free Trade Agreement

Background and Description

The China-ASEAN Free Trade Agreement (ACFTA) was China's first comprehensive FTA. It was signed in November 29, 2004, and entered into force July 1, 2005.[23] Reportedly, the main driver leading to the ACFTA was a common interest in regional economic integration, financial policy coordination, and political stability.[24] The ACFTA comprises discrete tariff concessions between China and individual ASEAN members within a framework of common elements addressing ROO, SPS rules, investment, and other issues.[25] Tariff concessions vary by member in scope and staging, as discussed below. The ACFTA will result in duty-free treatment for trade between signatories in products in 95 percent of HS subheadings for the "ASEAN-6" countries[26] and China by 2012.[27] Duty-free treatment will also apply to an average of 90 percent of the HS subheadings for Cambodia, Laos, and Burma and 93.7 percent for Vietnam.

Tariff and Tariff-Rate Quota Concessions

Tariff concessions are specific to the individual agreement between China and a particular ASEAN member country. Tariff concessions under the ACFTA are under three schedules: (1) an early harvest program (EHP); (2) a normal track; and (3) a sensitive track comprising a sensitive list and a highly sensitive list. Under the EHP, which was set forth in a framework agreement on economic cooperation in advance of the FTA, signatory countries agreed to immediately eliminate most tariffs on mutual trade in agricultural products in HS chapters 1 through 8.[28] This includes live animals and meat, dairy products and eggs, fruits, nuts, vegetables, and other animal and plant products. The EHP was amended to provide for the designation of specific products as well as the designation of product exclusions at the 8-digit HS level.[29] Countries opting for no exclusions include China, Brunei, Indonesia, Burma, Singapore, and Thailand. Malaysia has no exclusions for China, though it excludes eggs and certain fruit from other ASEAN members. Cambodia, Laos, and Vietnam exclude certain products from all ACFTA members, including certain live animals, poultry, eggs, and certain fruits and vegetables.[30] The EHP staging period varies by country. For China and the ASEAN-6 members EHP requires eliminating tariffs on relevant products between 2004 and 2006; for Cambodia, Laos, Burma, and Vietnam, it requires eliminating tariffs between 2004 and 2010.

The staging periods and annual reductions leading to tariff elimination for products under the second schedule—the normal track—vary by country.[31] One normal track required that relevant tariffs be eliminated no later than January 1, 2010, for the ASEAN-6 countries and China. The staging period is extended until January 1, 2015, for Cambodia, Laos, Burma and Vietnam, with a separate annual staging schedule for Vietnam. Another normal track provided for tariffs to be eliminated on certain products by January 1, 2012, for the ASEAN-6 and China or by January 1, 2018, for Cambodia, Laos, Burma, and Vietnam.[32] The Philippines began its normal-track tariff reductions on January 1, 2006.[33]

The ACFTA requires the inclusion of all products.[34] However, it allows each member country to designate sensitive products subject to lesser tariff reductions under the third schedule—the sensitive track.[35] These products are on either a sensitive list[36] or a highly sensitive list[37] (table 8.3). The ACFTA placed ceilings on the absolute number of tariff lines that could be designated as sensitive, as well as on the share of total tariff lines that could be so designated.[38] The ASEAN-6 countries and China must reduce tariffs on sensitive products to 20 percent ad valorem not later than January 1, 2012, and to between zero and 5 percent ad valorem no later than January 1, 2018. The other ASEAN countries have until January 1, 2015, to reduce tariffs on sensitive products to 20 percent ad valorem. The ASEAN-6 countries and China must reduce tariffs on highly sensitive products to between zero and 5 percent ad valorem by no later than January 1, 2015; other ASEAN countries have until January 1, 2018. Products subject to tariff-rate quotas (TRQs) are to be negotiated separately, between one or more ASEAN members and China, with the results to apply to all ACFTA members.[39]

Rules of Origin

The ACFTA requires specific ROOs be met in order for goods to qualify for preferential tariff treatment.[40] There are two possible criteria for origin under the ACFTA: either products are wholly obtained or produced in a member party,[41] or at least 40 percent of the content value originates in a member party. Product-specific rules (PSRs) are being developed to specify what degree of transformation is sufficient; these involve a change in tariff classification, a specific production process, a value criterion, or a combination thereof. Agricultural product PSRs have been agreed to for palm oil and ice cream.[42]

Sanitary and Phytosanitary Requirements

The ACFTA contained no specific SPS section. However, article 8.2 states that parties should identify nontariff barriers for elimination.[43] Article 12 (a) states that the ACFTA does not prevent countries from adopting measures "necessary to protect human, animal or plant life or health."[44] ACFTA members signed an SPS memorandum of understanding in 2007 and agreed to establish a consultative mechanism to facilitate the exchange of information, provide SPS notifications, collaborate on research, and assist lesser-developed parties.[45] As a result, ASEAN members and China have held two ministerial meetings on SPS issues.[46]

Investment

The ACFTA parties signed an investment agreement in August of 2009.[47] The stated objectives of the agreement are to liberalize investment regimes and promote the flow of investment among parties. The agreement provides for national treatment (except with regard to taxation and government procurement), compensation in case of government expropriation of an investment, repatriation of profits, and dispute settlement procedures. The agreement also provides for the institution of a permanent body to implement and administer the agreement.

Other Free Trade Agreements Involving China

A summary of provisions for China's other FTAs is presented in table 8.4.

Simulated Effects of Preferential Tariffs under China's Existing FTAs on U.S. Agricultural Exports

To estimate the effects of China's FTAs on U.S. agricultural exports, the Commission conducted an economic model simulation[48] in which bilateral tariff and TRQ preferences negotiated between China and its FTA partners were fully implemented for all merchandise trade, while Chinese tariffs on imports from its non-FTA trade partners, including the United States, remained unchanged. The resulting simulated trade flows were then compared to actual 2009 trade flows. The simulation results, represented by the difference between the two, are characterized as the effects of the full implementation of China's FTAs on U.S. exports to China. Results were obtained for 131 commodity sectors, a subset of which is presented in table 8.5. This subset includes (i) products with large actual 2009 U.S. exports to China (e.g., soybeans and animal hides), (ii) products for which the simulation showed large effects on U.S. exports to China (e.g., wheat and poultry), and (iii) products subject to TRQs in China (rice and sugar).

Table 8.3. China-ASEAN FTA sensitive and highly sensitive products, by member

Member	Sensitive track	Highly sensitive track
China	Coffee; dried pepper; wheat; rice; pineapples and pineapple juice; canned longan; coconut juice; tobacco	Corn seeds; certain rice; certain milled wheat, corn, and rice products; soybean, palm, rapeseed, and mustard oil; raw and refined sugar; cigars, cigarettes, smoking tobacco, and other tobacco products; carded or combed cotton
Brunei	None	None
Cambodia	Seaweed; pasta; potatoes; yeasts; truffles; protein concentrates; rum and tafia; raw cotton	Certain oil seeds; refined sugar; glucose; certain prepared or preserved vegetables; condiments; beer; vinegar; tobacco; cigarettes; dextrin
Indonesia	Tobacco	Certain corn; rice; rice flour; raw and refined sugar; ethanol
Laos	Live animals; meat; eggs; vegetables; nuts; fruit; dried chilies; rice	Alcoholic beverages; ethanol
Malaysia	None	Live chickens; poultry; milk and cream; eggs; cabbages; rice; cigars, cigarettes, tobacco, and tobacco products
Burma	Coffee; tea; dried chilies; rice; gum arabic; vegetable oils; margarine; insect waxes; meat products; raw and refined sugar; chewing gum; confectionery; cocoa products; pasta; biscuits and pastries; prepared or preserved fruit, vegetables, and nuts; alcoholic and non-alcoholic beverages; cigarettes and cigars; silkworm cocoons and raw silk; raw cotton	None
Philippines	Poultry; vegetables; provisionally preserved citrus and melon peel; dried chilies; ginger; corn and manioc starches	Meat; vegetables; rice; processed meat; raw and refined sugar; animal feed
Singapore	Certain alcoholic beverages	Beer
Thailand	Wheat flour; swine offal; processed tomatoes; fruit juices; pet food	Milk; vegetables; onions and garlic; certain fruit; coffee; tea; rice; soybeans; copra; vegetable seeds; vegetable oils; raw and refined sugar; coffee extracts; oil cake; tobacco; raw silk
Vietnam	Green tea; husked brown rice; certain processed meat products; beet sugar; chewing gum; certain alcoholic beverages	Hens' eggs; ducks' eggs; raw cane sugar; certain refined sugar; certain tobacco and tobacco products

Source: PRC, MOFCOM, China FTA Network, various schedules of tariff concessions.

Table 8.4. Summary of provisions for China's FTAs

Country	Background and description	Tariff and TRQ concessions	Rules of origin	SPS	Investment
Chile	The CCFTA was signed November 18, 2005, and entered into force on October 1, 2006. The CCFTA was the first non-Asian FTA signed by China. The CCFTA is a comprehensive FTA and includes provisions for institutional and administrative proceedings, market access, trade remedies, ROOs, SPS measures, technical barriers to trade, dispute settlement, and other cooperation.	The CCFTA will result in duty-free tariff treatment for 97 percent of HS subheadings during a 10-year staging period. China excluded 58 agricultural products at the 8-digit HS level, including wheat, corn, rice, certain milled grain products, certain vegetable oils, wool, and cotton. Chile excluded 24 HS subheadings, including wheat, rice, sugar, fructose, and blended syrups.	The CCFTA ROO provisions are similar to those in the ACFTA. Product-specific rules pertaining to agricultural products requires a change in chapter for HS chapters 1 through 22; a change in HS heading (4-digit HS level) for HS chapters 17 through 19; and a minimum 50 percent regional value content for the remaining products.	The CCFTA contains specific provisions regarding SPS measures. The stated objectives are to facilitate trade; protect human, animal, and plant health; and provide communication and cooperation mechanisms to implement and administer these measures. The CCFTA established an SPS committee to achieve these objectives within the general framework of the WTO SPS Agreement.	The CCFTA did not contain specific investment provisions. Rather, it established a Trade and Economic Mixed Commission that, among other duties, is responsible for dealing with investment issues between China and Chile.

Table 8.4. (Continued)

Country	Background and description	Tariff and TRQ concessions	Rules of origin	SPS	Investment
Pakistan	China and Pakistan agreed to an FTA in July 2007, and the agreement entered into force in July of the following year.	Both China and Pakistan complete their tariff reductions within five years of entry into force, with 15 percent of lines at the 8-digit level excluded from any reductions.	To be considered an originating good, at least 40 percent of the product (by content or value) must originate from within the partner country. There are also product-specific rules.	China and Pakistan have a separate agreement on SPS measures whereby the parties agree to accept the other party's measures as long as they achieve the appropriate level of procedure. They also agree to "cooperate for mutual recognition of SPS certificates."	Both parties agree that they will treat investors from the other country as fairly as investors from any third-party country and that any dispute concerning investment will be settled diplomatically. Meetings will also be held "from time to time" to review and discuss investment issues.
				Additionally, the agreement creates a committee made up of relevant Chinese and Pakistani officials responsible for consultation and cooperation on SPS issues.	

Country	Background and description	Tariff and TRQ concessions	Rules of origin	SPS	Investment
New Zealand	China and New Zealand agreed to an FTA on April 7, 2008. The agreement entered into force on October 1, 2008.	China will complete its tariff reductions by 2019 and New Zealand will complete its reductions by 2016. China maintains exceptions for grains, vegetable oils, beet and cane sugar, cotton, and wool. New Zealand does not retain any exceptions for agricultural imports from China.	Goods qualify as originating if the good is produced in the territory of one or both parties, using non-originating materials that conform to a change in tariff classification, a regional value content, a process requirement, or other requirements specified in annex 5.	China and New Zealand agree to work together to attempt to facilitate trade and improve communication. They established a joint management committee to help with market access, and each plans to expedite market access requests from the other country. The parties will jointly develop principles which will be used to determine equivalence of SPS measures.	The agreement provides important protections for investors, including national treatment (except with regard to taxation and government procurement), compensation in case of government expropriation of an investment, and dispute settlement procedures. The parties established a committee on investment that will consider any investment- related matters.
Peru	China and Peru agreed to an FTA on April 28, 2009. The agreement entered into force on March 1, 2010.	Both China and Peru complete their tariff reductions within 17 years of entry into force. Chinese exceptions to Peruvian imports include selected nuts, shellfish, livers and roes of fish, grains, cane and beet sugar, and vegetable oils.	To be considered an originating good, either the good must wholly originate from within the partner country, or the regional value content must be at least 40 percent, or the good must undergo a change in tariff classification as specified in the product-specific ROOs.	China and Peru agree to work together to attempt to achieve transparent and streamlined procedures related to SPS. They will also work on the harmonization of SPS measures and accept equivalent SPS measures used by the other party.	The agreement provides important protections for investors, including national treatment (except with regard to taxation and government procurement), compensation in case of government expropriation of an investment, and dispute settlement procedures.

Table 8.4. (Continued)

Country	Background and description	Tariff and TRQ concessions	Rules of origin	SPS	Investment
Costa Rica	China and Costa Rica signed an FTA in April of 2010. It is the first trade agreement between China and a Central American country.	China decreased tariffs on several goods, but maintained tariffs on imports of wheat, coffee, corn, rice, sugar, and certain vegetable oils. Costs Rica maintains tariffs on multiple sugars (including cane, maple, and beet), rice, and wheat.	Goods qualify as originating if the good is produced in the territory of one or both parties, using non-originating materials that conform to a change in tariff classification, a regional value content, a processing operation rule, a combination of these rules, or other requirements specified in annex 3.	China and Costa Rica formed a committee on sanitary and phytosanitary matters to convene at least every two years to communicate on SPS measures that affect the two parties. The two parties recognize the value of and will operate fully under the WTO SPS guidelines.	The agreement provides important protections for investors, including treatment no less favorable than a party would show its own suppliers. The parties agree to meet annually to review implementation of investment agreements.
Taiwan	China and Taiwan signed an economic cooperation frame-work agreement in June 2010.	China will cut tariffs on 539 products from Taiwan, a concession that is valued at $13.8 billion. Taiwan will cut tariffs on 267 products from China, a concession valued at $2.86 billion. Few agricultural products are initially included.	(a)	(a)	(a)

Country	Background and description	Tariff and TRQ concessions	Rules of origin	SPS	Investment
Asia-Pacific	The Asia-Pacific PTA involves China, India, Bangladesh, Laos, South Korea, and Sri Lanka.	The PTA included tariff reductions on the part of every signatory country, but tariffs on most goods were not eliminated completely.	To be considered an originating good, either the good must wholly originate from within the partner country, or processed products must contain no more than 55 percent value from non-participating countries, or the aggregate content originating in one of the participating states must be no less then 60 percent. Special rules apply for the least developed of the participants.	Every participating party will act in a transparent manner with one another and also act in accordance with all WTO nontariff SPS guidelines.	The PTA agreement is meant to spur economic activity, including increased investment, between the participants.
Africa	The Forum on China-Africa Cooperation (FOCAC) is an open forum between the African countries and China that promotes economic and social partnerships. There have been four summits between China and Africa, beginning in 2000 and with the most recent meeting held in 2009.	In 2007, to promote African exports to China, China exempted tariffs on 478 commodities for 31 least developed African countries.	(b)	(b)	With the 4th Ministerial Conference, it was recognized that China has been consistently increasing investment in Africa. There are currently mutual investment agreements between China and 31 African countries.

Source: MOFCOM, China FTA Network, various schedules of tariff concessions; MOFCOM, DWAA, "Give the Second Batch of African LDCs," November 3, 2010.

[a] The Taiwan-China economic cooperation framework agreement does not yet have published text outlining the specifics, such as ROO and SPS measures.

[b] There have been many summits between China and the African countries, but full trade agreements have not been agreed. The FOCAC is a forum for an economic and social relationship.

Table 8.5. China: China's tariffs, U.S. and Chinese trade statistics, and simulated effects of the full implementation of China's FTAs for selected agricultural products, 2009

Product	2009 trade-weighted AVE on U.S. exports to China	Actual 2009 U.S. exports to China	Range of simulated change in U.S. exports to China under full implementation of China's FTAs		2009 trade-weighted AVE on Chinese imports from FTA partner		Actual 2009 Chinese imports from FTA partners	Actual 2009 Chinese imports from the World	Range of simulated change in Chinese global imports under full implementation of China's FTAs	
					Before the implementation of FTAs	After the full implementation of FTAs				
	Percent	Million $	Million $	Percent	Percent		Million	Million	Million	Percent
Animal products Poultry	13	796	63–68	8–8	8	0	$ 21	$ 985	$ 95–107	10–11
Frozen pork	14	23	2–2	9–10	14	0	0	137	14–16	10–11
Bovine hides, other	7	646	0–1	0–0	13	0	35	1,398	13–16	1–1
Pork offal	13	52	(1)–1	(3)–1	14	0	3	391	(7)–7	(2)–2
Whey	6	75	(12)–(9)	(15)–(12)	11	0	10	284	(32)–(26)	(11)–(9)
Grains										
Rice	68	0	0–0	0–0	68	49	5	5	2–2	39–41
Corn	46	4	(1)–0	(24)–0	46	22	15	20	3–8	12–38
Wheat	68	84	(37)–(1)	(44)–(1)	68	0	0	205	(70)–1	(34)–0
Oilseeds and products Soybeans	2	6,993	15–15	0–0	2	0	0	18,787	33–38	0–0
Soybean oil	42	30	0–0	0–0	42	0	0	1,842	(16)–(12)	(1)–(1)
Horticultural products Almonds	15	87	0–0	0–0	15	1	0	111	0–0	0–0
Oranges	17	34	0–0	0–0	17	0	0	48	0–0	0–0

Product	2009 trade-weighted AVE on U.S. exports to China	Actual 2009 U.S. exports to China	Range of simulated change in U.S. exports to China under full implementation of China's FTAs		2009 trade-weighted AVE on Chinese imports from FTA partner		Actual 2009 Chinese imports from FTA partners	Actual 2009 Chinese imports from the World	Range of simulated change in Chinese global imports under full implementation of China's FTAs	
					Before the implementation of FTAs	After the full implementation of FTAs				
	Percent	Million $	Million $	Percent	Percent		Million	Million	Million	Percent
Pistachios, walnuts	14	107	(1)–0	(1)–0	14	0	0	305	(1)–0	0–0
Coconuts and other nuts	22	27	(1)–(1)	(4)–(3)	10	1	27	147	4–5	3–3
Apples	14	19	(3)–(2)	(14)–(11)	14	0	30	54	8–10	14–18
Grapes	18	59	(11)–(9)	(19)–(15)	18	0	107	189	35–45	18–24
Processed foods										
Processed rices	68	0	0–0	0–0	68	0	196	196	189–949	96–484
Foods prepared from vegetables	16	39	(1)–(1)	(4)–(3)	25	0	4	62	3–3	4–5
Foods prepared from fruits and nuts	18	63	(5)–(5)	(8)–(7)	21	0	21	128	12–13	9–10
Other prepared foods	18	94	(8)–(7)	(8)–(7)	21	0	81	469	44–46	9–10
Other										
Cotton	5	803	18–26	2–3	5	0	20	2,114	45–75	2–4

Table 8.5. (Continued)

Product	2009 trade-weighted AVE on U.S. exports to China	Actual 2009 U.S. exports to China	Range of simulated change in U.S. exports to China under full implementation of China's FTAs		2009 trade-weighted AVE on Chinese imports from FTA partner		Actual 2009 Chinese imports from FTA partners	Actual 2009 Chinese imports from the World	Range of simulated change in Chinese global imports under full implementation of China's FTAs	
					Before the implementation of FTAs	After the full implementation of FTAs				
	Percent	Million $	Million $	Percent	Percent		Million	Million	Million	Percent
Tobacco	16	104	12–13	12–12	16	0	13	742	114–122	15–16
Alcoholic beverages	29	137	2–4	2–3	29	0	3	737	16–26	2–3
Sugar and molasses	0	0	0–0	0–0	1	0	49	378	2–3	0–1
Seeds for planting and other plant parts	6	61	(1)–0	(1)–0	5	0	50	254	27–30	11–12
Pet food and other feeds	8	100	(1)–(1)	(1)–(1)	7	0	7	241	1–1	0–0
All other	14	504	(51)–(45)	(10)–(9)	8	0	9,557	17,757	1,321–1,623	7–9
Total	6	10,942	(21)–48	0–0	10	0	10,253	47,986	1,853–3,105	4–6

Source: Commission staff calculations with simulation framework discussed in appendix F.

Notes: (1) AVE stands for ad valorem equivalent. (2) Parenthesis ( ) indicates a negative number. (3) A range of simulated effects was obtained by varying the magnitude of trade elasticities to account for the degree of statistical uncertainty in the economic estimates of the elasticities.

Table 8.5 shows Chinese trade-weighted ad valorem tariff equivalent rates for the United States, for China's FTA partners (before and after full implementation of their FTAs), and for the world; actual 2009 U.S. exports to China, Chinese imports from FTA partners, and total Chinese imports; and the Commission's simulated effect of the fully implemented FTAs on U.S. exports and total Chinese imports of the selected agricultural products.[49] The results of the model simulation suggest that China's FTAs had a minimal effect on total U.S. food and agricultural exports to China in 2009. The effects, presented in ranges to account for the statistical uncertainty in key economic parameters, straddle zero, [50] including an overall contraction of U.S. food and agricultural exports to China of up to $21 million or an expansion in U.S. food and agricultural exports to China of up to $48 million in 2009. These small shifts in trade flows generated by the simulation total less than a 1 percent change in 2009 U.S. agricultural exports to China.

The effects of full implementation of China's FTAs on U.S. exports are small relative to the effects of the removal of China's tariffs and TRQs presented in chapter 7. This is primarily because the tariff effects presented in chapter 7 are the result of a simulation which removed China's tariffs and TRQs on imports from all sources, including the United States, thus directly affecting U.S. trade. The FTA effects presented here are the result of a simulation which either lowered or removed bilateral tariffs between China and its FTA partners, affecting U.S. trade only indirectly. In addition, the effects of China's FTAs on U.S. exports are relatively small because, for several food and agricultural products, the level of U.S. exports to China is either very large or very small relative to the level of China's imports from its FTA partners.[51] Thus U.S. exports of these products to China would not be expected to be affected to a large degree by China's FTAs.

Among U.S. agricultural exports negatively affected by China's FTAs were wheat (U.S. exports to China were reduced by $1 million–$37 million), whey (a reduction of $9– $12 million), grapes (a reduction of $9 million–$11 million), and apples (a reduction of $2 million–$3 million). Among U.S. agricultural exports positively affected by China's FTAs were poultry (an expansion of $63 million–$68 million), cotton (an expansion of $18 million–$26 million), and soybeans (an expansion of $15 million).

The simulations indicate that eliminating Chinese agricultural tariffs with FTA partners results in lower U.S. exports to China for some products as Chinese importers substitute the relatively cheaper products supplied by FTA partners for U.S. products. For example, the model simulation suggests that U.S. wheat exports would be as much as $37 million lower with implementation of China's FTAs. This is because, according to the simulation, the FTAs cause China to increase imports of rice from its FTA partners (e.g., Thailand and Malaysia). As a result, the general equilibrium effects in the model, in which factors of production move between sectors in response to the tariff changes in China and its FTA partners,[52] reflect the movement of resources in China out of rice production and into wheat production. The expansion of wheat production in China leads to a reduction in China's wheat imports from non-FTA countries, including the United States.

At the same time, the model simulation indicates that certain U.S. agricultural exports benefit from the tariff provisions of China's FTAs. For example, the model simulation shows that tariff preferences resulting from China's FTAs would lead to increases in U.S. exports of soybeans, cotton, tobacco, and poultry. In the case of poultry, U.S. exports are shown to increase by as much as 8 percent.

For these products, this seemingly anomalous outcome can be traced to the general equilibrium effects of the movement of factors of production among sectors in response to price changes. For example, according to the simulation, China's livestock production declines as resources are drawn to the expanding cotton, wheat, and fruits and vegetable sectors. As a result, China imports more meat products, including U.S. poultry.

In the case of cotton, the growth in U.S. exports to China, which increase by between $18 million and $26 million, can be attributed to the expansion in China's production of textiles as a result of its FTAs. To satisfy increased demand for cotton by its textiles industry, China would expand its domestic production and import more cotton from all sources, including the United States, which accounted for almost 40 percent of China's cotton imports in 2009. Tobacco represents a similar situation to cotton. Chinese imports of tobacco from the United States increase by between $12 million and $13 million because China's production of tobacco products, using U.S. tobacco as inputs, expands to satisfy increased demand for these products from its FTA partners.

The simulation also indicates an increase in U.S. exports of soybeans of $15 million as a result of China's FTAs. According to the simulation, Chinese paddy rice and cotton production expand at the expense of oilseed production, causing domestic oilseed prices to increase relative to those of imported oilseeds. Since Chinese demand for oilseeds does not change, China imports more oilseeds, including U.S. soybeans.

The simulated FTA effects are the marginal effects of the full implementation of the tariff and market access provisions for manufactured and agricultural goods in China's ratified FTAs with Hong Kong, ASEAN member countries, Chile, Pakistan, New Zealand, Peru, and Costa Rica. Note that the simulations incorporate only the marginal effects of the FTA provisions regarding tariffs and TRQs for goods trade and do not incorporate any other effects, such as those for rules of origin or investment. The model simulates trade policy effects on prices, trade, and supply and demand in the global economy with many countries, and many product markets, including markets for primary factors—that is land, labor and capital.

Bibliography

ASEAN-China Free Trade Area Business Portal. "Rules of Origin (ROO)," December 23, 2009. http://www.asean-cn.org/Item/1172.aspx.

Association of Southeast Asian Nations (ASEAN). *Agreement on Investment of the Framework Agreement on Comprehensive Economic Cooperation between the Association of Southeast Asian Nations and the People's Republic of China,* August 15, 2009. http://www.asean.org/22974.pdf.

———. *Agreement on Trade in Goods of the Framework Agreement on Comprehensive Economic Cooperation between the Association of Southeast Asian Nations and the People's Republic of China,* November 29, 2004. http://www.aseansec.org/16646.htm.

———. *Framework Agreement on Comprehensive Economic Co-operation between ASEAN and the People's Republic of China,* November 4, 2002. http://www.aseansec.org/13196.htm.

———. *Memorandum of Understanding between the Association of Southeast Asian Nations and the Government of the People's Republic of China on Strengthening Sanitary and Phytosanitary Cooperation,* November 20, 2007. http://www.aseansec.org/documents/ASEAN-

———. *Protocol to Amend the Framework Agreement on Comprehensive Economic Co-Operation between the Association of Southeast Asian Nations and the People's Republic of China,* October 6, 2003. http://www.aseansec.org/22196.pdf.

———. "ASEAN-China Free Trade Area." http://www.aseansec.org/19105.htm (accessed July 14, 2010).

———. "Joint Statement of 2nd ASEAN-China-Ministerial Meeting on Quality Supervision, Inspection and Quarantine (SPS Cooperation), Phnom Penh, 25 October 2010," October 25, 2010. http://www.aseansec.org/25399.htm.

Bilaterals.org. *Free Trade Agreement between the Government of the People's Republic of China and the Government of the Republic of Chile,* November 18, 2005. http://www.bilaterals.org/IMG/pdf/tlc chile china ing junio 2006.pdf.

Chile. National Customs Service (NCS). "Chile–China Free Trade Agreement." http://www.aduana.cl/prontus_aduana_eng/site/artic/20070227/pags/20070227172530.html (accessed November 1, 2010).

China Daily. "China, Costa Rica Ink Free Trade Agreement," April 8, 2010. *CNN World.* "Taiwan, China Sign Historic Trade Deal," June 29, 1010.

Macau Economic Services Directorate (DSE). "CEPA (October 17, 2003)," 2007. http://www.economia.gov.mo/web nfpb=true& pageLabel=Pg CEPA CEPA I&locale=en US.

People's Republic of China (PRC). Ministry of Commerce (MOFCOM). China FTA Network. http://fta.mofcom.gov.cn/english/index.shtml (accessed July 20, 2010).

———. China FTA Network. "China–Chile FTA." http://fta.mofcom.gov.cn/topic/enchile.shtml (accessed August 3, 2010).

———. China FTA Network. *Free Trade Agreement between the Government of the People's Republic of China and the Government of the Republic of Chile*, November 18, 2005. http://fta.mofcom.gov.cn/chile/xieyi/freetradexieding2.pdf.

———. China FTA Network. "Mainland and Hong Kong Closer Economic Partnership Arrangement." http://fta.mofcom.gov.cn/topic/enhongkong.shtml (accessed July 20, 2010).

———. China FTA Network. "Mainland and Macau Closer Economic Partnership Arrangement." http://fta.mofcom.gov.cn/topic/enmacau.shtml (accessed July 20, 2010).

———. China FTA Network. *Tariff Concession List of Chile.* http://fta.mofcom.gov.cn/chile/xieyi/chile offer.xls (accessed August 4, 2010).

———. China FTA Network. *Tariff Concession List of China.* http://fta.mofcom.gov.cn/chile/xieyi/china_offer.xls (accessed August 4, 2010).

———. Department of Western Asian and African Affairs (DWAAA). "The Second Batch of African LDCs Exports to China Began to Implement the Policy of Zero-tariff Treatment," July 5, 2007. http://xyf.mofcom.gov.cn/aarticle/ghlt/ckzcfg/200707/20070704859086.html.

Republic of China (ROC). Ministry of Economic Affairs (MEA). "Ministry of Economic Affairs Announced ECFA: Online English Text," September 21, 2010.

http://www.moea.gov.tw/Mns/populace/news/News.aspx?kind=1&menuid=40&newsid=19723.

Republic of the Philippines. Department of Trade and Industry (RPDTI). "ASEAN-China Free Trade Agreement (ACFTA)," 2008. http://www.dti.gov.ph/dti/index.php?p=688.

Republic of the Philippines Tariff Commission (RPTC). *Executive Order No. 485*, December 29, 2005. http://www.tariffcommission.gov.ph/eo 485.htm.

———. *Executive Order No. 487*, January 12, 2006. http://www.tariffcommission.gov.ph/eo_487.htm.

Shulong, Chu. "China's Approach to the Free-Trade Area." Presentation at the Roundtable Discussion on Free Trade and East Asia, Foreign Policy Research Institute and World Trade Center of Greater Philadelphia, October 6, 2006. http://www.fpri.org/pubs/20061102.chu.chinafreetradearea.pdf.

The Government of Hong Kong Special Administrative Region (HKSAR). Trade and Industry Department (TID). *Mainland and Hong Kong Closer Economic Partnership Agreement,* June 29, 2003. http://www.tid.gov.hk/english/cepa/legaltext/fulltext.html.

———. *Mainland and Hong Kong Closer Economic Partnership Agreement: Annexes,* September 29, 2003. http://www.tid.gov.hk/english/cepa/legaltext/fulltext.html.

———. *Goods Entitled to CEPA Zero Tariff Preference: Mainland 2010 Tariff Codes, Product Description and Rules of Origin,* May 31, 2010. http://www.tid.gov.hk/english/cepa/tradegoods/trade goods.html.

Universal Access to Competitiveness and Trade (UACT). *ASEAN-China Free Trade Agreement: A Primer,* May 4, 2009. http://www.uactphilippines.org/images/stories/uact/publications/pdf/aseanchina primer.pdf.

Wang, Vincent Wei-cheng. "The Logic of China-ASEAN Free Trade Agreement: Economic Statecraft of 'Peaceful Rise.'" Presentation at "China in the World, the World in China: Implications of a Transforming China; Domestic, Regional and Global Impacts." Institute of China Studies, University of Malaya, August 5–6, 2007. http://www.iadb.org/intal/intalcdi/PE/2008/01283.pdf.

World Trade Organization (WTO). *Framework Agreement on Comprehensive Economic Cooperation between the Association of Southeast Asian Nations and the People's Republic of China: Communication from the Parties to the Agreement; Addenum.* WT/COMTD/51/Add.6, October 16, 2007. http://www.wtocenter.org.tw/SmartKMS/do/www/readDoc?document id=88300.

Yue, Chia Siow. "ASEAN-China Free Trade Agreement." Paper presented at the AEP Conference, Hong Kong, April 12–13, 2004. http://www.hiebs.hku.hk/aep/Chia.pdf.

9. Chinese Agricultural Nontariff Measures

Overview

Nontariff measures (NTMs) include all "government measures other than ordinary tariffs that can potentially have an economic effect on international trade in goods, changing quantities traded, or prices or both."[1] Many laws and regulations have the potential to restrict international trade. Those that affect trade are considered to be barriers if there is a protectionist intent.[2] Without seeking to identify that intent, this chapter will focus on measures identified as having had an economic effect on existing or potential U.S. agricultural exports to China.

Economic simulations indicate that China's NTMs may have a greater impact on U.S. agricultural exports than do China's applied tariffs. In the absence of Chinese NTMs, it is estimated that total U.S. agricultural exports to China would have been \$2.6–\$3.1 billion higher in 2009. Economic simulations were conducted on 12 U.S. agricultural product sectors for which (1) Chinese import prices were higher than world prices and (2) Commission research indicated that specific NTMs were impeding U.S. agricultural exports. Unlike the tariff simulation, this simulation estimates the impacts of the removal of all known and unknown NTMs specific to these products, not the elimination of a specific policy or set of policies. The sectors included in this simulation were wheat, several horticultural products (potatoes, apples, and stone fruits), cotton, and meat products (beef, pork, and poultry). The products for which the model indicated the greatest change in trade flows (and therefore considered to be most affected by Chinese NTMs) were wheat, cotton, and pork.

Some of China's NTMs keep certain U.S. products out of the Chinese market completely. Others increase costs for traders, or increase uncertainty and therefore risk. Some of China's NTMs affect virtually all agricultural products, and can make U.S. products uncompetitive or dissuade U.S. exporters from entering the Chinese market. The value-added tax (VAT) exemption for Chinese primary agricultural producers, for instance, impacts all agricultural products by conferring a substantial cost advantage on domestically produced product. [3] Other NTMs are specific to a particular product. Table 9.1 summarizes the principal NTMs faced by U.S. agricultural products entering the Chinese market.

Sanitary and Phytosanitary Issues

The World Trade Organization (WTO) agreements, including the 1994 General Agreement on Tariffs and Trade (GATT), recognize the right of WTO member countries to maintain health and food safety measures to protect their plant, animal, and human populations. However, the WTO Agreement on the Application of Sanitary and Phytosanitary Measures (SPS Agreement) establishes a framework to ensure that these measures are not used merely to protect a domestic industry from import competition. Article 3 of that agreement allows a member to set standards other than the international standards, guidelines, or recommendations only when there is scientific justification for doing so, or if scientific evaluation of the international standards, guidelines, or recommendations reveals that they do not afford the level of safety the member determines to be appropriate.[4] The SPS Agreement explicitly recognizes three relevant international organizations that develop and review accepted standards, guidelines, and recommendations:[5] the OIE,[6] the Codex Alimentarius (Codex) Commission (food safety), and the organizations operating within the International Plant Protection Convention.

Article 5 of the SPS Agreement requires that SPS measures more restrictive than international standards or guidelines be based on an assessment of the risks to the relevant population, and that the risk analysis be based on scientific evidence.[7] In any risk assessment, WTO members are to take into account "relevant processes and production methods; relevant inspection, sampling and testing methods; prevalence of specific diseases or pests; existence of pest- or disease-free areas; relevant ecological and environmental conditions; and quarantine or other treatment."[8]

Table 9.1. Reported Chinese NTMs affecting imports of U.S. agricultural products

NTM	Description
H1N1 influenza restriction	U.S. pork has been denied access due to fears related to "swine flu." The World Organisation for Animal Health (OIE) has reported that there is no risk of influenza infection from consuming pork.
Ractopamine ban	China has a zero tolerance for ractopamine, a commonly used feed additive, in pork. This limits opportunities for farmers producing pork for other markets that could otherwise profitably export some cuts to China.
Zero tolerance for pathogens	Zero tolerance is unsupported by a scientific risk assessment. This policy can serve to limit imports of meat and poultry.
Bovine spongiform encephalopathy (BSE) restrictions	China stopped imports of U.S. beef following the discovery of BSE in the U.S. cattle herd in December 2003. This is contrary to OIE guidelines.
	Also related to BSE, China prohibits use of protein-free tallow ingredients derived from ruminants and imported ingredients in U.S. pet food exported to China, including ingredients that are themselves approved for import in China.
Low pathogenic avian influenza (LPAI) restrictions	China bans imports of poultry products from certain U.S. states in which LPAI has been detected. This is contrary to OIE guidelines.
Fire blight restrictions	Only two varieties of apples from three U.S. states are approved for import, and no pears. There is no known research demonstrating a risk of fire blight from symptomless commercial varieties of apples or pears.
Potato pest risk assessment	A Chinese pest risk assessment has been forthcoming for U.S. potatoes for a decade. Movement on this issue has reportedly been tied to U.S. movement on a variety of SPS issues affecting Chinese exports to the United States.
Strawberry ban	China does not allow imports of fresh strawberries, although there were no reported problems when this restriction was temporarily lifted during the Beijing Olympic Games.
Biotechnology regulations	All products containing generally modified organisms (GMOs) must be labeled, the registration process cannot begin in China until registration is completed in the exporting country, and registrations must be renewed every three years.
VAT policies	VAT policies provide a cost advantage to Chinese domestic agricultural producers and processers that purchase domestic agricultural products rather than imports.
Labeling requirements	Some products reportedly must be labeled entirely in Chinese or must have non-Chinese characters on their labels covered with a sticker.
Customs measures	Some imports are subject to reference pricing, classification is not consistent, and clearance may be delayed.
Multiplicity and duplication	Multiple ministries and agencies are involved in licensing, certification, and inspection and do not share information among themselves.
Provincial and local variation	Regulations, standards, and enforcement can vary by location.
Tariff-rate quota (TRQ) administration	Large allocations are reserved for state trading enterprises; only small allocations are available for private traders, and there is little reallocation.
Lack of transparency	Many Chinese ministries and regulatory agencies fail to follow agreed-upon comment and notification procedures. TRQ allocations and the identity of import license holders are not made public.

Source: Compiled by Commission staff.

Use of SPS restrictions to manage the flow of agricultural imports has been a consistent tool of the Chinese government. At the time of its accession to the WTO in December 2001, China agreed to conform its SPS measures to the terms of the SPS Agreement. However, China's SPS measures have shortcomings as a result of the structure of its regulatory system, including a lack of expertise on the part of the bureaucracy issuing the standards, a shortage of resources, and an apparent lack of national treatment in China's enforcement of its SPS standards. Furthermore, the restrictions are noticeably lessened when market shortages require China to admit imports that would otherwise be barred because of SPS concerns.[9] Because of these problems, China's SPS measures introduce an element of uncertainty that increases risks and therefore costs for exporters of agricultural products to China.[10]

Plant Health Standards

The International Plant Protection Convention (IPPC) governs regulations related to measures that protect plant health. Like the broader SPS Agreement, the IPPC stipulates that measures that restrict imports be no more restrictive than necessary to achieve the appropriate level of risk. The IPPC also requires that regulatory authorities take international standards into account and that measures more restrictive than international standards be justified by a pest risk analysis (PRA) or similar evaluation.[11]

The SPS standards that China enforces against some agricultural imports exceed commonly accepted international standards for the same products. For example, the United States has been unable to export pears or most varieties of apples to China because of concerns over fire blight.[12] U.S. negotiators have argued that commercial apples and pears that do not show symptoms of fire blight cannot transmit the disease and have provided peer-reviewed scientific confirmation of this, but regulators in China have not approved importation of U.S. pears or additional apple varieties.[13] Access for U.S. pears was first sought in 2007, and additional information was provided in 2009. China's Administration of Quality Supervision, Inspection, and Quarantine (AQSIQ) has begun, but not yet completed, a risk assessment on imported pears. Access for apples from California was first requested in 1999, but to date Chinese officials have not provided a PRA to substantiate the ban on apples from California.[14]

U.S. officials and producers have also sought access for U.S. fresh potatoes since 1999. In 2000, AQSIQ requested information on U.S. production practices and pests, and agreed to begin a PRA on potatoes from the Pacific Northwest. Chinese scientists visited the U.S. growing region in 2001, and additional information was requested in 2006, 2007, and 2008. Chinese officials reported that delays in initiating and completing the PRA were due to a lack of resources and the existence of other priorities. Unofficially, over the past decade, U.S. industry representatives have been told that access for U.S. potatoes to China's market has been linked to access for China's apples to the U.S. market.[15] To date, in the absence of a completed PRA, there is no market access for U.S. fresh potatoes.[16]

Phytosanitary regulations are designed to control the introduction or spread of pests and diseases that may infect plant species or harm domestic producers. Regulations may be in the form of an outright ban on imports from some countries or regions, tests for the presence of the disease or pest, or requirements for treatment to control the risks of infection. Requirements for imports may vary with the pest or disease status of the exporting country or region. Certain agricultural products require a phytosanitary or sanitary certificate for import, issued by the foreign government, which attests to the pest-free status of the product or the treatment performed. If the risk of introduction of diseases or pests is minimal or nonexistent, a requirement for a phytosanitary certificate can be viewed as an NTM (box 9.1).

U.S. wheat exports to China are constrained by China's standards for the presence of a fungus that can contaminate U.S. wheat exports. *Tilletia controversa Kuhn* (TCK) is a fungus that is found in some varieties of wheat—predominately soft white wheat—that lowers the yield of infected plants. It can only thrive under very specific climatic conditions, so it is found in only some regions and some classes of wheat. TCK reportedly is unable to thrive in wheat-growing areas of China.[17] In a 1999 agreement, China agreed to work toward a bilateral agreement on its tolerance level for TCK and, in the interim, to allow imports of wheat certified by an accredited laboratory as having no more than 30,000 TCK spores per sample. The agreement further stipulated that once a shipment was certified as meeting the tolerance level, there were to be no restrictions on port of delivery and no measures that prevented expeditious discharge.[18] In practice, China currently restricts imports of wheat that are subject to infection with TCK to a single facility in Guangdong province, even though shipments have been certified to be within the tolerance limit. The product is required to be retested, and may be treated, before it is allowed to be sold. Both the testing methods and the treatment method are unknown.[19] The uncertainty and additional expense involved with shipping soft white wheat to China, along with the restrictions attributable to China's tariff-rate quota (TRQ) administration (discussed below), have retarded the development of China as a market for U.S. wheat.[20]

BOX 9.1.

U.S. Phytosanitary Certification Policy Affects Trade

Some NTMs related to certification are reportedly more of a barrier for U.S. exports to China than for other countries' exports due to differences in government policies in the shipping countries. Specifically, the certification processes of U.S. agencies, including the U.S. Department of Agriculture (USDA), are reportedly less flexible than certifying agencies in some other major exporting countries. This is seen by some traders as a disadvantage, with U.S. regulators unwilling to adjust to a changing trade environment.[a] Some U.S. regulators see this as an advantage, since it could imply that a certification from a U.S. government agency means more than certification from an agency that is willing to bend its standards.[b] Recent events involving U.S. soybean oil trade show how the U.S. government has shown enough flexibility to allow trade to occur, while maintaining its stance that its trading partners adhere to international standards. In early 2010, regulators in China began requiring a phytosanitary certificate for imports of soybean oil.[c] A phytosanitary certificate specifies that the exported product has been inspected or tested "according to appropriate procedures," to control the risk of introducing injurious plant pests to the importing country.[d] However, edible oils such as soybean oil are processed to such an extent that they are incapable of being infested with plant pests and therefore international standards recommend that no sanitary measures or certifications be required.[e]

The USDA's Plant Protection and Quarantine, under the Animal Plant Health and Inspection Service (APHIS), program provides inspections and certifications for U.S. exporters of agricultural commodities that are based on international standards,[f] and their stated policy recognizes the need to work with other national plant protection authorities when there are differences in the perceived need for a phytosanitary certificate.[g]

In June 2010, USDA officials agreed to issue phytosanitary certificates for U.S. exports of soybean oil to China on an interim basis while China's Ministry of Agriculture conducts a PRA that would justify the need for a such a certificate.[h] Once the interim certificates accompanied shipments, considerable volumes of U.S. soybean oil began to be exported to China. In the two months between July 1 and August 31, 2010, U.S. exports of soybean oil to China totaled over 120,000 metric tons valued at over $100 million.[i]

[a] Industry official, interview by Commission staff, Shanghai, China, September 14, 2010.
[b] Government official, telephone interview by Commission staff, October 18, 2010.
[c] American Soybean Association, written submission to the USITC, June 22, 2010, 2.
[d] USDA, APHIS, Phytosanitary Certificate, PPQ Form 577, 2001.
[e] FAO, *International Standards for Phytosanitary Measures*, 2009, 427–28.
[f] USDA, APHIS, "International Services: International Role and Opportunities," accessed October 26, 2010.
[g] USDA, APHIS, "Frequently Asked Questions," accessed October 26, 2010.
[h] Government official, telephone interview by Commission staff, October 18, 2010.
[i] GTIS, Global Trade Atlas database.

Human and Animal Health Standards

Diseases

Some diseases that can infect humans or animals can be spread by trade in live animals or animal products. For this reason, the OIE has established standards for international trade in animals and animal products. These are described in its *Terrestrial Animal Health Code*. General obligations of the importing country include the following: regulations should be no more restrictive than necessary to meet the appropriate level of protection; if regulations are more restrictive than OIE standards, they should be based on a risk analysis; and regulations should not require a greater level of protection than measures applied to domestic products.[21] Many of China's regulations restricting imports of animal products that are stricter than OIE standards have been promulgated without notice of any scientific justification. Further, many appear to be stricter than standards for comparable domestic Chinese products.[22]

Avian influenza (AI) is a disease of poultry caused by infection with any of several influenza subtypes. AI primarily affects birds, but can infect humans as well. China continues to ban imports of poultry that are produced in, or shipped through, four U.S. states because of the presence of low pathogenic avian influenza (LPAI), [23] which the OIE distinguishes from highly pathogenic avian influenza (HPAI). Both must be reported to the OIE, but internationally accepted health standards allow imports of poultry from a country, zone, or compartment that is free of HPAI.[24]

U.S. and Chinese officials have been unable to reach agreement on requirements for trade in a variety of beef products, owing to China's regulations related to bovine spongiform encephalopathy (BSE).[25] In June 2006, China agreed to allow imports of boneless U.S. beef from cattle less than 30 months of age.[26] However, approval was subject to a number of stipulations, many unrelated to BSE risk, and an agreement has not been reached. The OIE classifies the United States as a controlled-risk country for BSE. OIE guidelines for a controlled-risk country allow trade in boneless and bone-in beef from cattle of any age, subject to removal of certain materials (specified risk materials) from the carcass.[27] Negotiators were able to reach agreement on trade in several other bovine products that present a low risk of BSE (bovine semen and embryos), but were unable to reach an agreement on trade in beef tallow. [28] OIE guidelines allow trade in tallow with a maximum of 0.15 percent insoluble impurities by weight from any country, regardless of BSE status.[29] China's regulations require that tallow be "protein-free," and thus appear to be inconsistent with internationally accepted standards.[30]

U.S. exports of pet food to China are also restricted by SPS measures. Under a veterinary protocol signed in 2004 between the USDA and China's Ministry of Agriculture (MOA), U.S. producers that wish to export pet food to China must first receive certification from USDA that the facilities meet the requirements set by MOA, and then be granted a license by MOA. The Pet Food Institute (PFI), a U.S. association, contends that China's restrictions on pet food imported from the United States violate China's commitments under the SPS Agreement on several points. The SPS Agreement Annex C, paragraph 1c, states that "information requirements are limited to what is necessary for appropriate control, inspection and approval procedures." The PFI reports that foreign manufacturers of pet food and ingredients attempting to register with MOA are being required to include information on such items as floor plans of production facilities, equipment specifications, and detailed descriptions of the production process. [31] This type of information would not be needed for a scientific risk assessment of the imported product.[32]

The PFI also alleges that China's restrictions on pet food from the United States (allegedly related to concerns over BSE) violate the national treatment provision of the SPS Agreement by requiring that all ingredients used in the manufacture of pet food in the United States be of U.S. origin. This means that ingredients available for use by firms in China, such as lamb from Australia and New Zealand, are not allowed to be used in U.S.-manufactured pet food for export to China.[33]

China maintains a zero tolerance policy for bacteria such as salmonella, E. coli, and listeria in imports of pork and poultry without having presented supporting scientific risk assessments. The zero tolerance levels appear to be enforced only intermittently. Further, China does not appear to apply this standard to its domestic production.[34]

In 2009, following outbreaks of H1N1 influenza in North America, China banned imports of pork from the United States.[35] In May 2009, the OIE, jointly with the WTO, the World Health Organization, and the UN Food and Agricultural Organization, issued a statement noting that "pork and pork products, handled in accordance with good hygienic practices ... , will not be a source of infection."[36] The OIE *Terrestrial Animal Health Code* recommends no restrictions on trade in pork due to the presence of H1N1 influenza. In March 2010, China agreed to reopen its market to U.S. pork, and limited exports began in April 2010, but for nearly a year, U.S. pork was denied access to the Chinese market.

China bans all imports of sheep meat because of concerns regarding scrapie, a degenerative neurological disease that affects sheep and goats.[37] Scrapie has not been found to pose a human health risk; moreover, the OIE guidelines allow for imports of sheep meat from any country, regardless of the risk status of the exporting country.[38]

On January 19, 2010, the government of China notified the WTO that it planned to change the requirements for its dairy import certificate, and that the new requirements would go into force on March 1, 2010. China's new dairy certificate requirements, as reported by AQSIQ, contain significant differences from its old certificate, including new requirements that countries or areas exporting dairy products to China be free from tuberculosis, paratuberculosis, and anthrax during the previous 12 months.[39] These new requirements for dairy products do not adhere to international standards.[40]

During the spring of 2010, major non-U.S. dairy suppliers to China successfully negotiated new dairy import certificates with China despite the fact that one or more of the prohibited diseases are present at some level in most major dairy-exporting countries.[41] There are some significant differences between the various certificates, and between the certificates and the AQSIQ notification. For example, Australia's new health certificate for dairy product exports to China certifies that the products are from animals free of clinical signs of diseases on the OIE "List A."[42] The List A nomenclature is no longer used by the OIE, and when it was used it did not include tuberculosis, paratuberculosis, or anthrax.[43] Other countries' certificates focus on the specific product rather than the disease status of the country. Argentina's new dairy export certificate, for example, certifies that the animals from which the milk was derived are subject to an ongoing control program for tuberculosis, and does not mention anthrax or paratuberculosis.[44] Thus, these certificates do not address some of the factors that AQSIQ has identified as important.

On April 21, 2010, the Chinese government notified the U.S. government that it would stop accepting U.S. dairy products on May 1, 2010, if the new certificate regulations were not met. Subsequently, that deadline was postponed to June 1, 2010. During May 2010, U.S. government officials reportedly worked with their Chinese counterparts to resolve this issue by proposing an alternative to China's new dairy certificate based on international guidelines. On May 31, 2010, the government of China informed USDA's Foreign Agricultural Service that China's borders would be kept open to U.S. dairy products while bilateral negotiations continued. During September and October 2010, USDA and USTR requested a response from the government of China regarding the latest U.S. dairy certificate proposal. As of December 2010, no resolution had been reached. [45]

U.S. dairy exports to China have not been halted during these bilateral negotiations. However, the ongoing uncertainty increases business risk for U.S. exporters, since the sudden closure of China's borders would not only result in future lost sales but also cause financial losses on dairy shipments en route to China. In addition, the uncertainty has reduced export opportunities for U.S. exporters, as many Chinese purchasers are unwilling to enter into new business deals with U.S. suppliers out of concern that the issue could end up not being resolved, leaving them without needed supplies.[46]

Residues

Imports of all food and feed products are subject to restrictions related to residue limits. As with measures related to diseases, regulations related to residue limits should take into account an evaluation of the risk to the subject population. For residues, this analysis includes an evaluation of the amount of the product that will be consumed. Maximum residue limits (MRLs) have been established for a variety of products and cover a wide range of substances, including metals, mycotoxins, veterinary drugs, herbicides, and insecticides. The Codex Commission, the international food standards-setting body recognized in the SPS Agreement, has established a set of MRLs widely accepted in international trade.

The veterinary drug ractopamine is a feed additive that is commonly added to swine feed in the United States and other countries to promote the gain of lean meat over fat. In 2004, the Joint FAO/WHO Export Committee on Food Additives (JECFA), a committee that provides independent scientific expert advice to Codex, recommended the establishment of an MRL for ractopamine in both pork and beef.[47] Two subsequent studies by the JECFA have confirmed this recommendation. In 2009, Codex paused at the eighth step in its eight-step process of establishing an MRL for ractopamine in order to give the JECFA time to evaluate new data provided by China. China maintains a zero tolerance policy for ractopamine and has taken such a position in the Codex proceedings to establish an MRL. Data subsequently submitted by China relating to residue levels in muscle, fat, and kidney tissue supported the JECFA recommendation for an MRL. The JECFA found that data relating to ractopamine residues in lung tissue were both inconclusive and inapplicable to the ongoing assessment. [48] To date, an MRL for ractopamine has yet to be established by Codex.

China has not provided a scientific risk assessment to support its position with respect to ractopamine.[49] In the United States ractopamine was first approved for use in swine by the FDA in 1999, following a scientific review of possible health effects. Ractopamine is approved for use in 26 countries.[50] Because China's standard differs from the standards of many other countries, its zero tolerance for ractopamine limits opportunities for producers to maximize returns by shipping specific cuts to markets with the greatest demand.[51]

Genetically Modified Organisms

Regulations regarding GMOs may fall under the WTO's SPS Agreement or its Technical Barriers to Trade (TBT) Agreement.[52] As indicated above, per the SPS Agreement, any restriction of trade in GMOs on SPS grounds could be justified only by a risk assessment. Restrictions on GMOs imply that products produced through biotechnology are different from or are less safe than comparable products. Requirements for special treatment of GMO products impose costs on both exporters and consumers.[53]

China's rules on GMOs require that any product developed outside of China must first be approved for sale in the exporting country. Only after approval is secured in the exporting country can an application be made for sale within China. This results in an automatic delay in the approval process. [54] Moreover, while China has approved imports of genetically modified varieties of soybeans, corn, cotton, and canola, approval must be sought for each specific "event." This includes each trait developed for each crop, and each "stack" or combination of traits, even though individual traits have already been approved. A shipment of U.S. corn was recently denied entry into China because of a trait that has been approved in the United States and has been incorporated into "stacks" of traits in commercial production, but has not yet been approved for use in China on its own or as part of a stack of traits.[55]

Each certificate for a GMO of approval is valid for three years. Although there has been no reported disruption of trade because of failure to renew a certificate of approval, the process adds to uncertainties in the market. China's authorities further require that a separate safety certificate be issued for each individual shipment of genetically modified soybeans, even though the safety of the trait or event has already been certified. According to the

American Soybean Association, this requirement imposes an additional cost of RMB 3,000 ($400) per shipment on U.S. soybean exports to China.[56]

Licensing and Certification

China has been reducing the number of imports subject to licensing requirements since the 1990s. In its WTO accession agreement, China agreed to give all enterprises within China the right to trade in all goods within three years, except those specifically reserved for state trading enterprises (STEs). At the same time, China committed to increasing the transparency and predictability of its licensing procedures.[57] Currently, importers of all agricultural products are required to obtain a Quarantine Import Permit (QIP), except meat and poultry importers, who are required to obtain a Meat Quarantine Import Permit, and for some products, importers are required to obtain an "automatic license" for each shipment.[58]

Automatic Licensing

The Chinese government has established a monitoring mechanism for some imports, in the form of an automatic registration form (ARF). The ARF was first applied to poultry imports. More recently, imports of soybeans, soybean oil, and pork have been subject to ARF requirements. For these products, an importer must obtain an ARF from the Ministry of Commerce (MOFCOM), and only certain entities in China are eligible to apply for an ARF. These entities can then either import the product themselves or sell the ARF to an importer. Theoretically, the ARF serves solely as an information-gathering device, enabling the Chinese government to keep track of the volume of these products as they are imported.[59] However, the ARF does not always appear to be "automatic," and reportedly has been used to slow the entry of imports during periods of political tension.[60]

The existence of the ARF system itself serves as another complication and cost for exporters looking for a consistent market.[61] A separate "automatic" license must be obtained by an eligible importer in order to apply for a permit to import a set volume.[62] For instance, an eligible poultry importer must first receive a meat QIP from AQSIQ and an ARF from MOFCOM before placing an order with a foreign supplier. The import permit would allow importation of 400 metric tons of one product from one supplier and is valid only for a limited time. Further, the implementation of the ARF is opaque. For instance, when the ARF system was first applied to poultry imports, only about 73 entities within China were given the right to apply for an ARF. Many of these were not themselves poultry importers, but entities that instead sold the import permits to traders, imposing an additional cost on imports. An unknown number of additional entities have since been added to the list. It is believed that approximately 200 entities are eligible for an ARF to import poultry, but the updated list of eligible importers has not been made public.[63]

Other Licenses and Certifications

In addition to the ARF system, trading with China requires a wide array of licenses and certifications. Imports of cotton, for instance, require an invoice, a bill of lading, a plant quarantine certificate, a quality certificate, a certificate of origin, a packing list, a specification of weight, and a nonwood packing certificate.[64] As discussed above, U.S. producers wishing to export pet food to China must receive certification from the USDA that products meet the requirements of a protocol negotiated between the USDA and China's MOA, and then obtain a separate license from the MOA. Certifications often involve more than one agency and are often duplicative, serving to increase costs for importers.[65] Reportedly, as of April 1, 2011, importers will be additionally be required to report imports of beef, pork, mutton, and related products to the China Chamber of Commerce of Import and Export of Foodstuffs, Native Produce and Animal By-Products.

As previously noted, for imports of most agricultural products, importers must obtain a QIP from AQSIQ. Originally, the QIP system was reportedly used to ensure that an adequate number of inspectors was available to inspect imported goods at the point of entry.[66] Reportedly, AQSIQ may slow down or suspend the issuance of QIPs without notice, and the suspension of QIPs is used to limit the volume of imports during peak Chinese harvest periods.[67] Shipments that are subject to a contract, or that are purchased when prices for a specific commodity are low, may be shipped before a QIP is received. Cargoes arriving without a QIP are not allowed to be unloaded, but are held until a QIP is obtained, resulting in demurrage charges.[68]

Tax Treatment

As noted in chapter 4, China's farmers and other agricultural producers are exempt from paying China's value-added tax (VAT) on sales of their agricultural products. In contrast, the VAT must be paid on the full import value of the vast majority of imported agricultural goods.[69] This practice confers a cost advantage on Chinese domestic primary agricultural products. Additionally, the VAT on imported products is charged not only on the customs value of an import, but also on the duty assessed on that customs value.[70] Some U.S. industry groups have alleged that this practice affects most U.S. agricultural exports to China and violates Article III of the General Agreement on Tariffs and Trade.[71]

Chinese processors that purchase domestic agricultural products are able to take substantial deductions from the VAT they must pay. In China, a VAT is assessed on each transaction; it is recorded by the seller as a sales VAT and by the purchaser as a purchase VAT. When selling a processed agricultural product, a processor is liable for the VAT on its sales (typically 17 percent of the value), less the purchase VAT paid on inputs. Although no VAT is actually paid on domestically produced agricultural inputs, agricultural processors are still able to deduct 13 percent of the purchase price of the raw agricultural inputs when calculating the VAT to be paid on the processed product. Processors of Chinese domestic agricultural products are thus able to take a deduction for a tax not actually paid, thereby lowering their effective VAT rate.[72]

State-owned agricultural enterprises also are exempted from the VAT. On June 1, 1999, the central government established that authorized state-owned grain companies, such as the China National Cereals, Oils and Foodstuffs Import and Export Company (COFCO), are exempt from VAT charges for both domestically produced and imported grains, including wheat, rice, and feed grains.[73] Given the large share of import TRQs controlled by STEs for wheat, corn, and rice, this provides STEs with a distinct advantage over nonSTE importers. For private grain traders, there are no deductions of the 13 or 17 percent VAT applied at the time of import.

Labeling and Quality Standards

A primary purpose of product labeling "is to give consumers information so that they can better choose products that match their individual preferences."[74] Governments use labeling requirements to protect consumers from deceptively labeled products and producers from unfair competition.[75] But diverse and confusing labeling rules can hinder trade in agricultural products.[76] Specific labeling requirements in a given market can restrict trade by differing from generally accepted international labeling norms, creating a burdensome label acquisition and approval process, or constructing a perception that one product is inferior to a competing product.[77] Examples of each type exist in China's market for U.S. agricultural exports.

Labeling

Labeling requirements in China are established by the central government and may involve several ministries and agencies.[78] Some labeling requirements, such as those that require that labels be placed on individual containers within bulk packages, that labels be only in Mandarin, and that notices be integrated with the packaging as opposed to being affixed with a sticker, add to producers' cost of delivery.[79] Labels in China must also identify the distributor or distributors. This can be costly for firms that deal with multiple distributors.[80]

According to industry representatives, the scope of China's labeling rules has unintended effects on certain agricultural products that differ from everyday packaged foods, such as bulk products for sale through food service distributors or quick service restaurants, and bottled wine and spirits. Reportedly, U.S. exporters of bulk products have had shipments delayed because, while bulk packages were labeled, inner packages did not have the labels that would be required for packages to be sold at retail.[81] Such deviations from internationally accepted norms are particularly problematic in the case of products with a long shelf life and producers that export to multiple countries, as the destination may not be known when the product is packaged. Labels for bottled wine and spirits sold in China are required to indicate a bottling date. This is not the industry standard, and is an additional expense for U.S. exporters who must handle exports to China separately from product bound for other destinations, or may face delays while the product's label is brought into compliance.[82]

Standards

The Standardization Administration of China (SAC), a government body administered under AQSIQ, is responsible for setting national standards, administering the standards system, and ensuring that China's standards conform with international standards and fulfill China's commitments under the TBT Agreement. Additionally, the Certification and Accreditation Administration of China (CNCA), also under AQSIQ, is responsible for administering and implementing China's conformity assessment regime.[83] Since 2002, China has been in the process of reviewing its technical regulations to ensure that deviations from international standards are justified.[84]

China's national standards, which are developed by the SAC, are known as "GB" standards. According to the WTO, only about 46 percent of China's national standards are equivalent to international standards.[85] The SAC has not established national standards for all products in all applications. Where no national standard exists, relevant authorities are permitted to develop sectoral or local standards.[86] In practice, this means that in the absence of a national "GB" standard, any central government ministry or agency may develop and implement a nationwide standard in the field in which it has responsibility. Additionally, a provincial or local governmental body may develop and implement a standard within its jurisdiction. This creates the possibility of overlapping and/or conflicting standards. In the area of food safety, local standards may be more stringent than sectoral standards, and both may be more stringent than national standards for the same product.[87] As a result, in addition to the national standards, there is a patchwork of sectoral, provincial, and local standards that increase the complexity and costs of exporting agricultural products to China.

Differences between Chinese and U.S. inspection standards and sampling methods sometimes lead to delays in unloading shipments and in increased costs for traders.[88] However, in general, traders experience more problems involving products and applications for which the SAC has not established national standards.[89] For instance, labels for packaged foods require a complete list of ingredients and additives, but China has not established standards for some ingredients in some applications. There are additives used in potato products for which China has a standard for shelf-stable products, such as potato chips, but has no standard for frozen products, such as frozen french fries. Therefore, use of such an additive in frozen french fries could be interpreted as out of conformity with a Chinese food standard. In such a case, the product may be allowed to be imported in one instance and denied entrance at another port or time, depending on a particular official's view.[90] Reportedly, if an ingredient or additive is not widely used in China, the SAC places a low priority on establishing a standard for its use, even if international standards or guidelines exist.[91]

Customs Procedures

The import of agricultural goods into a country involves several discrete components, any one of which could hinder smooth, timely, and accurate importation. Customs procedures alleged to constitute NTMs in their application include customs valuations methods in which the actual import valuation differs from the valuation used for customs purposes; inconsistent customs classification procedures; and customs clearance procedures involving inspections and documentation that create additional cost. According to the U.S. Trade Representative (USTR), China has what appears to be a modern customs clearance process, which should enhance speedy and low-cost importation, but the process in actual practice often results in delays and unexpected expenses. Problems with customs valuation or classification and clearance delays vary from port to port, with fewer problems at ports with more traffic and, therefore, more experience.[92]

Customs Valuation

Generally, Chinese duties on imports are based on the actual transaction value, including insurance and freight. However, the Customs Administration audits the reported transaction value of every shipment in order to evaluate the accuracy of declared transaction values. For this purpose, customs officials make use of a commodity pricing information database.[93] If customs staff has a doubt about the accuracy of the transaction value, additional supporting documentation is requested. If customs staff does not accept the accuracy of the declared transaction value, the customs value can be calculated by customs staff based on the value of identical or similar goods, or the value may be estimated on a "reasonable" basis.[94] Some traders have complained that, in some cases, customs valuation is based on a reference price, rather than the actual transaction price.[95]

Reportedly, for some products, customs value is always based on a reference price. The reference price is routinely announced and the prices are widely available, even though the methodology used to calculate the reference prices is unknown.[96] In other cases, problems with customs valuation arise because the declared value of the shipment is lower than that for similar products and transactions. While sometimes this may reflect an attempt to under-invoice the goods (in order to save money on VAT and tariff), at other times the actual transaction price may legitimately be lower because of a sales promotion, an attempt to introduce a lower-cost alternative product, or seasonality. Similar to problems with customs classification, a trader's ability to convince customs officials of the validity of the valuation may at times depend on the reputation of the trader and the trader's relationship with the officials.[97] Reportedly, the problems with customs valuation tend to be seen as a legitimate effort by customs officials to deal with problems of under-invoicing, but the uncertainty adds to the risk of trading with China.[98]

Customs Classification

China's customs officials apparently have wide latitude in determining the customs classification of imported products, which may have a significant effect on the level of tariff. Problems with customs classification add to the uncertainty of trade with China and increase risks for traders. As with customs valuations, customs classifications can vary by port. Reportedly, the difficulty stems from a lack of expertise with specific products in some ports, as well as the lack of national classification standards for these products. Again, the success of a trader's attempt to convince customs officials to classify a product into a category with a lower tariff rate may depend on the reputation of the trader and the trader's relationship with customs officials.[99]

Delays in Customs Clearance

Although China instituted reforms to its customs clearance process in 2005,[100] many agricultural exporters continue to report delays or wide variations in the time needed for customs clearance. Industry representatives indicated that customs clearance varies from port to port and can take up to eight weeks.[101] Delays increase costs and discourage imports of time-sensitive products. Delays in the time it takes to clear customs is one of the reasons why almost all U.S. pork exports to China are frozen, rather than fresh or chilled.[102] Often, traders that deal in perishable products will routinely contract with customs clearance companies to speed the process of customs clearance.[103] These firms maintain close ties with customs officials and can often smooth the clearance process. The use of such "customs clearance specialists" increases the cost in delivery for traders of these commodities.

Tariff-Rate Quota Administration

China maintains TRQs on wheat, cotton, corn, rice, wool, and sugar. In any country, the implementation of a TRQ necessarily increases the complexity of the trading environment. Such implementation can be considered an NTM if the volumes allowed under the TRQ are less than commercially feasible, or if the complexity of the TRQ administration itself significantly restricts trade or adds significantly to the costs of trade. China's TRQ administration does all three. A low fill rate is an indication that such an NTM may exist.

According to industry representatives, China has not been a consistent market for U.S. wheat, partly because of the administration of the wheat TRQ. China's TRQ for wheat has a very low fill rate in most years. Under the terms of its WTO accession, China agreed to open its market to a TRQ of 9.64 million metric tons of wheat per year from all import sources, with an in-quota duty rate of 1 percent ad valorem and an out-of-quota duty rate of 65 percent ad valorem. The volumes of wheat that may be imported under China's TRQ are divided between COFCO, a state trading enterprise, and multiple private traders. According to the agreement, any quota volume that remains unfilled in the first three quarters is to be reallocated to any firms wishing to import. The volumes allocated to individual private traders are unknown, as the information is not made public, but according to the U.S. Wheat Associates, in practice, 90 percent of the within-quota volume is allocated to COFCO, and unused within-quota volumes are not reallocated.[104] Therefore, if COFCO does not use a portion of its allocation, that volume is lost.

According to industry sources, the volume offered to any one wheat trader under the wheat TRQ is often less than what would be commercially feasible. Since the information on specific allocation volumes is not public, it is difficult for traders to pool within-quota volumes. Additionally, containerized cargo vessels, which are used for smaller (i.e., not bulk) shipments of wheat are not readily available in some parts of the United States. U.S. producers and exporters in these areas cannot profitably ship the small volumes allocated to these Chinese traders.[105]

The allocation of import volume under China's cotton TRQ is also not transparent. Cotton imports enter China in one of three categories. In 2010, the within-quota volume of 894,000 metric tons (mt) was subject to a tariff of 1 percent ad valorem. An additional 1 million mt is subject to a sliding tariff based on the price of cotton. A further 400,000 metric tons may be imported duty-free by export-oriented mills, subject to export of the finished products.[106] The Chinese government operates a reserve system to control the domestic price of cotton, and the volumes and functioning of the reserve system are opaque. Taken together, these policies discourage long-term relationships with Chinese importers and increase uncertainty for U.S. cotton exporters.[107] Nonetheless, China's total cotton imports during August 2009–July 2010 were more than double the within-quota volume.

Transparency

Article X of the GATT 1994 requires that laws and regulations that affect trade be administered in "a uniform, impartial, and reasonable manner." A consistent, clearly understandable, and fully participatory system of regulations affecting international trade encourages increased trade by, among other things, removing the risk of unexpected or unexplained government action. Both the SPS Agreement and the TBT Agreement use the term "transparency" in describing requirements for regulatory systems put in place by member countries. Within the SPS Agreement, the basic requirements for transparency are spelled out in Annex B, "Transparency of Sanitary and Phytosanitary Regulations." In the TBT Agreement, the requirements for transparency are found throughout the document, including in Annex 3—the "Code of Good Practice for the Preparation, Adoption, and Application of Standards."

Notice and Comment Procedures

Both the SPS and TBT Agreements require that a member develop clear and consistent notice and comment procedures, and establish an inquiry point to respond to reasonable questions and provide copies of relevant documents. "Notice and comment" is a term used to describe official processes by which a government regulatory agency publishes a new or changed regulation and receives comments from entities outside the government, such as private citizens, companies, or other governments, expressing opinions about the regulation before it is to take effect. A number of countries, including the United States, have expressed concerns about China's irregular and occasionally abbreviated process for notifying its trading partners of legal or regulatory changes that affect the import of agricultural goods. The process commonly consists of three parts—notice, comment, and implementation period—and U.S. government officials as well as industry representatives have cited areas for improvement at each step.

According to the WTO, China has not established consistent criteria for publishing regulations.[108] Two methods of notice, each serving specific purposes, are publication in China's official journal (the *Foreign Economic and Trade Gazette*) and notification to the WTO. Publication in the journal generally provides for public comment and establishes a deadline for comment submission, if any. Notification to the WTO satisfies China's commitments to allow its trading partners an opportunity to review proposals or actions that affect trade and comment on their adherence to international trade agreements. Reportedly, while changes to regulations are usually published, changes in enforcement or implementing regulations often are not.[109] Moreover, although notification procedures have improved, the SPS Agreement requires notification "at an early stage" so that trading partners have time to comment on proposed regulations.[110] Similarly, the TBT Agreement requires notification "at an early appropriate stage," but recommends notification at least 60 days before a regulation is enacted.[111] Industry representatives have reported that the notification process in China is often too abbreviated to allow time for comment.[112]

Chinese authorities typically notify the WTO of the title of the proposed change or regulation in English, but publish the text of documents only in Mandarin. Both the SPS and TBT Agreements require that developed countries provide translations into one of the official WTO languages (English, French, or Spanish), but both make exceptions for developing countries like China.[113] However, under the terms of its accession to the WTO, China agreed to publish translations of relevant documents in one of the official WTO languages within 90 days of implementation. Many ministries and agencies reportedly do not comply with this requirement.[114]

Additional Problems with Transparency

According to the USTR, "China remains among the least transparent and predictable of the world's major markets for agricultural markets, largely because of selective intervention in the market by China's regulatory authorities." [115] As noted, China's national laws are enforced by many different ministries and agencies,[116] and provincial and local jurisdictions may impose separate regulations, especially in the area of food safety. The necessity of dealing with multiple regulators increases complexity and costs for traders of agricultural products.

The enforcement of regulations requiring zero tolerance for pathogens in meat products appears to be selective. The speed of China's ARF system is reported to vary with geopolitical relationships between China and its trading partners.[117] The QIP system administered by AQSIQ appears to slow at times. Risk assessments that are reportedly based on scientific testing are instead said to be linked to market access for China's exports.[118] Enforcement of regulations reportedly varies by location and according to personal relationships with regulators.[119]

Simulated Effects of China's Nontariff Measures on U.S. Agricultural Exports

The Commission conducted an economic model simulation[120] to determine the effects of China's NTMs on U.S. agricultural exports. The assumption that guides this model simulation is that NTMs increase the cost of imported products and lead to price gaps— higher prices in the market with the NTM than in other markets—that can be observed. Price gaps can be observed for product groups with significant trade flows. However, for products with little or no trade, and those known to be commonly traded in the grey market, price gaps cannot be observed; therefore, quantity gaps were estimated by adjusting 2009 trade flows to reflect the levels they would have reached if the United States captured the same share of China's import market as the U.S. share of global trade. Both price and quantity gaps attempt to measure the discrepancy between observed trade and estimated trade flows in the absence of NTMs.

The Commission's model simulation was conducted by removing the estimated price and quantity gaps, considered here to be tariff equivalents.[121] The simulation results are the Commission's estimate of the effects of the removal of China's NTMs on these products.[122]

For the NTM model simulation, the Commission considered products for which prices on imports into China were higher than world prices and for which Commission staff research indicated that NTMs were impeding U.S. agricultural exports. Commission staff also considered products for which U.S. exports to China were particularly low or effectively zero. Among those products for which U.S. exports were low or zero, the list included only those products for which Commission staff was able to document Chinese NTMs that raise prices and/or restrict imports. Twelve agricultural product groups met all of these criteria.

The results of the Commission's economic model simulation of the removal of China's NTMs on the 12 product groups are presented in table 9.2. The estimated increase in U.S. exports of these agricultural products to China following the simultaneous removal of these NTMs, relative to a 2009 baseline, is between \$2.6 and \$3.1 billion.[123] The majority of this increase is attributable to export expansions in wheat (U.S. exports to China would have been between \$1.5 billion and \$1.7 billion higher), cotton (\$524–\$630 million), pork offal (\$305–\$363 million), and beef (\$156–\$213 million).[124] Since this simulation removed NTMs for only 12 agricultural products, the Commission's model simulation suggests that NTMs may have a greater impact on U.S. agricultural exports than do China's applied tariffs.[125]

Table 9.2. China: U.S. and Chinese trade statistics and simulated effects of removal of China's NTMs for selected agricultural products, 2009

Product	Actual 2009 U.S. exports to China	Range of simulated change in U.S. exports to China in the absence of China's NTMs		Actual 2009 Chinese imports from the World	Range of simulated change in total Chinese global imports in the absence of China's NTMs	
	Million $	Million $	Percent	Million $	Million $	Percent
Wheat	84	1,452–1,704	1,722–2,022	205	1,702–1,956	830–954
Cotton	803	524–630	65–79	2,114	450–533	21–25
Pork offal	52	305–363	586–697	391	898–1,010	229–258
Fresh beef	0	75–96	*	8	80–103	1,004–1,293
Frozen pork	23	49–56	215–245	137	223–229	163–168
Beef offal	0	45–72	*	9	360–380	3,997–4,214
Poultry	796	35–40	4–5	985	(6)–(3)	(1)–0
Frozen beef	0	38–45	*	37	258–281	699–761
Preserved pork	0	29–42	*	1	30–44	4,484–6,472
Potatoes	1	27–31	3,202–3,678	1	33–38	2,288–2,632
Apples	19	15–18	79–96	54	10–12	18–22
Stone fruits	5	1–1	12–16	21	0–1	0–3
Sum	1,782	2,595–3,098	146–174	3,962	4,037–4,583	102–116

Source: Commission staff calculations with simulation framework discussed in appendix F.

Notes: (1) An asterisk identifies products with no U.S. exports to China in 2009. (2) Parenthesis ( ) indicates a negative number. (3) A range of simulated effects was obtained by employing low and high values in the estimated price gaps.

The simulation considers the elimination of all possible measures that may increase prices and/or restrict trade. This includes the aggregate effect of all known and unknown NTMs specific to these products, as well as issues affecting importation of U.S. agricultural products generally, such as customs procedures, notice and comment procedures, and tax policies. Decomposing the effect of particular measures is problematic. The results, therefore, do not represent the impacts of the elimination of a specific policy or set of policies. The simulated removal of NTMs differs from that for tariffs in that regard. For instance, it may be that if one NTM that currently limits imports is eliminated, another measure continues to be restrictive.

Bibliography

American Soybean Association. Written submission to the U.S. International Trade Commission in connection with inv. no. 332-518, *China's Agricultural Trade: Competitive Conditions and Effects on U.S. Exports*, June 22, 2010.

Australia. Department of Agriculture, Fisheries, and Forestry. Quarantine and Inspection Service. Health Certificate, ZDCNO2-11/07. http://jckspaqj.aqsiq.gov.cn/ybzspjyjy/ggshrzpzsyb/200803/P020100201415935357804.pdf (accessed November 24, 2010).

Biotechnology Industrial Organization. Commercial Status of Certain Agricultural Biotechnology Products database. June 25, 2010. http://www.biotradestatus.com/ (accessed November 2, 2010).

Bloomberg News. "China Agrees to Reopen Market to Argentine Soybean Oil Imports, People Say," October 11, 2010. http://www.bloomberg.com/news/2010-10-11/china-agrees-to-reopen-marketto-argentine-soybean-oil

Center for Environmental Risk Assessment, GM Crop database. http://www.cera-gmc.org/?action=gm_crop (accessed November 2, 2010).

China Daily. "China Said to Seek Brazil, U.S. Soybean Oil Supply," May 5, 2010.

Codex Alimentarius Commission. "Report on the Nineteenth Session of the Codex Committee on Residues of Veterinary Drug Residues in Food," September 2010.

Cotton International. *China Sets Limit on Cotton Imports for 2011*, December 30, 2010. http://www.cotton247.com/news/ci/?storyid=1756

Food and Agricultural Organization of the United Nations (FAO). "Categorization of Commodities According to their Pest Risk." *International Standards for Phytosanitary Measures*, (ISPM) no. 32. Secretariat of the International Plant Protection Convention, 2009.

Global Trade Information Service, Inc. (GTIS). Global Trade Atlas database (accessed November 15, 2010).

International Trade Centre. "Nontariff Requirements in the Domestic Market: Chapter 6; Market Profiles, China." http://www.cottonguide.org/chapter-6/china/non-tariff- (accessed October 6, 2010).

Joint FAO/WHO Expert Committee on Food Additives (JECFA). "Summary of Evaluations Performed by the Joint FAO/WHO Expert Committee on Food Additives," January 28, 2006. http://www.inchem.org/documents/jecfa/jeceval/jec_2080.htm.

———. "Residue Evaluation of Certain Veterinary Drugs." Rome: FAO, 2010.

Mattson, Jeremy W., Won W. Koo, and Richard D. Taylor. *Non-Tariff Trade Barriers in Agriculture*. Agribusiness & Applied Economics Report, no. 531, March 2004.

http://ageconsearch.umn.edu/bitstream/23501/1/aer531.pdf.
National Potato Council. "U.S. Fresh Potato Market Access to China." Submitted to the International Trade Commission in connection with inv. no. 332-518, *China's Agricultural Trade: Competitive Conditions and Effects on U.S. Exports*, June 22, 2010.
OIE, see the World Organisation for Animal Health.
Organisation for Economic Co-operation and Development (OECD). "Analysis of Non-Tariff Measures: The Case of Labelling." TC/TC/WP(2002)40/FINAL, November 13, 2003. http://www.olis.oecd.org/olis/2002doc.nsf/LinkTo/NT00004A22/$FILE/JT00153610.PDF.
Pet Food Institute. "Dossier on Chinese Barriers to U.S. Pet Food Products," December 2008.
People's Republic of China (PRC). General Administration of Customs. "The Rules Regarding Determination on Customs Value of Imported and Exported Goods." Decree no. 148, March 28, 2006.
———. Ministry of Commerce (MOFCOM). Decree no. 538 of the State Council of the People's Republic of China, November 10, 2008. http://www.fdi.gov.cn/pub/FDI EN/Laws/GeneralLawsandRegulations/AdministrativeRegulations/P020081121386712812518.pdf.
———. General Administrative of Quality Supervision, Inspection, and Quarantine (AQSIQ). "Mission." http://english.aqsiq.gov.cn/AboutAQSIQ/Mission (accessed December 7, 2010).
———. Ministry of Finance and State Administration of Taxes. *Circular on VAT Exemption for Staple Crops Enterprises*. Caishui No. 198, 1999.
Republic of Argentina. Ministry of Agriculture, Livestock and Fisheries. National Service of Agrifood Health and Quality. Health Certificate. http://jckspaqj.aqsiq.gov.cn/ybzspjyjy/ggshrzpzsyb/201011/P020101102594710865033. pdf (accessed November 30, 2010).
Reuters. "China Quarantine Bureau Rejects U.S. Corn Cargo," November 2, 2010.
Standards Portal. "Standardization Organizations in China: Standardization Administration of China," October 15, 2010. http://www.sac.gov.cn/templet/english/ShowArticle_zwpt_en.jsp?id=4112&cluster_node_id=zw pten.
U.S. Department of Agriculture (USDA). Animal Plant Health and Inspection Service (APHIS). "International Services: International Role and Opportunities." http://www.aphis.usda.gov/international safeguarding/ index.shtml (accessed October 26, 2010).
———. "National Scrapie Eradication Program; Fiscal Year 2009 Report," January 15, 2010.
———. Plant Export. "Frequently Asked Questions." http://www.aphis.usda.gov /import_export/plants/plant_exports/faqs.shtml (accessed October 26, 2010).
U.S. Department of Agriculture (USDA). Foreign Agricultural Service (FAS). *China, Peoples Republic of: Exporter Guide; China's Customs Valuation Operation, 2008*, by Zhang Lei. GAIN Report no. CH8090, October 6, 2008.
———. *China, Peoples Republic of: Cotton and Products Annual*, by Chanda Beckman and Wu Xinping. GAIN Report no. CH10019, May 1, 2010.
———. *China, Peoples Republic of: Poultry and Products; Challenges to Increasing U.S Sales of Chicken Paws to China, 2006*, by Casey Bean and Suzanne Ryan. GAIN Report no. CH6117, December 15, 2006.
———. *China, Peoples Republic of: Food and Agricultural Import Regulations and Standards, 2007,* by Mark Petry, Joseph Jacobson, Jennifer Schriber, and Zhang Jian Ping. GAIN Report no. CH7043, June 13, 2007.

U.S. Department of Agriculture (USDA). Food Safety and Inspection Service (FSIS). "Bovine Spongiform Encephalopathy: 'Mad Cow Disease.'" Fact Sheet, March 2005. http://www.fsis.usda.gov/Fact_Sheets/Bovine_Spongiform_Encephalopathy_Mad_Cow_Disease/ index.asp.

———. "Export Requirements for the People's Republic of China," December 23, 2010. http://www.fsis.usda.gov/Regulations_&_Policies/China (accessed January 4, 2011).

U.S. Department of State. Office of the Spokesman."U.S. Concerned That EU's Biotech Labeling Could Be Burdensome," July 4, 2003.

U.S. Wheat Associates. Written submission to the U.S. International Trade Commission in connection with inv. no. 332-518, *China's Agricultural Trade: Competitive Conditions and Effects on U.S. Exports*, September 16, 2010.

U.S. Trade Representative (USTR). *National Trade Estimate Report*. Washington, DC: USTR, March 2009.

———. *National Trade Estimate Report*. Washington, DC: USTR, March 2010.

———. "Agreement on U.S.-China Agricultural Cooperation." Washington, DC: USTR, 1999.

———. *2010 Report to Congress on China's WTO Compliance*. Washington, DC: USTR, December 2010.

———. *2010 Report on Sanitary and Phytosanitary Measures*. Washington, DC: USTR, March 2010.

———. *2010 Report on Technical Barriers to Trade*. Washington, DC: USTR, March 2010.

Wells, G. A. H., S. A. C. Hawkins, R. B. Green, R. A. Austin, I. Dexter, Y. I. Spencer, M. J. Chaplin, M. J. Stack, and M. Dawson. "Preliminary Observations on the Pathogenesis of Experimental Bovine Spongiform Encephalopathy (BSE); An Update." *The Veterinary Record* 142 (1998): 103–6.

World Organisation for Animal Health (OIE). "Joint WTO/OIE/WHO/FAO Statement on A/H1N1 Influenza." News release, May 2, 2009.

———. "Old Classification of Diseases Notifiable to the OIE," January 25, 2005. http://www.oie.int/en/animal-health- (December 14, 2010).

———. *Terrestrial Animal Health Code*. Pairs: OIE, 2010. http://www.oie.int/eng/normes/mcode/en sommaire.htm.

———. World Animal Health Information database. http://www.oie.int/wahis/public.php?page=home (accessed November 24, 2010).

World Trade Organization (WTO). *Accession of the People's Republic of China*. WT/L/432. WTO, November 23, 2001.

———. *Agreement on the Application of Sanitary and Phytosanitary Measures (SPS Agreement)*. Geneva:WTO; 1994.

———. *Agreement on Technical Barriers to Trade*. Geneva: WTO, 1994.

———. *"Current Issues, 8.1: Genetically Modified Organisms."* In *SPS Agreement Training Module*. WTO, n.d. http://www.wto.org/english/tratope/sps_e/sps_agreement_cbt_e /c8s1p1_e.htm.

———. *Trade Policy Review: China*. Geneva: WTO, April 26, 2010.

———. Committee on Sanitary and Phytosanitary Measures. "Notification G/SPS/N/CHN/203," January 19, 2010. http://www.wtocenter.org.tw/SmartKMS /do/www/readDoc?document id=104678.

———. Secretariat. "Data Day at the WTO: Trade and Market Access Data for Policy Makers," May 18– 19, 2009. http://www.wto.org/english/res e/statis e/data day may09 e/brochure datadaymay09 e.pdf.

APPENDIX A. REQUEST LETTER

United States Senate
COMMITTEE ON FINANCE
WASHINGTON, DC 20510-6200

DOCKET NUMBER
2724
Office of the Secretary
Int'l Trade Commission

April 1, 2010

The Honorable Shara L. Aranoff
Chairman
U.S. International Trade Commission
500 E Street, S.W.
Washington, DC 20436

Dear Chairman Aranoff,

We are writing to request that the U.S. International Trade Commission conduct an investigation under section 332(g) of the Tariff Act of 1930 (19 U.S.C. 1332(g)) regarding competitive factors affecting agricultural trade between China and the United States.

Since it joined the World Trade Organization in 2001, China's imports of U.S. agricultural products have grown substantially. China is now the fourth largest market for U.S. agricultural exports. Yet sales are highly concentrated in a few products—soybeans, cotton, poultry, and hides and skins accounted for more than 85 percent of Chinese imports of U.S. agricultural products in 2009. Chinese imports of several globally competitive U.S. agricultural products, such as certain meat, feedgrains, and processed food, are limited. With rapidly rising per capita income and resource constraints on domestic production growth, China has the potential to provide greater opportunities for expanding U.S. agricultural exports.

At the same time, several factors threaten the ability of U.S. agricultural exporters to realize these opportunities. Chinese government policies aimed at boosting domestic production and curbing imports, non-tariff measures, including sanitary/phytosanitary measures and technical trade barriers, and increased competition from third-country suppliers, especially those with which China has negotiated trade agreements, are important factors that could weaken the competitive position of U.S. agricultural products in the Chinese market.

The Commission's report should cover the period 2005-2009, or the period 2005 to the latest year for which data are available. In addition, to the extent possible, the report should include the following:

- an overview of China's agricultural market, including recent trends in production, consumption, and trade;
- a description of the competitive factors affecting the agricultural sector in China, in such areas as costs of production, technology, domestic support and government programs related to agricultural markets, foreign direct investment policies, and pricing and marketing regimes;

RECEIVED
APR 01 2010
OFFICE OF THE SECRETARY
U.S. INTL. TRADE COMMISSION

- an overview of China's participation in global agricultural export markets, particularly in the Asia-Pacific region and in those markets with which China has negotiated trade agreements;
- a description of the principal measures affecting China's agricultural imports, including tariffs and non-tariff measures such as sanitary and phytosanitary measures and technical barriers to trade; and
- a quantitative analysis of the economic effects of China's MFN tariffs, preferential tariffs negotiated under China's free trade agreements, and China's non-tariff measures on U.S. agricultural exports to China and on imports from the rest of the world.

The Commission should submit its final report no later than eleven months from the receipt of this request. As we intend to make the report available to the public, we request that it not contain confidential business information.

Sincerely,

Max Baucus
Chairman

Charles E. Grassley
Ranking Member

Appendix B. Federal Register Notice

Federal Register / Vol. 75, No. 87 / Thursday, May 6, 2010 / Notices 24969

62815) and determined on March 8, 2010 that it would conduct an expedited review (75 FR 13779, March 23, 2010).

The Commission transmitted its determination in this review to the Secretary of Commerce on April 30, 2010. The views of the Commission are contained in USITC Publication 4148 (April 2010), entitled *Crepe Paper Products from China: Investigation No. 731–TA–1070A (Review).*

By order of the Commission.

Issued: May 3, 2010.

Marilyn R. Abbott,

Secretary.

[FR Doc. 2010–10691 Filed 5–5–10; 8:45 am]

BILLING CODE 7020–02–P

INTERNATIONAL TRADE COMMISSION

[Investigation No. 731–TA–1178 (Preliminary)]

Glyphosate From China

AGENCY: United States International Trade Commission.

ACTION: Notice of withdrawal of petition in antidumping investigation.

SUMMARY: On April 29, 2010, the Department of Commerce and the Commission received letters on behalf of the petitioner in the subject investigation (Albaugh, Inc., Ankeny, IA) withdrawing its petition. Commerce has not initiated an investigation as provided for in section 732(c) of the Tariff Act of 1930 (19 U.S.C. 1673a(c)). Accordingly, the Commission gives notice that its antidumping investigation concerning glyphosate from China (investigation No. 731–TA–1178 (Preliminary)) is discontinued.

DATES: *Effective Date:* April 29, 2010.

FOR FURTHER INFORMATION CONTACT: Amy Sherman (202–205–3289), Office of Investigations, U.S. International Trade Commission, 500 E Street, SW., Washington, DC 20436. Hearing-impaired individuals are advised that information on this matter can be obtained by contacting the Commission's TDD terminal on 202–205–1810. Persons with mobility impairments who will need special assistance in gaining access to the Commission should contact the Office of the Secretary at 202–205–2000. General information concerning the Commission may also be obtained by accessing its Internet server (*http://www.usitc.gov*). The public record for this investigation may be viewed on the Commission's electronic docket (EDIS) at *http://edis.usitc.gov.*

By order of the Commission.

Issued: April 30, 2010.

Marilyn R. Abbott,

Secretary to the Commission.

[FR Doc. 2010–10649 Filed 5–5–10; 8:45 am]

BILLING CODE 7020–02–P

INTERNATIONAL TRADE COMMISSION

[Investigation No. 332–518]

China's Agricultural Trade: Competitive Conditions and Effects on U.S. Exports

AGENCY: United States International Trade Commission.

ACTION: Institution of investigation and scheduling of hearing.

SUMMARY: Following receipt on April 1, 2010, of a request from the United States Senate Committee on Finance (Committee) under section 332(g) of the Tariff Act of 1930 (19 U.S.C. 1332(g)), the U.S. International Trade Commission (Commission) instituted investigation No. 332–518, *China's Agricultural Trade: Competitive Conditions and Effects on U.S. Exports.*

DATES: *May 25, 2010:* Deadline for filing requests to appear at the public hearing.

June 3, 2010: Deadline for filing prehearing briefs and statements.

June 22, 2010: Public hearing.

June 29, 2010: Deadline for filing posthearing briefs and statements.

September 15, 2010: Deadline for filing all other written submissions.

March 1, 2011: Transmittal of Commission report to the Committee.

ADDRESSES: All Commission offices, including the Commission's hearing rooms, are located in the United States International Trade Commission Building, 500 E Street, SW., Washington, DC. All written submissions should be addressed to the Secretary, United States International Trade Commission, 500 E Street, SW., Washington, DC 20436. The public record for this investigation may be viewed on the Commission's electronic docket (EDIS) at *http://www.usitc.gov/secretary/edis.htm.*

FOR FURTHER INFORMATION CONTACT: Project leader Joanna Bonarriva (202–205–3312 or *joanna.bonarriva@usitc.gov*) or deputy project leader Marin Weaver (202–205–3461 or *marin.weaver@usitc.gov*) for information specific to this investigation. For information on the legal aspects of this investigation, contact William Gearhart of the Commission's Office of the General Counsel (202–205–3091 or *william.gearhart@usitc.gov*). The media should contact Margaret O'Laughlin, Office of External Relations (202–205–1819 or *margaret.olaughlin@usitc.gov*). Hearing-impaired individuals may obtain information on this matter by contacting the Commission's TDD terminal at 202–205–1810. General information concerning the Commission may also be obtained by accessing its Internet server (*http://www.usitc.gov*). Persons with mobility impairments who will need special assistance in gaining access to the Commission should contact the Office of the Secretary at 202–205–2000.

Background: As requested by the Committee, the Commission will conduct an investigation and prepare a report on the conditions of competition in China's agricultural market and trade and their effect on U.S. agricultural exports. As requested, to the extent possible, the report will include—

(1) An overview of China's agricultural market, including recent trends in production, consumption, and trade;

(2) A description of the competitive factors affecting the agricultural sector in China, in such areas as costs of production, technology, domestic support and government programs related to agricultural markets, foreign direct investment policies, and pricing and marketing regimes;

(3) An overview of China's participation in global agricultural export markets, particularly in the Asia-Pacific region and in those markets with which China has negotiated trade agreements;

(4) A description of the principal measures affecting China's agricultural imports, including tariffs and non-tariff measures such as sanitary and phytosanitary measures and technical barriers to trade, and;

(5) A quantitative analysis of the economic effects of China's MFN tariffs, preferential tariffs negotiated under China's free trade agreements, and China's non-tariff measures on U.S. agricultural exports to China and on imports from the rest of the world.

The Committee asked that the Commission's report cover the period 2005–2009, or the period 2005 to the latest year for which data are available. The Committee requested that the Commission deliver its report by March 1, 2011.

Public Hearing: The Commission will hold a public hearing in connection with this investigation at the U.S. International Trade Commission Building, 500 E Street, SW., Washington, DC, beginning at 9:30 a.m. on Tuesday, June 22, 2010. Requests to appear at the public hearing should be filed with the Secretary no later than

24970 Federal Register / Vol. 75, No. 87 / Thursday, May 6, 2010 / Notices

5:15 p.m., May 25, 2010, in accordance with the requirements in the "Submissions" section below. All prehearing briefs and statements should be filed with the Secretary not later than 5:15 p.m., June 3, 2010; and all posthearing briefs and statements responding to matters raised at the hearing should be filed with the Secretary not later than 5:15 p.m., June 29, 2010. All hearing-related briefs and statements should be filed in accordance with the requirements for filing written submissions set out below. In the event that, as of the close of business on May 25, 2010, no witnesses are scheduled to appear at the hearing, the hearing will be canceled. Any person interested in attending the hearing as an observer or nonparticipant may call the Office of the Secretary (202–205–2000) after May 25, 2010, for information concerning whether the hearing will be held.

Written Submissions: In lieu of or in addition to participating in the hearing, interested parties are invited to file written submissions concerning this investigation. All written submissions should be addressed to the Secretary, and all such submissions (other than pre- and post-hearing briefs and statements) should be received not later than 5:15 p.m., September 15, 2010. All written submissions must conform with the provisions of section 201.8 of the Commission's Rules of Practice and Procedure (19 CFR 201.8). Section 201.8 requires that a signed original (or a copy so designated) and fourteen (14) copies of each document be filed. In the event that confidential treatment of a document is requested, at least four (4) additional copies must be filed, in which the confidential information must be deleted (see the following paragraph for further information regarding confidential business information). The Commission's rules authorize filing submissions with the Secretary by facsimile or electronic means only to the extent permitted by section 201.8 of the rules (*see* Handbook for Electronic Filing Procedures, *http://www.usitc.gov/secretary/fed_reg_notices/rules/documents/handbook_on_electronic_filing.pdf*). Persons with questions regarding electronic filing should contact the Office of the Secretary (202–205–2000).

Any submissions that contain confidential business information must also conform with the requirements of section 201.6 of the Commission's Rules of Practice and Procedure (19 CFR 201.6). Section 201.6 of the rules requires that the cover of the document and the individual pages be clearly marked as to whether they are the "confidential" or "nonconfidential" version, and that the confidential business information be clearly identified by means of brackets. All written submissions, except for confidential business information, will be made available for inspection by interested parties.

In its request letter, the Committee stated that it intends to make the Commission's report available to the public in its entirety, and asked that the Commission not include any confidential business information in the report it sends to the Committee. Any confidential business information received by the Commission in this investigation and used in preparing this report will not be published in a manner that would reveal the operations of the firm supplying the information.

By order of the Commission.

Issued: April 30, 2010.

Marilyn R. Abbott,

Secretary to the Commission.

[FR Doc. 2010–10650 Filed 5–5–10; 8:45 am]

BILLING CODE 7020–02–P

INTERNATIONAL TRADE COMMISSION

[USITC SE–10–014]

Government in the Sunshine Act Meeting Notice

AGENCY HOLDING THE MEETING: United States International Trade Commission.

TIME AND DATE: May 14, 2010 at 10 a.m.

PLACE: Room 101, 500 E Street, SW., Washington, DC 20436, Telephone: (202) 205–2000.

STATUS: Open to the public.

MATTERS TO BE CONSIDERED:

1. Agenda for future meetings: none.
2. Minutes.
3. Ratification List.
4. Inv. Nos. 701–TA–475 and 731–TA–1177 (Preliminary) (Certain Aluminum Extrusions from China)—briefing and vote. (The Commission is currently scheduled to transmit its determinations to the Secretary of Commerce on or before May 17, 2010; Commissioners' opinions are currently scheduled to be transmitted to the Secretary of Commerce on or before May 24, 2010.)
5. Inv. Nos. 731–TA–770–773 and 775 (Second Review) (Stainless Steel Wire Rod from Italy, Japan, Korea, Spain, and Taiwan)—briefing and vote. (The Commission is currently scheduled to transmit its determinations and Commissioners' opinions to the Secretary of Commerce on or before May 28, 2010.)
6. Outstanding action jackets: None.

In accordance with Commission policy, subject matter listed above, not disposed of at the scheduled meeting, may be carried over to the agenda of the following meeting.

By order of the Commission:

Issued: April 26, 2010.

William R. Bishop,

Hearings and Meetings Coordinator.

[FR Doc. 2010–10801 Filed 5–4–10; 4:15 pm]

BILLING CODE 7020–02–P

DEPARTMENT OF JUSTICE

Federal Bureau of Investigation

[Docket No. FBI 122]

FBI Records Management Division National Name Check Program Section User Fees

AGENCY: Federal Bureau of Investigation (FBI), Justice.

ACTION: Notice.

SUMMARY: This notice establishes the user fee schedule for federal agencies requesting name-based background checks of the FBI's National Name Check Program for noncriminal justice purposes. These checks of the Central Records System are performed by the Records Management Division.

DATES: *Effective Date:* June 7, 2010.

FOR FURTHER INFORMATION CONTACT: FBI, RMD, National Name Check Program Section, 170 Marcel Drive, Winchester, VA 22602, *Attention:* Michael Cannon, 540 868–4400.

SUPPLEMENTARY INFORMATION: Pursuant to the authority in Public Law 101–515 as amended, the FBI has established user fees for federal agencies requesting noncriminal name-based background checks of the Central Records System (CRS) through the National Name Check Program (NNCP) of the Records Management Division (RMD). The final rule, to be codified under 28 CFR 20.31 (f), is set out elsewhere in today's issue of the **Federal Register**.

The following fee schedule provides the user fees for name-based CRS checks by the NNCP through the FBI's RMD.

NAME-BASED NNCP CHECKS

If the check is a/an	*The fee is*
Electronic transaction:	
Batch Process Only	$1.50
Batch + File Review	29.50
Manual Submission	56.00
Expedited Submission	56.00

APPENDIX C. HEARING WITNESSES

CALENDAR OF PUBLIC HEARING

Those listed below appeared as witnesses at the United States International Trade Commission's hearing:

Subject:	China's Agricultural Trade: Competitive Conditions and Effects on U.S. Exports
Inv. No.:	332-518
Date and Time:	June 22, 2010 - 9:30 a.m.

A session was held in connection with this investigation in the Main Hearing Room (room 101), 500 E Street, S.W., Washington, D.C.

ORGANIZATION AND WITNESS:

National Potato Council
Washington, D.C.

John Keeling, CEO and Executive Vice President

Almond Board of California
Modesto, CA

David C. Norris, President, Jessup, Norris & Trunick, Inc.

Kristi Mika Saitama, Associate Director, Asia-Pacific, Almond Board of California

Tuttle Taylor & Heron
Washington, D.C.
on behalf of

Blue Diamond Growers

Julian B. Heron) – OF COUNSEL

ORGANIZATION AND WITNESS (continued):

American Soybean Association
St. Louis, MO

Randy Mann, Vice President, American Soybean Association; Chair of Trade Policy & International Affairs Committee, American Soybean Association; and Member of the Board of Directors of the U.S. Soybean Export Council

R-CALF United Stockgrowers of America
Billings, MT

Bill Bullard, CEO

Stewart & Stewart Law Offices
Washington, D.C.

Elizabeth J. Drake, Esq.

-END-

APPENDIX D. SUMMARY OF VIEWS OF INTERESTED PARTIES

Summary of Views of Interested Parties

The Commission held a public hearing in relation to its investigation on China's agricultural trade on June 22, 2010, in Washington, DC. Interested persons were also invited to file written submissions for the investigation. This appendix summarizes the views expressed to the Commission via testimony, written submission, or both, and reflects the principal points made by the particular party. The views expressed in the summarized materials should be considered to be those of the submitting parties and not the Commission. In preparing this summary, Commission staff did not undertake to confirm the accuracy of, or otherwise correct, the information summarized. For the full text of hearing testimony, written submissions, and exhibits, see entries associated with investigation no. 332-518 at the Commission's Electronic Docket Information System (https://edis.usitc.gov/edis3-internal/app).

Almond Board of California[1]

According to its hearing statement, the Almond Board of California administers a federal marketing order under the U.S. Department of Agriculture (USDA), with a broad range of programs funded by a mandatory assessment of almond growers. The Almond Board stated that the California almond industry is export-dependent, with the majority of almonds being sold abroad. According to the Almond Board, China lacks commercial domestic production and thus relies on almond imports. It also claimed that growth in demand for almonds in China has been dramatic but that prices, inflated by duties and valued-added tax (VAT), limit increased consumption. The Almond Board asserted that because marketing channels for the imports of almonds are fragmented, Chinese custom statistics tend to undercount almond imports; nonetheless, its exports there have rapidly increased. It said that a duty reduction on inshell almonds from 24 percent to 10 percent (the same as shelled almonds)—or, better, a cut in the duty on all almonds to 5 percent— would add transparency. The Almond Board expressed concern that China's free trade agreement negotiations with Australia and Chile could result in a competitive disadvantage on tariffs for U.S. almonds in China. According to the Almond Board, a tariff reduction on inshell almonds would help China by leading to greater direct investment and better economic opportunities with the expansion of almond processors.[2]

American Dehydrated Onion and Garlic Association[3]

In a written submission, the American Dehydrated Onion and Garlic Association (ADOGA), a group of companies operating dehydration, packaging, distribution, and warehousing facilities and representing the vast majority of U.S. production of dehydrated (dehy) onions and garlic in the United States, stated that Chinese dehy garlic has a competitive advantage over U.S.-produced dehy garlic in all markets because of lower production costs. In the submission, ADOGA stated that China is the largest global producer of garlic for all uses, that its large supplies can drive U.S. and world prices down, and that U.S. exports are unsuccessful selling into any other market because of China's predatory pricing in those markets. In addition, ADOGA stated in the submission that Chinese dehy garlic is being undervalued for U.S. Customs valuation purposes and that Chinese traders have been circumventing the large U.S. import tariff on Chinese dehy garlic by shipping product through a number of other countries.

American Soybean Association[4]

In its submission, the American Soybean Association (ASA), an advocate and representative of U.S. soybean farmers on national and international policy issues, highlighted the importance of the U.S. soybean trade with China and expressed concern over its future. It noted that China is the largest soybean export market for the United States and that soybeans comprise over half the value of total U.S. agricultural exports to China. In 2009, the United States exported \$9.2 billion worth of soybeans to China, and ASA predicted that Chinese consumption of soybeans will increase approximately 8 percent annually over the next three to four years, a good prospect for U.S. exports. Although ASA stated that there are no major impediments to the soybean trade currently, it noted some concerns about future trade. According to ASA, China has shown concern over discolored U.S. soybeans, believing that the discoloration might indicate the presence of pesticide residue instead of natural discoloration. It also stated that the United States and China have agreed to communicate over such issues to improve understanding about things like testing methods. ASA asserted that China's required certification for soybean oil is not necessary under international standards. It characterized China's regulatory approval process for commodities containing biotech traits as opaque and, given the need to re-register traits every three years, unpredictable in nature. According to ASA, this is a potential threat to trade. ASA also asserted that the Chinese registration system has redundancy, owing to the need to register stacked traits both individually and together. It hopes that China's greater use of biotechnology may help resolve some of these issues. Finally, ASA reported that Argentina employs differential export taxes (DETs) as a way to lower the price of soybean oil exports to China, a practice that negatively affects the United States in the Chinese market.

Blue Diamond Growers[5]

A statement submitted by Blue Diamond Growers, a 100-year-old nonprofit marketing cooperative composed of almond growers, outlined U.S. almond production and its role in the Chinese market as follows: U.S. almonds are grown exclusively in California; over $2 billion worth of almonds were exported from California in 2009; and over 100 million pounds worth $200 million were exported to China. Blue Diamond reported three main issues with current U.S. almond trade with China. First, it stated that the Chinese word for "almonds" is the same as the Chinese word for "bitter apricot kernels," often causing the two commodities to be confused both during trade negotiations and in marketing outlets. Blue Diamond claimed that this confusion may be the biggest obstacle it faces in seeking tariff reduction. Second, according to Blue Diamond, the 24 percent Chinese tariff on inshell almonds was negotiated when China joined the World Trade organization because the industry incorrectly thought China would be a "shelled" market. Blue Diamond stated that China has become a major inshell market, but with the tariff and a VAT of 13 percent, the total duties on inshell almonds are 37 percent. Blue Diamond asserted that a reduction for almonds is in order, given the 2008 duty reduction for inshell pistachios. Third, according to Blue Diamond, the majority of California almonds are being imported through Hong Kong and then shipped to mainland China through "grey channels" to avoid the high duties. Blue Diamond asserted that they would like to legitimize China's importation process through reduced tariffs, which would increase its shipments and increase the duty revenue China collects.

California Cling Peach Board[6]

In a written submission, the California Cling Peach Board (Board), a nonprofit, quasi-governmental association representing all 600 cling peach producers in California, stated that the U.S. canned cling peach industry has been greatly diminished by unfairly priced imports into the U.S. market of Chinese canned peaches, both in institutional-size metal containers and in single-serve plastic cups. In the submission, the Board stated its industry efforts to compete more favorably have included the initiation of a tree-pull program to reduce market supply and the use of new thinning technology to reduce costs, but that such efforts have been unsuccessful in restoring U.S. industry profitability because of rising grower costs, static U.S. demand, increasing imports, and increased competition in all global markets from Chinese production. In addition, the Board stated that increased U.S. imports of Chinese canned peaches and canned fruit mixtures are displacing U.S. industry sales and are putting downward pressure on prices of U.S. products in the retail and institutional sales sectors.

National Cotton Council[7]

In a written submission, the National Cotton Council of America (Cotton Council) stated that China is the dominant force in cotton textiles and apparel and Chinese policies significantly distort world cotton production markets. Specifically, the Cotton Council claimed that U.S. textile manufacturing has declined as China's has risen, thus increasing U.S. cotton producers' dependence on Chinese purchases of cotton. According to the Cotton Council, China uses policies including seed subsidies, a variable levy, and an internal reserve system to support Chinese cotton farmers in a fashion inconsistent with China's World Trade Organization (WTO) accession documents. Additionally, the Cotton Council alleged that the Chinese government tends to give import licenses to companies that agree to use imported cotton for production of textiles and apparel to be exported rather than domestically consumed. The Cotton Council asserted that these policies restrict access to the China market, making it difficult to find a place to sell U.S. cotton on the international market.

National Milk Producers Federation and U.S. Dairy Export Council[8]

The National Milk Producers Federation (NMPF) and the U.S. Dairy Export Council (USDEC) submitted a joint written submission. NMPF is a national farm commodity organization that represents dairy farmers and the dairy cooperative marketing associations they own and operate throughout the United States. USDEC is a nonprofit, independent membership organization that represents the export trade interests of U.S. dairy producers, proprietary processors and cooperatives, ingredient suppliers, and export traders.

The joint submission addresses China's possible revision of its dairy health certificate requirements for U.S. dairy products. NMPF/USDEC stated that the new requirements appear to impose stricter requirements on U.S. food-grade dairy exports than on exports of the major dairy suppliers of other nations, without any indication that U.S. dairy products pose a greater risk to human health. NMPF/USDEC asserted that U.S. dairy products are manifestly safe, given that production is done under a robust U.S. oversight and inspection system. U.S. exports are made from the same milk source as dairy products destined for U.S. consumers, and U.S. dairy products are recognized by millions of foreign consumers around the world for their quality and safety. NMPF/USDEC stated that the U.S. government has offered several certificate-language options to China to address this issue in a way that they view as consistent with sound science and international norms for dairy products, but that the government of China has not approved any of these options. NMPF/USDEC reports that they have urged the U.S. government to work hard to resolve this situation and have urged China to pursue a resolution that reflects international scientifically informed guidelines and complies with China's SPS requirements under the WTO.

NMPF/USDEC made the following additional points:

- U.S. dairy exports to China averaged $168 million over the last three years, making China the third largest export destination after NAFTA partners Canada and Mexico;
- While U.S. exports are currently permitted entry to China, no guarantee exists as to how long this will last, given the ongoing bilateral negotiations regarding China's health certificate;

- The uncertainty surrounding the dairy health certificates has led to a decline in U.S. dairy exports to China in the months since May 2010, the last month for which acceptance of U.S. dairy products in China was officially guaranteed in advance;
- The market in China for dairy products has been expanding over the past few years and expectations for future demand growth for imported dairy products is high, if the question of China's health certificate for U.S. dairy products can be resolved; and
- If resolution is not forthcoming, opportunities in China will be forfeited to U.S. competitors, who have all managed to secure an agreed-upon health certificate for access to China's market, and thus it is critical to ensure that U.S. suppliers are able to supply much-needed quality dairy products to China at a time of rapidly growing demand for dairy products in that country.

National Potato Council[9]

In two written submissions, the National Potato Council (NPC), representing the U.S. potato industry, identified China as an important market for U.S. fresh potato exports and identified obtaining market access to China for U.S. fresh potatoes as one of its highest international priorities. In its June 22 submission, the NPC listed a number of events involving negotiators from the United States and China since July 2000 concerning market entry for U.S. fresh potato exports, especially in the area of plant quarantine. In its June 24 submission, the NPC stated that because of China's free trade agreements (FTAs) with Australia, New Zealand, and Thailand, buyers in China have shifted their purchases away from the United States, and the U.S. potato industry is risking the loss of important markets in Asia and Oceania.

Northwest Horticultural Council[10]

In a written submission, the Northwest Horticultural Council (Council), representing growers, packers, and shippers of tree fruit in the Pacific Northwest, stated that conditions of competition in trade between the United States and China are of great interest to their industry, particularly the supply and demand situation in China for apples, pears, and cherries. The Council requested that the Commission focus its study on examining current and future consumption of apples, pears, and cherries in China, including how current increases in labor costs and constraints on water and land might affect a reallocation of agricultural resources in China, so as to assist U.S. exporters in their marketing efforts both in China and in other export markets in which China currently competes.

Ranchers-Cattlemen Action Legal Fund, United Stockgrowers of America[11]

In its written submission, the Ranchers-Cattlemen Action Legal Fund, United Stockgrowers of America (R-CALF USA) stated that it represents thousands of live cattle producers in 46 states. R-CALF USA stated that it does not support further opening of beef trade with other countries until imbalances between the cattle industry and the beef-packing industry are resolved. Regarding trade with China, R-CALF USA maintains that the United States should have accepted China's offer to allow shipments of U.S. beef from cattle under 31 months of age instead of holding out for unrestricted access for all beef. According to R-CALF USA, it is "indefensible" to reject this offer on the grounds that the deal is not compliant with World Organization for Animal Health (OIE) guidelines for three reasons: first, if the United States believes the Chinese ban is causing monetary losses to the beef industry, then rejecting a partial opening exacerbates the loss where accepting it could mitigate the loss; second, if the United States thinks exports are necessary to revitalize rural areas, then depriving the beef industry of even limited export opportunities hurts rural areas; third, all countries except the United States are rejecting what R-CALF USA claimed are inadequate OIE guidelines on bovine spongiform encephalopathy (BSE).[12]

R-CALF USA asserted that Chinese beef consumption has not risen along with income because other factors are significantly affecting beef consumption, primarily China's undervalued currency. R-CALF USA also added that if China expands its beef production and/or exports, the U.S. cattle industry would be hurt. R-CALF USA claimed that opening up the beef trade will, on balance, hurt the U.S. cattle industry, because the meat-packing industry will benefit from exports while the cattle industry will face increased competition in imports. R-CALF USA made a number of recommendations; those specific to China are to accept China's offer to import U.S. beef from cattle under 31 months and to counteract the "tariff" caused by an undervalued renminbi.

Stewart and Stewart[13]

In two written submissions by Stewart and Stewart, a law firm with over 50 years' experience representing U.S. industries, farmers, ranchers, and workers in international trade matters, the firm laid out information regarding China's agricultural future as well as what they perceive as China's tariff and nontariff barriers to trade. According to Stewart and Stewart, China has numerous support programs and plans for agriculture. For example, under China's 11th Five Year Plan, the Chinese government is incentivizing production of both grains and high-value crops by subsidizing agricultural inputs such as fertilizer and water, exempting farmers from paying taxes, and maintaining an undervalued currency so as to keep exports competitively priced. Stewart and Stewart claimed that several of these measures contribute to the environmental degradation facing China. According to the law firm, China also employs various SPS regulations that act as barriers, such as excluding poultry produced in four U.S. states and banning all but two types of U.S. apples due to unsupported SPS concerns. Stewart and Stewart acknowledged that China has made attempts to lower tariffs, but argued that other barriers, including domestic support programs and SPS regulations, continue to make it difficult for U.S. industries to export to China.

U.S. Wheat Associates[14]

In a written submission, U.S. Wheat Associates asserted that China violates its WTO commitments for wheat in a way which harms U.S. wheat exports to China. According to U.S. Wheat Associates, its industry is export-dependent, with about half of production being exported every year. It maintained that China currently has certain policies, based on a desire to be self-sufficient in grains, which result in few wheat imports. In particular, U.S. Wheat Associates asserted that there are four major WTO commitments that China is violating. It stated that the tariff-rate quota for wheat is not transparent, which causes difficulties for both importers and exporters of wheat to China; that China has failed to comply with a 1999 agreement resolving *Tilletia controversa Kuhn* (TCK) spores found in some U.S. wheat—among other things, by not accepting U.S. certification and by creating barriers that de facto ban U.S. soft white wheat; that China is discriminatory in applying a 13 percent value-added tariff against imports but not domestic wheat; and that China is giving more support, via subsidies, to domestic wheat producers than is allowed under its WTO commitments.

U.S. Wheat Associates recommended that a multilateral coalition be formed to address China on a number of commodity issues and to correct these violations, in order to avoid Chinese retaliation directly toward the United States. U.S. Wheat Associates asserted that the resolution of these issues would allow U.S wheat to compete fairly in a market where significant new sales could be gained.

Western Pistachio Association[15]

According to its written submission, the Western Pistachio Association (WPA) represents commercial growers, ranchers, and processors of pistachios in the United States. The WPA stated that greater China (the mainland plus Hong Kong) is a large market for U.S. pistachios. According to WPA, Iranian pistachios are its main competition in greater China. WPA claimed there are two main problems which affect its pistachio exports to China. The first is the mislabeling and bleaching of Iranian product in order to pass it off as high-quality, safe U.S. product. WPA asserted that, in addition to infringing on the U.S. brand, bleaching is a potential health hazard. The second issue is the tariff level of 10 percent, a reduction which WPA asserted would increase U.S. market access. WPA also made four arguments concerning competitive factors in China. First, WPA contended that China's subsidies for tree planting could lead to future pistachio competition for U.S. exports. Second, according to WPA, China maintains arbitrary and nontransparent food safety and SPS rules. WPA asserted that while the U.S. has dealt with these barriers, they decrease the competitiveness of U.S. pistachios with grey market shipments. Third, WPA stated that distribution channels in China are currently inadequate and that improvements are needed to increase country-wide competitiveness. Fourth, WPA maintained there is a lack of intellectual and trademark protection, which diminishes the "U.S. pistachios" generic brand, as many Iranian pistachios are relabeled U.S. pistachios.

Wine Institute et al.[16]

In a written submission, the Wine Institute (the largest advocacy and public policy association for California wineries and affiliated businesses), the California Wine Grape Growers Association (representing the interests of California wine grape farmers), and Wine America (the national association of American wineries outside of California) (together "Wine Institute et al.") commented that China, as the world's largest emerging market, represents unlimited potential for U.S. wine exports. The Wine Institute et al. stated that U.S. wine producers have the capacity to supply the Chinese market with quality U.S. wine in all price categories to meet growing Chinese consumption. However, it said that China's high tariffs, nontariff measures, intellectual property rights infringement, nontransparent regulations, and government support from third-country wine suppliers are substantial obstacles to increased exports of U.S. wine to China.

The Wine Institute et al. made the following additional points:

- That Chinese wine consumption is increasing by 17 percent per year, making it the world's fastest-growing wine market;
- That wine production in China has increased to meet growing demand, but that domestic production is currently not "at par" with imported wine;
- That a number of factors are limiting U.S. export growth to China, particularly Chinese tariffs and taxes on U.S. wine that raise the price of U.S. bottled wine by 48 percent and bulk wine by 56 percent;
- That competitor wine suppliers Chile and New Zealand benefit from FTAs that have resulted in lower wine duties for those suppliers; that other leading wine exporters, including Australia, benefit from lower shipping costs owing to proximity to China; and that French producers benefit from domestic subsidies;
- That nontransparent SPS regulations and technical barriers "are becoming more of an impediment as China moves its economy into the mainstream"; and
- That U.S. wine is increasingly subject to counterfeiting (labeling Chinese wine as California wine) as U.S. brands become better known in the Chinese market.

APPENDIX E. COMPETITIVE CONDITIONS ANALYTICAL FRAMEWORK

A Framework for Analyzing the Competitive Conditions Affecting China's Agricultural Trade

Several recent Commission factfinding investigations concern competitive conditions affecting U.S. agricultural markets.[1] In these studies, "competitive conditions" refer mostly to factors that determine costs of production, such as natural resource base, input costs, government subsidies, technology, transportation costs, marketing infrastructure, and exchange rates.[2] However, discussions with industry representatives in connection with this investigation suggest that the competitive conditions affecting Chinese agricultural trade go far beyond costs of production and include a wide range of market, institutional, and regulatory factors.[3] Presented in this section is an analytical framework that reflects a broad definition of competitive conditions. It defines the analytical assumptions, parameters, and structure regarding competitive conditions in order to provide a common context and consistent perspective of analysis.[4]

Competitive conditions in agriculture refer to the economic, institutional, and regulatory environment in which firms compete. Differences between countries in their competitive conditions provide opportunities and incentives for agricultural trade to take place. The competitiveness of a country's agricultural sector is defined as the ability of its farmers and food processors to sell their products in domestic and overseas markets.[5] The ability of suppliers to sell agricultural products is determined by purchasers that base their buying decisions on a set of desired product characteristics, such as low cost, product differentiation, and reliability of supply. Competitive factors are defined as direct and indirect determinants of the ability of suppliers to meet the desired product characteristics of buyers. The remainder of this section discusses these definitions in more detail, leading to a framework for analyzing competitive conditions affecting China's agricultural trade.

Porter's Framework for Competitive Advantage

Summary of the Framework

Michael Porter provides a useful starting point from which to develop a framework for analyzing competitive conditions affecting agricultural trade.[6] According to Porter, there are two basic types of competitive advantage—low cost and differentiation. From these, Porter describes three generic strategic approaches firms can employ to achieve a competitive advantage in an industry—overall cost leadership, differentiation, and focus. Firms can pursue a competitive advantage within an industry by becoming the low-cost producer. Cost leadership strategies for firms involve aggressively pursuing preferential access to low cost inputs, seeking economies of scale, investing in cost-saving technologies, and minimizing costs associated with research and development, advertising, marketing, and sales. Porter notes that firms seeking cost leadership typically supply generic, undifferentiated, no-frills products and "place considerable emphasis on reaping scale or absolute cost advantage from all sources."[7]

Alternatively, firms may seek a competitive advantage in a market through product differentiation. With this strategy, firms create an advantage in the market by offering a product perceived by purchasers as being special or unique. Porter identifies several forms of differentiation, including product design or brand image, special features, customer service, and dealer networks. Product differentiation creates brand loyalty among customers, who respond by being less sensitive to price in making their purchasing decisions.

Cost leadership and product differentiation strategies are employed by firms competing for a broad range of consumers in many segments of the market. However, Porter describes a third strategy whereby firms seek a competitive advantage by focusing on a narrow market segment or consumer type. Under a "focus" strategy, firms target a narrow segment of the market (e.g., a certain demographic or income level, consumers with unusual or specific needs, consumers for which a specific delivery system better suits their needs) and aim to provide products and services better than firms trying to satisfy many consumers in a broader market segment. A focus strategy assumes that the needs of the target market are not well served by firms serving the entire market.[8]

Porter points out that the three strategies are not mutually exclusive. Firms looking for a competitive advantage through cost leadership must not ignore product quality and customer service. Similarly, a differentiation strategy does not allow firms to ignore costs and the importance of maintaining costs close to those of their competitors. Further, Porter describes two types of focus strategies—cost focus (firms aim to be the cost leader in the target market) and differentiation focus (firms seek differentiation in the target market).

Critique of the Framework

Although Porter's theory of competitive advantage still remains quite influential in business management circles, it is often criticized in the academic literature for oversimplication or the lack of empirical evidence. Some critics have called his analysis a tautology because the term competitive advantage is never properly defined. Porter, in effect, argues that companies are successful because of factors that make them successful. Less analysis is provided on how companies set up business conditions to attain advantageous competitive factors.

Because Porter's ideas on how firms can achieve competitive advantage were developed in the late 1970s and early 1980s, changing economic conditions have called into question the relevance of the framework to firms' decision making today.[9] One view that Porter's competitive forces framework is outdated is referred to as the "resource-based approach." Supporters of this approach analyze internal (also called "intrinsic" or "core competence") factors to determine a firm's competitiveness, including corporate culture, worker morale, team communication, corporate wide technologies, integrating production skills and new technology, and management leadership.[10] Porter either neglects these factors in his analysis or disregards them altogether. In assessing Porter's work, Downes lists three "new forces" which require a new competitiveness framework and new analytical tools to evaluate their impact on businesses: Digitalization, symbolized by the Internet and electronic financial transactions; Globalization, best characterized by improvements in worldwide distribution logistics and communications; and Deregulation, resulting in the decline of government influence over many industries. [11]

Sharp and Dawes argue that Porter's strategies are oversimplified, at times recommending two or more mutually exclusive business strategies to achieve success, and not drawing sharp distinctions between the differentiation or focus strategies—thereby rendering his recommendations nearly impossible to implement. [12] In addition, the theory of competitive advantage is ambiguous about whether Porter's prescriptions apply to firms or products. Porter views differentiation as something that firms undertake, but consumers purchase products, and the attributes of those products create customer value. So, while Porter discusses the competitiveness of firms, his competitiveness factors, such as low costs and differentiation, are really product qualities.[13]

Another common criticism of Porter is his inattention to national and cultural dimensions in any detail. He explains industrial success on a disaggregated, industry-by-industry basis, but it is also reasonable to take a more macroeconomic approach and ask why certain cultures give rise to industrial excellence or why some people (or cultures) are more hardworking, flexible, and entrepreneurial. Going further, Porter fails to explore how these cultural determinants of national competitive success are produced, and the role of government or civil society in sustaining or replicating such conditions.[14] Porter himself acknowledges that non-market forces, such as government regulation, may need to be factored into his competitiveness model, but he argues that non-market factors do not necessarily require an additional "force" in his "five forces model".[15]

Several criticisms of Porter's framework have been made on the basis that there is little empirical evidence to support the notion that the success of a firm is based on either a cost leadership or product differentiation strategy. Porter warns against the dangers companies face in not pursuing either a purely low cost strategy or a differentiated one, thereby being "stuck in the middle" of two competing strategies. In support of this premise, he discusses a

U-shaped relationship between return on investment and market share, a relationship used to illustrate the pitfalls of a "middle" strategy. But Porter cites only two empirical examples for this phenomenon: U.S. fractional horsepower electric motors and the global automobile market, and in both cases fails to provide a convincing assessment on whether the data support his conclusions.[16]

A Framework for Agriculture

In spite of the weaknesses identified in the Porter approach, it still provides a useful starting point from which to develop a framework for agriculture. Porter developed this framework for individual firms competing in an industry. However, it is possible to apply it at a more macro level to analyze how individual country agricultural sectors (made up of several firms) compete in a global marketplace.[17] Porter identified low costs and differentiation as drivers of competitive advantage for firms in a market. Similarly, within global agricultural markets, delivered cost (low costs) and product characteristics (differentiation) form the essential criteria upon which importers and buyers of agricultural products make their purchasing decisions. For many bulk, unprocessed agricultural products such as wheat, corn, and soybeans, success in global markets is determined largely by whichever supplier is able to offer buyers the lowest delivered cost (assuming that minimum quality standards and other basic product specifications of the buyer are met). For other agricultural products—especially value-added, processed products such as infant formula, alcoholic beverages, and snack foods—buyers are less sensitive to delivered cost and choose among suppliers more on the basis of product differentiation (assuming cost is not prohibitive).[18]

In addition to delivered cost and product differentiation identified by Porter, importers and buyers of agricultural products view the *reliability of the supplier* as a further factor in their selection among competing suppliers.[19] Reliability refers to the ability of a supplier to deliver a product in the desired form, at the desired place and time, and in sufficient volume on a consistent basis. Although reliability of supply is important for nonagricultural products, the inherent riskiness of agricultural production (because of weather and disease, for example) and the economic and political importance of agricultural production in many countries means that reliability of supply for agricultural products takes on a particularly strong significance.

Porter's theoretical framework of competitiveness, in combination with practical knowledge of how agricultural products are traded internationally, provides the building blocks for an analytical framework to address the competitive conditions and factors affecting global agricultural trade. This framework assumes that purchasers base their buying decisions on three main criteria: delivered cost, product differentiation, and reliability of supply. Buyers evaluate the importance they place on these criteria and then make purchasing decisions based on which competing supplier is best able to meet their requirements.

This framework highlights several important aspects of agricultural competitiveness:

- Competitiveness is a relative, not absolute, concept, and not all products are competitive in all markets. For example, consider the market for soybean oil in India. Buyers in the Indian market are highly price-sensitive and base their purchasing decisions largely on delivered cost. Even though the United States is a highly efficient producer and one of the world's lowest-cost suppliers of soybean oil, Indians buy this product almost exclusively from Argentina and Brazil. One reason is that these countries may have even lower costs, after adjusting for exchange rates.[20] In addition, Argentina's use of differential export taxes on soybeans, compared to soybeans oil, has the effect of significantly reducing the price level at which Argentina soybean processors can sell oil profitably to other countries. In this case, being a low-cost producer is not enough to make the United States competitive in India's soybean oil market.[21]

- As with all efficiently functioning markets, buyers of agricultural products are the ultimate arbiters of which suppliers are competitive, not the suppliers themselves. However, suppliers can make their products attractive to a buyer by offering low-cost, differentiated products and reliable delivery in order to entice a buyer to select their product.
- Competition among agricultural suppliers takes place in two markets—domestic and export. Domestic competitiveness is the ability of local suppliers to sell goods in the domestic market with better delivered cost, product differentiation, and/or reliability than other domestic suppliers and competing import suppliers. Export competitiveness is the ability of local suppliers to sell goods in foreign markets with better delivered cost, product differentiation, and/or reliability than competing domestic and foreign producers.
- The relative importance of delivered cost, product characteristics, and reliability of supply in determining competitiveness depends on the type of agricultural product traded. Many agricultural products are highly heterogeneous, differing, for example, in terms of the level of processing, branding, by type of purchaser (food processor/food consumer), or whether the products are used in food or industrial applications.[22] For bulk, undifferentiated products, purchasers typically buy based largely on delivered cost. In fact, for some products, cost may be the only consideration. But as we move along the marketing chain to semiprocessed, highly processed, and branded products, purchasers increasingly consider specific product characteristics, in addition to cost and reliability, in making their buying decisions.

Bibliography

Aktouf, Omar. "The False Expectations of Michael Porter's Strategic Management Framework," *Revista Gestao e Planejamento*, Vol. 6:11, January/June 2005.

Downes, Larry. "Beyond Porter." *Context Magazine*, December 1997. http://www.libyaforum.org/archive/index2.php?option=com_content&do_pdf=1&id=4121 (accessed December 17, 2010).

Klein, Jeremy. "A Critique of Competitive Advantage." Paper presented at Strategy Stream Critical Management Studies Conference, Manchester, UK, July 2001. http://www.mngt.waikato.ac.nz/ejrot/cmsconference/2001/Papers/Strategy/Klein-ACritiqueofCompetitiveAdvantage.pdf (accessed December 17, 2010).

O'Shaugnessy, Nicholas J. "Michael Porter's Competitive Advantage Revisited." *Management Decision*, Vol. 34/6 (2006). http://www.kouvola.lut.fi/files/download/Michael Porters CompetitiveAdvantagerevisited.pdf (accessed December 17, 2010).

Porter, Michael E. *Competitive Advantage*. New York: The Free Press, 1985.

———. *The Competitive Advantage of Nations*. New York: The Free Press, 1990.

———. *Competitive Strategy*. New York: The Free Press, 1980.
President's Commission on Industrial Competitiveness. *Global Competition—New Reality*, Vol. 1, Washington, DC: Government Printing Office, 1985.
Recklies, Dagmar. "Beyond Porter: A critique of the critique of Porter," *TheManager.org*, 2001 http://www.themanager.org/strategy/BeyondPorter.htm (accessed December 17, 2010).
Sharp, Byron and John Dawes. "Is Differentiation Optional? A Critique of Porter's Generic Strategy Typology." *Management, Marketing and the Competitive Process*. Peter Earl, ed. London: Edward Elgar, 1996.
Speed, Richard J. "Oh Mr. Porter! A Reappraisal of Competitive Strategy." Marketing Intelligence and Planning. Vol. 7, May/June 1989.
U.S. International Trade Commission (USITC). *Canned Peaches, Pears, and Mixtures: Conditions of Competition Between U.S. and Principal Foreign Supplier Industries*. USITC Publication 3972. Washington, DC: USITC, 2007.
———. *Conditions of Competition for Certain Oranges and Lemons in the U.S. Fresh Market*. USITC Publication 3863. Washington, DC: USITC, 2006
———. *Conditions of Competition for Milk Protein Products in the U.S. Market*. USITC Publication 3692. Washington, DC: USITC, 2004.
———. *Guidelines for Developing an Economic Framework for an USITC Study*. Office of Economics. Unpublished, 2008.
———. *India: Effects of Tariffs and Nontariff Measures on U.S. Agricultural Exports*. USITC Publication 4107. Washington, DC: USITC, 2009.
———. *Sub-Saharan Africa: Effects of Infrastructure Conditions on Export Competitiveness*. USITC Publication 4071. Washington, DC: USITC, 2009.

Appendix F. Analytical Framework and Assumptions for Economic Simulations

Simulation Framework

In this study, the effects of China's food and agriculture tariffs and nontariff measures (NTMs) on U.S. food and agricultural exports have been simulated with a framework that has a general equilibrium (GE) model component and a partial equilibrium (PE) model component. The effects of China's free trade agreements (FTAs) on U.S. food and agricultural exports have been simulated with the same framework.

A GE model simulates the effects of trade policy on prices, trade, supply, and demand in the global economy with many countries and many product markets, including markets for primary factors— that is, land, labor, and capital. In contrast, a PE model simulates the effects of trade policy on the prices of a few products, assuming that the prices of all other products remain constant. Applied GE models specify products only in the aggregate, whereas PE models can specify products in detail. The advantage of linking a PE model to a GE model is that the PE model accounts for differences in bilateral trade policies at the product level, while the GE model reflects intersectoral linkages within the United States and all other economies.

The GE model comprises 57 sectors covering all aspects of the economies under consideration, including trade among China, the United States, China's FTA partners, and the rest of the world (RoW). The PE model is linked to the GE model and focuses on food and agricultural trade among China, the United States, China's FTA partners, and the RoW for 139 product categories, 131 of which are food and agricultural products— the focus of the simulations.

Both the PE model and the GE model contain 18 economies: the United States, China, Hong Kong, Cambodia, Indonesia, Laos, Malaysia, Burma, the Philippines, Singapore, Thailand, Vietnam, Chile, Pakistan, New Zealand, Peru, Costa Rica, and a RoW region representing all other economies.

The General Equilibrium Model

The GE model is the Global Trade Analysis Project (GTAP) model of world trade.[1] The GTAP framework consists of a database containing global data on international trade, inter-industry relationships, and national income accounts, together with a simulation model. In the GTAP model, domestic products and imports are consumed by firms, governments, and households. Product markets are assumed to be perfectly competitive, implying zero economic profits for firms, with imports viewed as imperfect substitutes for domestic products and with sectoral production equal to global demand.

In addition to the data on bilateral trade in each of the 57 GTAP sectors, the database includes data on domestic production and use for each sector, including intermediate use in the production of other commodities and services, as well as data on use of land, capital, and labor employment by sector. An additional component of the data is a set of parameters that, in the context of the model's equations, determine economic behavior. These parameters are principally a set of elasticities that determine, among other things, the extent to which imports and domestically produced goods are substitutes for one another. The GTAP data used in the simulation framework have been updated from their 2004 base year to 2009 (the most recent year for which statistics are available), using statistics on trade and on gross domestic product.[2]

The Partial Equilibrium Model

The PE model contains 139 product categories, which are organized into 23 product groups. Each product group corresponds to one of 23 sectors in the GTAP model that contain food and agricultural products. Table F.1 lists these 23 GTAP sectors. Of the 139 product categories in the PE model, 131 are the food and agricultural products which are the focus of the simulations. Table F.2 lists the 131 products and the U.S. Harmonized Tariff System 6-digit items which have been aggregated into each one of these product categories.

Simulations Performed

Three simulations were performed to estimate three sets of economic effects:

1) The effects of China's tariffs on U.S. food and agricultural exports to China and to the RoW were estimated by simulating the effects of the absence of China's applied tariffs (and tariff equivalents of tariff-rate quotas (TRQs)) on food and agricultural imports from all sources;

2) The effects of China's nontariff measures (NTMs) on U.S. food and agricultural exports to China and to the RoW were estimated by simulating the effects of the absence of China's NTMs on food and agricultural imports from all sources; and
3) The effects of China's preferential tariffs and TRQs (negotiated under China's FTAs) on U.S. food and agricultural exports to China, to China's FTA partners, and to the RoW were estimated by simulating the effects of the full implementation of China FTAs for all goods. The Chinese FTAs considered in this simulation are those with Hong Kong, ASEAN,[3] Chile, Pakistan, New Zealand, Peru, and Costa Rica.

Table F.1. GTAP model sectors containing the 131 products in the partial equilibrium model

GTAP sector and description	
Paddy rice	Wheat
Cereal grains n.e.c.	Vegetables, fruits, and nuts
Oil seeds	Sugarcane, and sugar beet
Plant-based fibers	Crops n.e.c.
Cattle, sheep, goats, horses	Animal products n.e.c.
Wool, silkworm cocoons	Forestry
Fishing	Meat
Meat products n.e.c.	Cattle
	Sheep
	Goats
	Horses
Dairy products	Vegetable oils and fats
Sugar	Processed rice
Beverages and tobacco products	Food products n.e.c.
Chemical, rubber, and plastic products	Textiles

Sources: Hertel, Global Trade Analysis, 1997; Narayanan and Walmsley, Global Trade Assistance, and Production, 2008.
Note: The acronym n.e.c. means not elsewhere classified.

A PE-GE simulation of, for example, the absence of China's import tariffs consists of three steps. First, the absence of China's food and agricultural tariffs is simulated using the PE model at the 131-product level (table F.2) to provide tariff shocks for the 23 GTAP sectors (table F.1). Second, the absence of China's food and agricultural tariffs is simulated using the GE model to obtain GE effects. Third, certain GE effects are incorporated into a second-round PE simulation of the absence of China's food and agricultural tariffs at the 131-product level.

Table F.2. Food and agricultural products in the partial equilibrium model and the HTS 6-digit items contained in each product

Product no.	Product name	HTS6 codes						
1	Rice, husked and unhusked	100610,	100620					
2	Wheat	100110,	100190					
3	Corn grain	100510,	100590					
4	Other grains	100200,	100300,	100400,	100700,	100810,	100820,	
		100830,	100890					
5	Potatoes	070110,	070190,	071420				
6	Tomatoes	070200						
7	Onions	070310,	070320,	070390				
8	Cabbages	070410,	070420,	070490				
9	Lettuces	070511,	070519,	070521,	070529			
10	Carrots	070610,	070690,	070700				
11	Other vegetables	070810,	070820,	070890,	070910,	070920,	070930,	
		070940,	070951,	070952,	070959,	070960,	070970,	
		070990,	071331,	071332,	071333,	071339,	071340,	
		071350,	071390,	071410,	071490			
12	Dried peas	071310						
13	Chickpeas	071320						
14	Walnuts and almonds in the shell	080211,	080231,	080232				
15	Pistachios and shelled almonds	080212,	080250					
16	Coconuts and other nuts	080110,	080111,	080119,	080221,	080222,	080290	
17	Brazil nuts	080120,	080121,	080122				
18	Cashews	080130,	080131,	080132				
19	Chestnuts	080240						
20	Macadamia nuts	080260						
21	Oranges	080510						
22	Mandarins	080520						

Table F.2. (Continued)

Product no.	Product name	HTS6 codes						
23	Grapefruits and other citrus fruits	080540,	080590					
24	Lemons and limes	080530,	080550					
25	Grapes	080610,	080620					
26	Apples	080810,	081330					
27	Cherries	080920						
28	Strawberries	081010						
29	Other berries	081020,	081030,	081040				
30	Other fresh and dried fruits	080300,	080410,	080420,	080430,	080440,	080450,	
		080720,	081050,	081060,	081090,	081340,	081350	
31	Stone fruits	080910,	080940,	081310,	081320			
32	Melons	080710,	080711,	080719				
33	Pears	080820						
34	Peaches	080930						
35	Soya beans	120100						
36	Sunflower seeds	120600						
37	Peanuts	120210,	120220					
38	Other oilseeds	120300,	120400,	120500,	120590,	120710,	120720,	
		120730,	120740,	120750,	120760,	120791,	120792,	120799
39	Rapeseed	120510						
40	Sugar beet and sugarcane	121291,	121292,	121299				
41	Cotton	520100						
42	Flax and hemp	530110,	530210					
43	Tobacco	240110,	240120,	240130				
44	Flowers and plants	060110,	060120,	060210,	060220,	060230,	060240,	
		060290,	060291,	060299,	060310,	060311,	060312,	
		060313,	060314,	060319,	060390			
45	Spices	090111,	090220,	090240,	090300,	090411,	090412,	

Product no.	Product name	HTS6 codes					
		090420,	090500,	090610,	090611,	090619,	090620,
		090700,	090810,	090820,	090830,	090910,	090920,
		090930,	090940,	090950,	091010,	091020,	091030,
		091040,	091050,	091091,	091099		
46	Seeds for planting and other plant parts	120910,	120911,	120919,	120921,	120922,	120923,
		120924,	120925,	120926,	120929,	120930,	120991,
		120999,	121010,	121020,	121110,	121120,	121130,
		121140,	121190,	121210,	121300,	121410,	121490,
		140310,	140390,	180100,	230810,	230890	
47	Live cattle, sheep, goats, and horses	010110,	010111,	010119,	010120,	010190,	010210,
		010290,	010410,	010420			
48	Bovine semen	051110					
49	Other live animals	010310,	010391,	010392,	010511,	010512,	010519,
		010591,	010592,	010593,	010594,	010599,	010600,
		010611,	010612,	010619,	010620,	010631,	010632,
		010639,	010690,	020820			
50	Eggs	040700					
51	Honey	040900					
52	Bovine hides and skins and	041000,	050210,	050290,	050300,	050400,	050510,
	miscellaneous animal products	050590,	050610,	050690,	050710,	050790,	051000,
		051199,	152190,	410110,	410120,	410121,	410122,
		410129,	410130,	410140,	410150,	410190	
53	Sheep skins	410210,	410221,	410229			
54	Other hides and skins	410310,	410320,	410330,	410390,	430120,	430130,
		430140,	430150,	430160,	430170,	430180,	430190
55	Mink furskins	430110					
56	Wool	510111,	510119				

Table F.2. (Continued)

Product	Product name	HTS6 codes					
57	Silkworm cocoons and other animal						
	fibers	500100,	510210,	510211,	510219,	510220	
58	Fresh beef	020110,	020120,	020130			
59	Frozen beef	020210,	020220,	020230			
60	Beef offal	020610,	020621,	020622,	020629		
61	Lamb and sheep meet	020410,	020421,	020422,	020423,	020430,	020441,
		020442,	020443				
62	Pork offal	020630,	020641,	020649,	020900,	150100	
63	Goat and horse meat and other edible						
	offal	020450,	020500,	020680,	020690,	150200,	150500
64	Fresh pork	020311,	020312,	020319			
65	Frozen pork	020321,	020322,	020329			
66	Preserved pork	021011,	021012,	021019			
67	Poultry	020710,	020711,	020712,	020713,	020714,	020721,
		020722,	020723,	020724,	020725,	020726,	020727,
		020731,	020732,	020733,	020734,	020735,	020736,
		020739,	020741,	020742,	020743,	020750	
68	Other meats and animal fats	020810,	020830,	020840,	020850,	020890,	021020,
		021090,	021091,	021092,	021093,	021099,	150300,
		150430,	150510,	150590,	150600,	230110	
69	Processed meats	160100,	160220,	160231,	160232,	160239,	160241,
		160242,	160249,	160250,	160290		
70	Soybean oil	150710,	150790				
71	Corn oil	151521,	151529				
72	Cotton linters	140420					
73	Peanut oil	150810,	150890				
74	Olive oil and other oilseed oils	150910,	150990,	151000,	151211,	151219,	151221,

Product	Product name	HTS6 codes						
		151229,	151311,	151319,	151511,	151519,	151530,	
		151540,	151550,	151560,	151590,	151610,	151620,	
		151710,	151790,	152110,	152200			
75	Palm oil	151110,	151190,	151321,	151329,	151410,	151411,	
		151419,	151490,	151491,	151499			
76	Soybean meal	120810,	230400					
77	Other oilseed meals	120890,	230500,	230610,	230620,	230630,	230640,	
		230641,	230649,	230650,	230660,	230670,	230690	
78	Milk and cream	040110,	040120,	040130,	040210,	040221,	040229,	
		040291,	040299					
79	Yogurt	040310,	040390					
80	Whey	040410,	040490					
81	Butter	040500,	040510,	040520,	040590			
82	Cheese	040610,	040620,	040630,	040640,	040690		
83	Lactose and lactose syrup	170210,	170211,	170219				
84	Ice cream	210500						
85	Casein	350110						
86	Caseinates	350190						
87	Processed rice (rice, semi- or wholly							
	milled)	100630,	100640					
88	Sugar and molasses	170111,	170112,	170191,	170199,	170220,	170310,	170390
89	Prepared eggs	040811,	040819,	040891,	040899			
90	Frozen potatoes	071010						
91	Frozen vegetables	071021,	071022,	071029,	071030,	071040,	071080,	071090
92	Preserved vegetables	071110,	071120,	071130,	071140,	071151,	071159,	071190
93	Dried vegetables	071210,	071220,	071230,	071231,	071232,	071233,	
		071239,	071290					

Table F.2. (Continued)

Product	Product name	HTS6 codes						
94	Preserved fruits	081110,	081120,	081190,	081210,	081220,	081290,	081400
95	Coffee and tea	090112,	090121,	090122,	090130,	090140,	090190,	
		090210,	090230					
96	Cereals and starches	110100,	110210,	110220,	110230,	110290,	110311,	
		110312,	110313,	110314,	110319,	110320,	110321,	
		110329,	110411,	110412,	110419,	110421,	110422,	
		110423,	110429,	110430,	110510,	110520,	110610,	
		110620,	110630,	110811,	110812,	110813,	110814,	
		110819,	110820,	110900,	121230			
97	Vegetable saps and extracts	130211,	130212,	130213,	130214,	130219,	130220,	
		130231,	130232,	130239				
98	Processed sugar and confectionery	170230,	170240,	170250,	170260,	170290,	170410,	170490
99	Cocoa products	180200,	180310,	180320,	180400,	180500,	180610,	
		180620,	180631,	180632,	180690			
100	Infant formula	190110						
101	Mixes, doughs, breads, and pastries	190120,	190190,	190510,	190520,	190530,	190531,	
		190532,	190540,	190590				
102	Pasta	190211,	190219,	190220,	190230,	190240,	190300	
103	Foods prepared from cereals, flour,							
	and starches	190410,	190420,	190430,	190490			
104	Foods prepared from vegetables	200110,	200120,	200190,	200210,	200290,	200310,	
		200320,	200390,	200410,	200490,	200510,	200520,	
		200530,	200540,	200551,	200559,	200560,	200570,	
		200580,	200590,	200591,	200599,	200600		
105	Foods prepared from fruits and nuts	200710,	200791,	200799,	200811,	200819,	200820,	
		200830,	200840,	200850,	200860,	200870,	200880,	
		200891,	200892,	200899				

Product	Product name	HTS6 codes						
106	Other fruit juices	200911,	200912,	200919,	200920,	200921,	200929,	
		200930,	200931,	200939,	200940,	200941,	200949,	
		200950,	200960,	200961,	200969,	200970,	200980,	200990
107	Apple juice	200971,	200979					
108	Other prepared foods	210110,	210111,	210112,	210120,	210130,	210210,	
		210220,	210230,	210310,	210320,	210330,	210390,	
		210410,	210420,	210610,	210690,	220900		
109	Bran, sharps and other residues	230210,	230220,	230230,	230240,	230250,	230310,	
		230320,	230800					
110	Pet food	230910,	230990					
111	Other food products	160210,	350210,	350211,	350219,	350220,	350510	
112	Water and non-alcoholic beverages	110710,	220110,	220190,	220210,	220290		
113	Wine	220410,	220421,	220429,	220430			
114	Alcoholic beverages	110720,	220300,	220510,	220590,	220600,	220710,	
		220720,	220810,	220820,	220830,	220840,	220850,	
		220860,	220870,	220890,	230330,	230700		
115	Tobacco products	240210,	240220,	240290,	240310,	240391,	240399	
116	Raw silk	500200,	500300,	500310,	500390			
117	Processed wool	510121,	510129,	510130,	510310,	510320,	510330	
118	Cotton waste	520210,	520291,	520299,	520300			
119	Other plant based fibers	530121,	530129,	530130,	530290			
120	Plant materials	060410,	060491,	060499,	130110,	130120,	130190,	
		140110,	140120,	140190,	140200,	140210,	140290,	
		140291,	140299,	140300,	140410,	140490		
121	Chemically modified animal or							
	vegetable fats and oils	151800						
122	Glycerol	152000,	152010,	152090,	290545			
123	Mannitol and sorbitol	290543,	290544					

Table F.2. (Continued)

Product	Product name	HTS6 codes					
124	Essential citrus fruit oils	330111,	330112,	330113,	330114,	330119	
125	Other essential oils	330121,	330122,	330123,	330124,	330125,	330126,
		330129					
126	Resinoids	330130,	330190				
127	Albumins and mixtures	330210,	350290				
128	Gelatins	350300					
129	Peptones	350400					
130	Glues and finishing agents	350520,	380910				
131	Fatty acids and alcohols	151911,	151912,	151913,	151919,	151920,	151930,
		382310,	382311,	382312,	382313,	382319,	382320,
		382330,	382340,	382360,	382370,	382390,	382460

Source: Compiled by Commission staff.

To illustrate how a PE-GE simulation is generated, figure F.1 specifies supply, demand, and trade linkages for a theoretical GE product group that contains only two of the 139 products in the PE model. In the figure, the world is divided into three regions: China, the United States, and the RoW. The quantities of total demand for the ith product group in China, $QD_{i,China}$, and domestic supply, $QO_{i,China}$, are exogenous in the PE model. In a first-round PE simulation of the absence of tariffs, the model simulates trade changes for each PE product under the restriction that demand and supply for the GE product group are not affected by the simulation. In a second-round PE simulation of the absence of tariffs, demand and supply for the GE product group are changed to reflect GE effects obtained from the GE simulation of the absence of tariffs.[4] Market-clearing conditions in the PE model ensure that supplies are equal to total demand for each of the 139 products identified in the model.

Producers determine the optimal supply for PE products ($QO_{i,1,China}$ and $QO_{i,2,China}$ in figure F.1) by maximizing revenues subject to a constant elasticity of transformation (CET) production possibilities frontier with an elasticity of transformation, ETG.[5] Optimal demand for PE products ($QD_{i,1,China}$ and $QD_{i,2,China}$) is determined by minimizing expenditures subject to CES (constant elasticity of substitution) trade-offs with an elasticity of substitution, ESG.[6] Optimal demand for the domestic varieties (QDSi,1,China) and total imports ($QIM_{i,1,China}$) of the PE products are determined by a CES function with an elasticity of substitution, ESD. Finally, the quantities of PE products imported from the United States and the RoW ($QXS_{i,1,USA,China}$ and $QXS_{i,1,RoW,China}$) are determined by a CES function with an elasticity of substitution, ESM. The PE model specifies that the domestic product is differentiated from imports, and consumers, whether final or intermediate, view imports of a particular product from a specific region as different from imports from all other regions. These two specifications constitute the Armington specification of product differentiation by country of origin.[7]

Data for Partial Equilibrium Model

The PE model requires certain statistics and economic parameters. The statistics are the dollar value of bilateral trade and demand for the domestic variety of a product and the corresponding bilateral import tariffs. Bilateral trade statistics were obtained from the United Nations Commodity Trade Statistics (UN Comtrade) database for 2009, the latest year at the time of this work, with complete and consistent statistics for China, the United States, and the other economies in the simulation framework.[8]

Statistics for demand for the domestic variety of a product are not currently available for the 131 food and agricultural products in the simulation framework. Thus, statistics from two sources were used to construct domestic demand statistics. The Food and Agriculture Organization of the United Nations (FAO) FAOSTAT databases provide export/production and import/demand shares for specific products and certain product groups. The GTAP database also provides these statistics for 23 GE product groups. Trade shares from the FAO and GTAP data were applied to the 131-product trade statistics to construct domestic supply and demand statistics.[9]

Applied ad valorem tariff equivalents were obtained from the United Nations Conference on Trade and Development (UNCTAD) Trade Analysis and Information System (TRAINS) data and the MAcMapHS6 data.[10] China's food and agricultural tariffs are current as of 2009.

The elasticities ESG, ESD, ESM (figure F.1) are based on the GTAP database. Estimated values for ESM are shown in table F.3. Employing the "rule of two," values for ESD and ESG are computed from ES_M as follows: $ES_D = ES_M/2$ and $ES_G = ES_D/2$.[11] Table F.3 also shows the values of ET_G, the parameter determining the supply response at the PE product level (figure F.1). For most product groups, ET_G is assigned the value –1. For product groups that may require relatively more resources and time to adjust to price changes, ET_G is assigned the value -0.7.

Model Limitations

Simulated effects from this PE-GE model are based on established U.S. export patterns which may exist for such reasons as the distance between countries and the presence or absence of transport infrastructure. These factors are imperfectly captured by the model. Furthermore, the model does not directly account for historical or cultural factors as determinants of trade patterns. The model assumes that these factors are unaffected by the economic policy changes.

Economic models capture the most important factors for the question under consideration: existing trade flows and trade policies, and the degree of consumer demand sensitivity to price changes. However, economic models are limited in their ability to reflect the degree of complexity evident in the real world.[12] Despite these limitations, the simulations performed here can be quite useful in providing insights on the effects of economic policies. The model presents a unified framework in which the likely effects of the policy can be assessed.

Table F.3. Values of partial equilibrium model parameters ES_M and ET_G

Product group	ESM	ETG
Paddy rice	10.1	-1.0
Wheat	8.9	-1.0
Cereal grains n.e.c.	2.6	-1.0
Vegetables, fruit, nuts	3.7	-0.7
Oil seeds	4.9	-1.0
Sugarcane, sugar beet	5.4	-1.0
Plant-based fibers	5.0	-1.0
Crops n.e.c.	6.5	-0.7
Cattle, sheep, goats, horses	4.0	-0.7
Animal products n.e.c.	2.6	-0.7
Wool, silkworm cocoons	12.9	-1.0
Forestry	5.0	1.0
Fishing	2.5	-1.0
Meat: cattle, sheep, goats, horses	7.7	-1.0
Meat product n.e.c.	8.8	-0.7
Vegetable oils and fats	6.6	-1.0
Product group	ESM	ETG
Dairy products	7.3	-0.7
Processed rice	5.2	-1.0
Sugar	5.4	-1.0
Food products n.e.c.	4.0	-1.0
Beverages and tobacco products	2.3	-0.7
Textiles	7.5	-1.0
Chemical, rubber, and plastic products	6.6	-1.0

Source: Estimated values for ESM are from Hertel et al., "How Confident Can We Be of CGE-Based Assessments of Free Trade Agreements?" 2007.

Note: The acronym n.e.c. means not elsewhere classified.

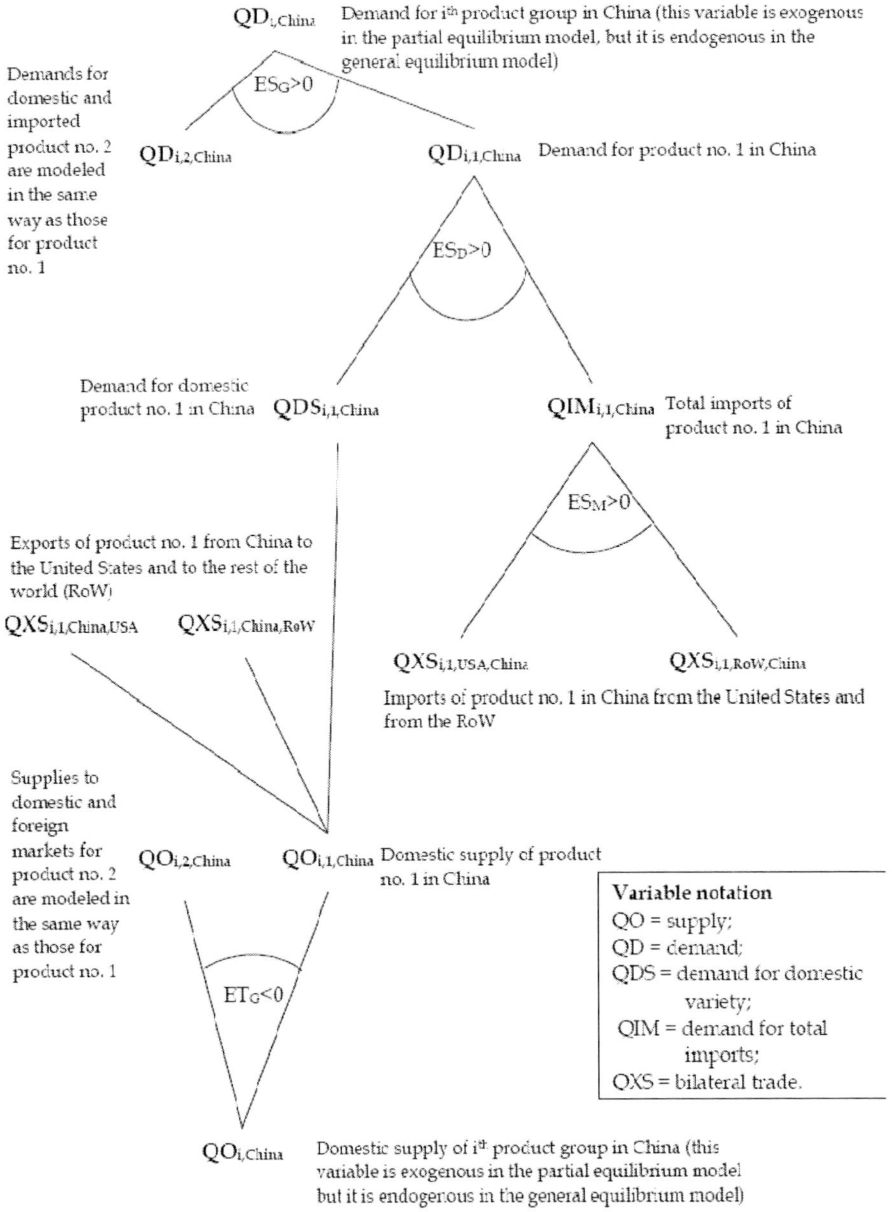

Source: Developed by Commission staff.

Figure F.1. Partial equilibrium (PE) model and linkages to general equilibrium (GE) model: China's supply, demand, and trade for a GE product group containing two hypothetical PE model products.

Estimation of Price Gaps for Nontariff Measures

Estimation of Nontariff Measure Price Gaps

The quantification of NTMs using the method of price gaps, or "tariff equivalents," has been frequently used in Commission studies on NTMs.[13] The NTM analysis in this study estimates supplier-specific gaps in a way that allows for both quality differences and the possibility that the NTMs may have a greater or lesser impact on Chinese prices for imports from different sources.[14] Price gaps in this study for products with known NTMs are presented in table F.4. Separate price gaps are estimated for Chinese imports from the United States and Chinese imports from the RoW, by comparing the price in China of an imported variety (i.e., a good from a particular source) with the price of that same variety in the world market.[15] The effects of removing these tariff equivalents are then analyzed using the modeling framework described above.

Table F.4. Known NTMs and estimated ad valorem equivalents (percent)

Product number	Product description	Estimated NTM price gap	
		U.S.	ROW
2	Wheat	119	61
5	Potatoes	6,658	1,720
26	Apples	45	0
31	Stone fruits	6	0
41	Cotton	24	4
61	Fresh beef	n.a.	n.a.
62	Frozen beef	n.a.	n.a.
63	Beef offal	n.a.	n.a.
65	Pork offal	n.a.	n.a.
68	Frozen pork	n.a.	n.a.
69	Preserved pork	n.a.	n.a.
70	Poultry	5	0

Source: Commission staff calculation.
Note: Products with non-applicable (n.a.) were analyzed with quantity gaps.

The estimation procedure uses unit values or "average prices" information from both bilateral and global trade statistics to estimate price gaps for all agricultural goods at the HS-6 subheading level. For U.S. goods, price gaps are estimated by comparing unit values obtained from Chinese imports from the U.S. on a cost-insurance-freight (c.i.f.) basis with unit values obtained from U.S. exports to the world on a free on board (f.o.b.) basis. These U.S. price gaps were then adjusted for transportation costs by subtracting transportation margins. These were obtained from the GTAP database and represent transportation margins for China's imports from the U.S. at the GTAP sector level.

Price gaps for Chinese imports of non-U.S. goods are estimated in a similar fashion for each non-U.S. supplier, by comparing Chinese c.i.f. unit values by supplier with the various suppliers' f.o.b. unit values to the world. These non-U.S., supplier-specific gaps are then aggregated into a RoW price gap for each good. This aggregation uses quantities imported into China by supplier as weights, to adjust for systematic quality differences among different suppliers to the Chinese market. The RoW price gaps were adjusted for transportation costs by using transportation margin information on China's imports from RoW, obtained from the GTAP database.

Import and export statistics were taken from the UN Comtrade database.[16] The analysis considered price data for three recent years (2006–08) to account for variable effects of

NTMs under different market conditions. Either the median price gap for the three years or the most representative price gap out of these, based on available industry information, was used as the estimate. Price gap estimates at the HS-6 level were aggregated to the 131 product categories used for the partial equilibrium model using a method that holds the implied NTM rents for the aggregate equal to the sum of implied NTM rents at the HS-6 level.[17] Some HS-6 products that presented data difficulties were excluded.[18]

Negligible Trade and Quantity Gaps

The NTM analysis considered the possibility that some measures might prohibit U.S. agricultural exports to China completely or almost completely over the period considered. For these products, beef being the most notable, obtaining a price of imports on which to base price gaps was infeasible or problematic. In these cases, quantity gaps were estimates. These gaps attempt to measure the discrepancy between observed trade and estimated trade flows in the absence of NTMs.

To estimate the quantity gaps, the trade data for these goods were first adjusted to account for a significant amount of imports into mainland China, particularly from Hong Kong, not reflected in the data. The amount of trade redirected from Hong Kong into China was approximated by using Singaporean imports for these goods as a reference point. Both Chinese imports from the U.S. and Chinese imports from the RoW were revised for a more complete depiction of the market for imported goods. Both in themselves and according to other information gathered, these unreported imports are a reflection of existing import barriers for these goods in China. Using these revised data, a simulation was then run to obtain the effect of NTMs under the assumption that, in the absence of such measures, U.S. exports to China would represent the U.S. global share of these goods. The overall effects of NTMs are assessed by comparing these simulation results with the original trade data.

Bibliography

Armington, P.S. "A Theory of Demand for Products Distinguished by Place of Production." *IMF Staff Papers* 16 (1969): 159–76.

Arrow, K., H. Chenery, B. Minhas, and R. Solow. "Capital-labor Substitution and Economic Efficiency." *Review of Economics and Statistics* 43 (1961): 225-250.

Boumellassa, H., D. Laborde, and C. Mitaritonna. *A Consistent Picture of the Protection across the World in 2004: MAcMapHS6 version 2*. IFPRI Discussion Paper 903, 2009.

Deardorff, Alan V., and Robert M. Stern. *Measurement of Non-Tariff Barriers*. Ann Arbor: University of Michigan Press, 1998.

FAOSTAT. http://faostat.fao.org/default.aspx (accessed July 1, 2010).

Ferrantino, Michael. "Quantifying the Trade and Economic Effects of Nontariff Measure," OECD. OECD Trade Policy Working Paper No. 28, January 2006.

Grant, J., T. Hertel, and T. Rutherford. "Extending General Equilibrium to the Tariff Line: U.S. Dairy in the Doha Development Agenda." Paper presented at the Tenth Annual Conference on Global Economic Analysis, W. Lafayette, Indiana, June 2007.

Hertel, T. W. (editor). *Global Trade Analysis: Modeling and Applications*, New York: Cambridge University Press, 1997.

Hertel, T., D. Hummels, M. Ivanic, and R. Keeney. "How Confident Can We Be of CGE-based Assessments of Free Trade Agreements?" *Economic Modelling* 24 (2007): 611–635.

Jansson, T., M. Kuiper, M. Banse, T. Heckelei, and M. Adenaeuer. "Getting the Best of Both Worlds? Linking CAPRI and GTAP for an Economywide Assessment of Agriculture." Paper presented at the Eleventh Annual Conference on Global Economic Analysis, Helsinki, June 2008.

APPENDIX G. SIMULATED EFFECTS OF CHINESE TARIFFS AND FREE TRADE AGREEMENTS: TABLES

Table G.1. China: China's tariffs, U.S. and Chinese trade statistics, and simulated effects on U.S. exports to China and Chinese imports from the world in the absence of Chinese tariffs for agricultural products, 2009

Product	2009 trade-weighted AVE on U.S. exports to China	Actual 2009 U.S. exports to China	Range of simulated change in U.S. exports to China absent Chinese tariffs				2009 trade-weighted AVE on Chinese imports from World	Actual 2009 Chinese imports from the World	Range of simulated change in Chinese global imports absent Chinese tariffs			
	Percent	Million $	Million $		Percent		Percent	Million $	Million $		Percent	
Rice, rough	68	0	0	0	0	0	68	5	0	20	119	378
Wheat	68	84	489	1,192	580	1,415	68	205	1,233	3,269	601	1,595
Corn	46	4	1	4	33	100	46	20	6	16	28	79
Other grains	2	0	0	0	0	0	1	450	(14)	(3)	(3)	(1)
Potatoes	11	1	0	0	0	0	11	1	0	0	0	0
Tomatoes	13	0	0	0	0	0	13	0	0	0	0	0
Onions	13	0	0	0	0	0	13	1	0	0	0	0
Cabbages	12	0	0	0	0	0	13	0	0	0	0	0
Lettuces	11	0	0	0	0	0	12	0	0	0	0	0
Carrots	13	0	0	0	0	0	13	0	0	0	0	0
Other vegetables	9	0	0	0	0	0	8	924	69	75	8	8
Dried peas	3	4	0	0	0	0	3	107	(1)	0	(1)	0
Chickpeas	4	0	0	0	0	0	4	0	0	0	0	0
Almonds	15	87	9	11	10	12	15	111	11	14	10	12
Pistachios, walnuts	14	107	10	22	17	21	14	305	53	64	17	21
Coconuts and other nuts	22	27	10	13	30	40	10	147	37	47	25	32
Brazil nuts	10	0	0	0	0	0	10	2	0	0	0	0
Cashews	10	0	0	0	0	0	17	90	10	12	10	12
Chestnuts	26	0	0	0	0	0	26	18	7	0	40	51
Macadamia nuts	22	0	0	0	0	0	15	0	0	0	0	0
Oranges	17	34	8	9	22	27	17	48	11	13	23	28
Mandarins	18	6	2	2	25	31	18	11	3	3	24	30
Grapefruits and other citrus fruits	19	2	0	0	0	0	21	5	1	1	15	18
Lemons and limes	11	8	1	1	12	15	11	10	1	1	12	15
Grapes	18	59	13	16	22	28	18	189	46	57	24	30
Apples	14	19	3	4	17	21	14	54	10	12	18	22
Cherries	14	10	1	1	9	11	14	37	3	4	9	11
Strawberries	24	0	0	0	0	0	24	0	0	0	0	0
Other berries	30	0	0	0	0	0	30	0	0	0	0	0
Other fresh and dried fruits	21	1	0	0	0	0	20	893	235	283	26	32
Stone fruits	21	5	1	2	31	39	19	21	5	7	26	32
Melons	12	0	0	0	0	0	24	34	10	12	30	35
Pears	15	0	0	0	0	0	15	0	0	0	0	0

Product	2009 trade-weighted AVE on U.S. exports to China	Actual 2009 U.S. exports to China	Range of simulated change in U.S. exports to China absent Chinese tariffs				2009 trade-weighted AVE on Chinese imports from World	Actual 2009 Chinese imports from the World	Range of simulated change in Chinese global imports absent Chinese tariffs			
	Percent	Million $	Million $		Percent		Percent	Million $	Million $		Percent	
Peaches	14	0	0	0	0	0	14	0	0	0	0	0
Soybeans	2	6,993	0	0	0	0	2	18,707	(70)	(63)	0	0
Sunflower seeds	8	11	(7)	0	11	15	7	22	2	3	10	15
Peanuts	0	3	1	2	0	0	9	6	1	1	14	20
Other oilseeds	7	3	0	0	0	0	5	486	22	29	4	6
Rapeseed	7	0	0	0	0	0	5	1,385	32	45	2	3
Sugar beet and sugar cane	17	4	0	0	32	32	17	24	8	8	33	31
Cotton	5	803	1	1	3	9	5	2,114	70	185	3	9
Flax and hemp	6	0	28	71	0	0	6	0	0	0	0	0
Tobacco	16	104	0	0	19	22	16	742	147	160	20	22
Flowers and plants	5	3	20	22	0	0	5	90	0	0	0	0
Spices	15	0	0	0	0	0	13	60	13	15	22	25
Seeds for planting and other plant parts	6	61	0	0	5	5	5	254	3	3	1	1
Live cattle, sheep, goats and horses	0	0	3	3	0	0	2	80	(1)	1	-1	1
Bovine semen	0	3	0	0	0	0	1	4	0	0	0	0
Other live animals	5	31	0	0	3	3	5	60	2	2	4	4
Eggs	18	0	1	1	0	0	18	1	0	0	0	0
Honey	17	0	0	0	0	0	17	5	1	1	18	22
Bovine hides, other	7	646	0	0	4	4	8	1,398	100	118	7	8
Sheep skins	8	10	24	26	0	0	8	257	11	14	4	5
Other hides and skins	12	18	0	0	7	9	12	110	9	11	8	10
Mink furskins	18	15	1	2	10	12	18	193	19	24	10	12
Wool	1	10	1	2	0	0	1	1,337	(33)	6	-2	0
Silk worm cocoons and other animal fibers	9	0	0	0	0	0	9	43	19	28	45	67
Fresh beef	19	0	0	0	0	0	19	8	8	13	96	168
Frozen beef	19	0	0	0	0	0	19	37	37	64	101	173
Beef offal	14	0	0	0	0	0	14	9	6	9	66	103
Lamb and sheep meet	14	0	0	0	0	0	15	139	151	263	109	189
Pork offal	13	52	0	0	90	162	14	391	365	636	90	163
Goat and horse meat and other edible offal	8	0	51	84	0	0	8	211	130	193	61	91
Fresh pork	20	0	0	0	0	0	20	0	0	0	0	0
Frozen pork	14	23	0	0	52	52	14	137	72	73	53	54
Preserved pork	26	0	12	12	0	0	26	1	1	1	124	142
Poultry	13	796	358	363	45	46	13	985	400	416	41	42

Table G.1. (Continued)

Product	2009 trade-weighted AVE on U.S. exports to China	Actual 2009 U.S. exports to China	Range of simulated change in U.S. exports to China absent Chinese tariffs				2009 trade-weighted AVE on Chinese imports from World	Actual 2009 Chinese imports from the World	Range of simulated change in Chinese global imports absent Chinese tariffs			
	Percent	Million $	Million $		Percent		Percent	Million $	Million $		Percent	
Other meats and animal fats	7	4	0	1	0	13	9	42	11	13	27	32
Processed meats	16	1	1	1	73	86	16	3	2	2	72	84
Soybean oil	42	30	3	3	10	11	42	1,842	1,277	1,627	69	88
Corn oil	10	0	0	0	0	0	10	1	0	0	0	0
Cotton linters	4	14	(5)	(4)	(38)	(32)	4	62	(8)	(6)	(12)	-10
Peanut oil	10	0	0	0	0	0	10	32	3	4	0	12
Olive oil and other oilseed oils	19	11	(2)	(2)	(21)	(16)	16	589	110	138	19	23
Palm oil	1	0	0	0	0	0	1	4,934	(973)	(794)	(20)	(16)
Soybean meal	8	0	0	0	0	0	5	47	(1)	(1)	(2)	(2)
Other oilseed meals	10	0	0	0	0	0	6	99	(6)	(5)	(6)	(5)
Milk and cream	13	13	2	2	12	15	14	604	72	93	12	15
Yogurt	25	0	0	0	0	0	24	4	1	1	23	20
Whey	6	75	4	5	6	7	7	204	16	20	6	7
Butter	23	0	0	0	0	0	23	66	14	18	21	27
Cheese	20	7	1	2	18	23	20	70	12	16	18	23
Lactose and lactose syrup	10	19	2	2	9	12	10	34	3	4	9	12
Ice cream	24	0	0	0	0	0	24	16	4	5	22	28
Casein	10	0	0	0	0	0	10	15	1	2	9	12
Caseinates	10	0	0	0	0	0	10	28	3	3	9	12
Processed rices	68	0	0	0	0	0	68	196	179	871	91	444
Sugar and molasses	0	0	0	0	0	0	0	378	(2)	(1)	(1)	0
Prepared eggs	22	0	0	0	0	0	22	0	0	0	0	0
Frozen potatoes	13	0	0	0	0	0	13	0	0	0	0	0
Frozen vegetables	11	11	2	2	19	19	11	13	2	3	19	20
Preserved vegetables	13	0	0	0	0	0	13	2	1	1	23	24
Dried vegetables	13	1	0	0	0	0	13	4	1	1	23	24
Preserved fruits	30	2	1	1	59	62	30	71	41	44	59	62
Coffee and tea	19	7	2	3	35	37	18	25	9	9	34	36
Cereals and starches	18	5	2	2	48	51	12	294	53	54	18	18
Vegetable saps and extracts	14	18	4	4	23	24	14	102	25	26	25	26
Processed sugar and confectionary	28	5	4	4	84	89	17	63	21	21	32	34
Cocoa products	10	25	2	2	8	8	12	209	36	37	17	18
Infant formula	15	4	1	1	16	17	15	605	79	82	13	13
Mixes, doughs, breads and pastries	19	11	4	4	37	39	16	349	91	95	26	27

Product	2000 trade-weighted AVE on U.S. exports to China	Actual 2009 U.S. exports to China	Range of simulated change in U.S. exports to China absent Chinese tariffs				2000 trade-weighted AVE on Chinese imports from World	Actual 2009 Chinese imports from the World	Range of simulated change in Chinese global imports absent Chinese tariffs			
	Percent	Million $	Million $		Percent		Percent	Million $	Million $		Percent	
Pasta	18	2	1	1	36	38	18	25	8	9	33	35
Foods prepared from cereals, flour, starches	26	5	2	3	50	53	27	25	13	13	51	53
Foods prepared from vegetables	10	39	10	11	26	27	18	62	20	21	33	34
Foods prepared from fruits and nuts	18	63	20	21	32	34	18	128	44	45	34	36
Other fruit juices	15	15	4	4	28	29	13	140	27	28	19	20
Apple juice	15	0	0	0	0	0	12	1	0	0	0	0
Other prepared foods	18	94	25	26	27	27	21	469	181	189	39	40
Bran, sharps and other residues	5	1	0	0	0	0	5	17	1	1	5	6
Pet food and other feeds	8	100	14	14	14	14	7	241	27	27	11	11
Other food products	11	32	5	5	15	15	12	181	27	28	15	15
Water and non-alcoholic beverages	27	4	1	1	24	32	24	59	10	14	18	23
Wine	25	25	3	4	12	15	26	457	58	75	13	17
Alcoholic beverages	29	137	32	43	24	31	28	737	153	202	21	27
Tobacco products	27	0	0	0	0	0	30	99	24	32	24	32
Raw silk	9	1	0	0	0	0	9	5	2	2	38	39
Processed wool	35	0	0	0	0	0	37	128	140	145	109	113
Cotton waste	9	1	0	0	0	0	7	36	6	6	17	18
Other plant based fibors	6	0	0	0	0	0	6	157	23	24	14	15
Plant materials	12	1	0	0	0	0	11	60	18	25	30	42
Chemically modified animal or veg. fats, oils	16	1	0	0	0	0	16	16	9	9	54	56
Glycerol	14	0	0	0	0	0	16	111	35	36	32	32
Mannitol and sorbitol	14	0	0	0	0	0	13	3	1	1	44	46
Essential citrus fruit oils	20	18	7	8	43	44	20	54	23	23	42	43
Other essential oils	10	10	7	7	67	69	10	44	26	27	59	61
Resinoids	20	3	2	2	67	69	20	14	9	10	65	68
Albumins and mixtures	22	46	26	27	56	58	22	149	82	84	55	57
Gelatins	13	2	1	1	45	46	13	7	3	3	44	45
Peptones	6	8	2	2	21	21	6	20	4	4	20	20
Glues and finishing agents	15	1	0	0	0	0	15	8	4	4	50	52
Fatty acids and alcohols	13	21	5	5	25	26	15	375	123	126	33	34
Total	6	10,942	1,251	2,000	11	19	9	47,986	5,466	9,781	11	20

Source: Commission staff calculations with simulation frame work discussed in appendix F.

Note: (1) AVE stands for ad valorem equivalent. (2) Parenthesis ( ) indicates a negative number. (3) A range of simulated effects was obtained by varying the magnitude of trade elasticities to account for the degree of statistical uncertainty in the economic estimates of the elasticities.

Table G.2. China: China's tariffs, U.S. and Chinese trade statistics, and simulated effects of the full implementation of China's free trade agreements for agricultural products, 2009

Product	2009 trade-weighted AVE on U.S. exports to China	Actual 2009 U.S. exports to China	Range of simulated change in U.S. exports to China under full implementation of China's FTAs				2009 trade-weighted AVE on Chinese imports from FTA partners		Actual 2009 Chinese imports from FTA partners	Actual 2009 Chinese imports from the World	Range of simulated change in Chinese global imports under full implementation of China's FTAs			
							Before the FTAs	After the FTAs						
	Percent	Million $	Million $		Percent		Percent		Million $	Million $	Million $		Percent	
Rice	68	0	0	0	0	0	68	49	5	5	7	2	39	41
Wheat	68	84	(37)	(1)	(44)	(1)	68	0	0	205	(70)	1	(34)	0
Corn	46	4	(1)	0	(24)	0	46	22	15	20	3	8	12	38
Other grains	2	0	0	0	0	0	1	0	4	450	4	5	1	1
Potatoes	11	1	0	0	0	0	12	0	0	1	0	0	0	0
Tomatoes	13	0	0	0	0	0	13	0	0	0	0	0	0	0
Onions	13	0	0	0	0	0	13	0	0	1	0	0	0	0
Cabbages	12	0	0	0	0	0	12	0	0	0	0	0	0	0
Lettuces	11	0	0	0	0	0	12	0	0	0	0	0	0	0
Carrots	13	0	0	0	0	0	13	0	0	0	0	0	0	0
Other vegetables	9	0	0	0	0	0	8	0	900	924	126	149	14	16
Dried peas	3	4	0	0	0	0	2	0	0	107	0	0	0	0
Chickpeas	4	0	0	0	0	0	4	0	0	0	0	0	0	0
Almonds	15	87	0	0	0	0	15	1	0	111	0	0	0	0
Pistachios, walnuts	14	107	(1)	0	(1)	0	14	0	0	305	(1)	0	0	0
Coconuts and other nuts	22	27	(1)	(1)	(4)	(3)	10	1	27	147	4	5	3	3
Brazil nuts	10	0	0	0	0	0	10	0	0	2	0	0	0	0
Cashews	19	0	0	0	0	0	17	0	99	99	11	14	11	14
Chestnuts	26	0	0	0	0	0	26	0	0	18	0	0	0	0
Macadamia nuts	22	0	0	0	0	0	14	0	0	0	0	0	0	0
Oranges	17	34	0	0	0	0	17	0	0	48	0	0	0	0
Mandarins	18	6	(1)	0	-8	0	18	0	3	11	1	1	11	14
Grapefruits and other citrus fruits	19	2	0	0	0	0	22	0	2	5	1	1	10	12
Lemons and limes	11	8	0	0	0	0	11	0	0	10	0	0	0	0
Grapes	18	59	(11)	(9)	(19)	(15)	18	0	107	189	35	45	18	24
Apples	14	19	(3)	(2)	(14)	(11)	14	0	30	54	8	10	14	18
Cherries	14	10	(2)	(2)	(23)	(19)	14	0	25	37	3	4	8	10
Strawberries	24	0	0	0	0	0	24	0	0	0	0	0	0	0
Other berries	30	0	0	0	0	0	27	0	0	0	0	0	0	0
Other fresh and dried fruits	21	1	0	0	0	0	20	0	880	893	296	366	33	41
Stone fruits	21	5	(1)	(1)	(23)	(19)	18	0	15	21	5	6	23	29
Melons	12	0	0	0	0	0	24	0	34	34	12	15	37	45
Pears	15	0	0	0	0	0	15	1	0	0	0	0	0	0
Peaches	14	0	0	0	0	0	14	0	0	0	0	0	0	0

Product	2009 trade-weighted AVE on U.S. exports to China	Actual 2009 U.S. exports to China	Range of simulated change in U.S. exports to China under full implementation of China's FTAs				2009 trade-weighted AVE on Chinese imports from FTA partners		Actual 2009 Chinese imports from FTA partners	Actual 2009 Chinese imports from the World	Range of simulated change in Chinese global imports under full implementation of China's FTAs			
							Before the FTAs	After the FTAs						
	Percent	Million $	Million $		Percent		Percent		Million $	Million $	Million $		Percent	
Soybeans	2	6,993	15	15	0	0	2	0	0	18,787	33	38	0	0
Sunflower seeds	8	11	0	0	0	0	8	0	4	22	1	1	3	5
Peanuts	8	3	0	0	0	0	14	0	1	6	0	1	0	13
Other oilseeds	7	3	0	0	0	0	5	0	45	486	6	7	1	1
Rapeseed	7	0	0	0	0	0	7	0	0	1,365	1	2	0	0
Sugar beet and sugar cane	17	4	(1)	(1)	(15)	(15)	17	0	20	24	0	0	31	32
Cotton	5	303	18	26	2	3	5	0	20	2,114	45	75	2	4
Flax and hemp	6	0	0	0	0	0	6	0	0	0	0	0	0	0
Tobacco	16	104	12	13	12	12	16	0	13	742	114	122	15	16
Flowers and plants	5	3	0	0	0	0	8	0	27	90	13	14	14	15
Spices	15	0	0	0	0	0	12	0	41	60	16	18	27	30
Seeds for planting and other plant parts	6	61	(1)	0	(1)	0	5	0	50	254	27	30	11	12
Live cattle, sheep, goats and horses	0	0	0	0	0	0	0	0	17	80	1	1	1	2
Bovine semen	0	3	0	0	0	0	5	0	0	4	0	0	0	0
Other live animals	5	31	0	0	0	0	6	0	15	60	1	1	2	2
Eggs	18	0	0	0	0	0	18	1	0	1	0	0	0	0
Honey	17	0	0	0	0	0	17	0	2	5	0	1	0	11
Bovine hides, other	7	646	0	1	0	0	13	0	35	1,390	13	16	1	1
Sheep skins	8	10	0	0	0	0	11	0	27	257	3	4	1	2
Other hides and skins	12	18	0	0	0	0	11	0	2	110	1	1	1	1
Mink furskins	18	15	0	0	0	0	18	1	0	193	1	1	0	0
Wool	1	10	0	0	0	0	1	0	101	1,337	69	83	5	6
Silk worm cocoons and other animal fibers	9	0	0	0	0	0	9	0	1	43	3	4	7	10
Fresh beef	19	0	0	0	0	0	20	1	0	8	0	0	0	0
Frozen beef	19	0	0	0	0	0	19	0	7	37	6	10	15	27
Beef offal	14	0	0	0	0	0	14	0	1	9	1	1	8	15
Lamb and sheep meet	14	0	0	0	0	0	14	0	82	139	42	71	30	51
Pork offal	13	52	(1)	1	(3)	1	14	0	3	391	(7)	7	(2)	2
Goat and horse meat and other edible offal	8	0	0	0	0	0	9	0	65	211	16	19	8	9
Fresh pork	20	0	0	0	0	0	20	0	0	0	0	0	0	0
Frozen pork	14	23	2	2	9	10	14	0	0	137	14	16	10	11
Preserved pork	26	0	0	0	0	0	26	0	0	1	1	1	109	150
Poultry	13	796	63	68	8	8	8	0	21	985	95	107	10	11
Other meats and animal fats	7	4	0	0	0	0	12	0	0	42	5	5	11	13

Table G.2. (Continued)

Product	2009 trade-weighted AVE on U.S. exports to China	Actual 2009 U.S. exports to China	Range of simulated change in U.S. exports to China under full implementation of China's FTAs				2009 trade weighted AVE on Chinese imports from FTA partners		Actual 2009 Chinese imports from FTA partners	Actual 2009 Chinese imports from the World	Range of simulated change in Chinese global imports under full implementation of China's FTAs			
							Before the FTAs	After the FTAs						
	Percent	Million $	Million $		Percent		Percent		Million $	Million $	Million $		Percent	
Processed meats	16	1	0	0	0	0	16	0	0	3	0	1	0	20
Soybean oil	42	30	0	0	0	0	42	0	0	1,842	(16)	(12)	(1)	(1)
Corn oil	10	0	0	0	0	0	10	0	0	1	0	0	0	0
Cotton linters	4	14	0	0	0	0	4	0	1	62	0	0	0	0
Peanut oil	10	0	0	0	0	0	10	0	0	32	0	0	0	0
Olive oil and other oilseed oils	19	11	(3)	(2)	(27)	(22)	18	0	226	589	125	164	21	28
Palm oil	1	0	0	0	0	0	1	0	4,548	4,934	(62)	(40)	(1)	(1)
Soybean meal	8	0	0	0	0	0	7	0	0	47	0	0	0	0
Other oilseed meals	10	0	0	0	0	0	7	1	34	99	5	6	5	7
Milk and cream	13	13	(7)	(6)	(53)	(45)	14	0	490	604	95	120	16	20
Yogurt	25	0	0	0	0	0	25	0	1	4	1	1	16	24
Whey	6	75	(17)	(9)	(15)	(12)	11	0	10	284	(32)	(26)	(11)	(9)
Butter	23	0	0	0	0	0	23	0	53	66	22	29	34	44
Cheese	20	7	(4)	(3)	(64)	(46)	20	0	32	70	7	10	10	14
Lactose and lactose syrup	10	19	(3)	(3)	(16)	(13)	10	0	2	34	(4)	(3)	(11)	(9)
Ice cream	24	0	0	0	0	0	24	0	1	16	0	0	0	0
Casein	10	0	0	0	0	0	10	0	7	15	1	1	3	5
Caseinates	10	0	0	0	0	0	10	0	13	28	3	3	11	11
Processed rices	68	0	0	0	0	0	68	0	196	196	189	949	96	484
Sugar and molasses	0	0	0	0	0	0	1	0	49	378	2	3	0	1
Prepared eggs	22	0	0	0	0	0	22	0	0	0	0	0	0	0
Frozen potatoes	13	0	0	0	0	0	13	0	0	0	0	0	0	0
Frozen vegetables	11	11	0	0	0	0	13	0	0	13	0	0	0	0
Preserved vegetables	13	0	0	0	0	0	13	0	0	2	0	0	0	0
Dried vegetables	13	1	0	0	0	0	13	2	0	4	0	0	0	0
Preserved fruits	30	2	0	0	0	0	30	0	11	71	9	10	13	14
Coffee and tea	19	7	0	0	0	0	18	0	1	25	0	0	0	0
Cereals and starches	18	5	(1)	(1)	(17)	(17)	12	0	254	294	56	59	19	20
Vegetable saps and extracts	14	18	(1)	(1)	(4)	(3)	13	1	13	102	4	4	4	4
Processed sugar and confectionary	28	5	0	0	0	0	15	0	19	63	7	8	12	12
Cocoa products	10	25	(3)	(3)	(12)	(11)	15	0	74	209	21	22	10	11
Infant formula	15	4	(1)	(1)	(25)	(24)	15	0	477	605	74	77	12	13
Mixes, doughs, breads and pastries	19	11	(1)	(1)	(13)	(12)	15	0	138	349	43	45	12	13
Pasta	18	2	0	0	0	0	18	0	7	25	3	3	13	14

Product	2009 trade-weighted AVE on U.S. exports to China	Actual 2009 U.S. exports to China	Range of simulated change in U.S. exports to China under full implementation of China's FTAs				Trade-weighted AVE on Chinese imports from FTA partners: Before the FTAs	After the FTAs	Actual 2009 Chinese imports from FTA partners	Actual 2009 Chinese imports from the World	Range of simulated change in Chinese global imports under full implementation of China's FTAs			
	Percent	Million $	Million $		Percent		Percent		Million $	Million $	Million $		Percent	
Foods prepared from cereals, flour, starches	26	5	(1)	(1)	(17)	(16)	26	0	10	25	6	6	23	25
Foods prepared from vegetables	16	39	(1)	(1)	(4)	(3)	25	0	4	62	3	3	4	5
Foods prepared from fruits and nuts	18	63	(6)	(6)	(8)	(7)	21	0	21	128	12	13	9	10
Other fruit juices	15	15	0	0	0	0	20	0	8	140	4	4	3	3
Apple juice	15	0	0	0	0	0	19	0	0	1	0	0	0	0
Other prepared foods	18	94	(8)	(7)	(8)	(7)	21	0	81	469	44	46	9	10
Bran, sharps and other residues	5	1	0	0	0	0	5	0	16	17	1	1	8	8
Pet food and other feeds	8	100	(1)	(1)	(1)	(1)	7	0	7	241	1	1	0	0
Other food products	11	32	(4)	(4)	(12)	(11)	12	0	90	181	17	18	10	10
Water and non-alcoholic beverages	27	4	0	0	0	0	25	0	5	59	3	4	5	8
Wine	25	25	(1)	(1)	(4)	(3)	28	0	63	457	22	32	5	7
Alcoholic beverages	29	137	2	4	2	3	29	0	3	737	16	26	2	3
Tobacco products	27	0	0	0	0	0	25	0	61	99	20	27	20	27
Raw silk	9	1	0	0	0	0	9	1	0	5	0	0	0	0
Processed woo	35	0	0	0	0	0	37	3	42	128	(1)	(1)	0	0
Cotton waste	9	1	0	0	0	0	9	0	24	36	6	7	18	19
Other plant based fibers	6	0	0	0	0	0	6	0	0	157	(3)	(3)	2	2
Plant materials	12	1	0	0	0	0	10	0	38	60	12	17	21	29
Chemically modified animal or veg. fats, oils	16	1	0	0	0	0	16	0	0	16	5	6	34	35
Glycerol	14	0	0	0	0	0	16	0	71	111	28	29	25	26
Mannitol and sorbitol	14	0	0	0	0	0	14	0	0	3	0	0	0	0
Essential citrus fruit oils	20	18	0	0	0	0	20	0	0	54	0	0	0	0
Other essential oils	18	10	(1)	(1)	(10)	(10)	18	0	7	44	6	6	13	13
Resinoids	20	3	0	0	0	0	20	0	1	14	1	1	6	7
Albumins and mixtures	22	46	(3)	(3)	(7)	(7)	22	0	13	149	21	23	14	15
Gelatins	13	2	0	0	0	0	13	0	0	7	0	0	0	0
Peptones	6	8	0	0	0	0	6	0	2	20	1	1	3	3
Glues and finishing agents	15	1	0	0	0	0	13	0	1	8	1	1	13	14
Fatty acids and alcohols	13	21	(8)	(8)	(39)	(38)	16	0	275	375	106	110	23	29
Total	6	10,942	(21)	48	0	0	10	0	10,253	47,986	1,853	3,105	4	6

Source: Commission staff calculations with simulation framework discussed in appendix F.

Notes: (1) AVE stands for ad valorem equivalent. (2) Parenthesis ( ) indicates a negative number. (3) A range of simulated effects was obtained by varying the magnitude of trade elasticities to account for the degree of statistical uncertainty in the economic estimates of the elasticities.

Jomini, P., R. McDougall, G. Watts, and P.S. Dee. *The SALTER Model of the World Economy: Model Structure, Database and Parameters.* Canberra, Australia: Industry Commission, 1994.

Linkins, Linda A., and Hugh M. Arce. "Estimating Tariff Equivalents of Non-Tariff Barriers." USITC. Office of Economics Working Paper EC1994-06-A(4), June 1994, rev. August 2002.

Liu, J., T.C. Arndt, and T.W. Hertel. "Parameter Estimation and Measures of Goodness of Fit in a Global General Equilibrium Model." *Journal of Economic Integration* 19, no. 3 (2004): 626–49.

Narayanan, B., and T. Walmsley (eds.). *Global Trade, Assistance, and Production: The GTAP 7 Data Base.* Center for Global Trade Analysis, Purdue University. 2008.

Narayanan, B., T. Hertel, and M. Horridge. "A Nested PE/GE Model for GTAP: Simulating the Disaggregated Impacts of Tariff-Liberalization on Automotive Industry in India." Paper presented at the Eleventh Annual Conference on Global Economic Analysis, Helsinki, June 2008.

Powell, A., and F. Gruen. "The Constant Elasticity of Transformation Production Frontier and Linear Supply System." *International Economic Review* 9 (1968): 315–28.

U.S. International Trade Commission (USITC). *Global Beef Trade: Effects of Animal Health, Sanitary, Food Safety, and Other Measures on U.S. Beef Exports*, USITC Publication 4033. Washington, DC: USITC, September 2008.

———. *India: Effects of Tariffs and Nontariff Measures on U.S. Agricultural Exports*, USITC Publication 4107. Washington, DC: USITC, November 2009.

———. Interactive Tariff and Trade Dataweb (Dataweb). http://dataweb.usitc.gov (accessed various dates).

———. *U.S.-Korea Free Trade Agreement: Potential Economy-wide and Selected Sectoral Effects*. USITC Publication 3949. Washington, DC: USITC, September 2007.

End Notes for Introduction

[1] Horticultural products include fruit crops, vegetable crops, ornamental and floral crops, and turf.

[2] Data are for 2009. FAO, FAOSTAT database; USDA, FAS, PSD Online database.

[3] Lohmar et al., *China's Ongoing Agricultural Modernization*, April 2009, 3–4.

[4] National Bureau of Statistics of China, *China Statistical Yearbook 2009*, 2009.

[5] IMF, *International Financial Statistics*.

[6] See chapter 3 of this report for further discussion of China's food consumption patterns and the factors affecting them.

[7] According to the U.S. Department of Agriculture, the U.S. agricultural sector is twice as reliant on overseas markets as the U.S. economy as a whole. USDA, FAS, *National Export Initiative: Importance of U.S. Agricultural Exports*, February 2010.

[8] USDA, ERS, "Rural and Natural Resource Indicators," June 2010.

[9] U.S. Census Bureau, International Data Base (accessed December 2, 2010).

[10] USDA, FAS, "India and China: Divergent Markets for U.S. Agricultural Exports," February 24, 2010.

[11] Comments by U.S. Department of Agriculture Secretary, Tom Vilsack, conference call briefing, December 1, 2010.

[12] USDA, FAS, "U.S.-China Trade," August 4, 2010.

[13] GTIS, Global Trade Atlas database (accessed February 18, 2011).

[14] The competitiveness of a country's agricultural sector can be defined as the ability of its farmers and food processors to sell their products in domestic and overseas markets.

[15] Industry official, interview by Commission staff, Shanghai, China, September 14, 2010.

[16] Appendix A contains the request letter for this investigation; appendix B contains the *Federal Register* notice; and appendix C contains the list of witnesses that appeared at the Commission's public hearing.

[17] Additional detail on these measures is found in appendix D, which summarizes the views of interested parties.

[18] Processed fish products classified in HS chapter 16 are also excluded from the WTO definition of agricultural products.

[19] In connection with this investigation, Commission staff traveled to the Chinese cities of Beijing, Shanghai, and Hong Kong, and the provinces of Shandong and Sichuan.

[20] The uncertain accuracy of China's official statistics has traditionally hampered analysis of China's economy. Gale, "China's Statistics: Are They Reliable?" 2002. However, the trustworthiness of Chinese data is considered to be improving, and Chinese statistics are now considered reliable enough to be used for analysis, especially for long-term growth trends. Improvements in data quality are attributed to better vetting efforts of the central government agencies responsible for compiling these data. Holtz, "OECD: The Institutional Arrangements for the Production of Statistics," January 19, 2005; Chow, *Are Chinese Official Statistics Reliable?* February 2006; U.S. government official, telephone interview by Commission staff, February 2, 2011.

[21] USITC, *India: Effects of Tariffs and Nontariff Measures*, November 2009.

[22] A price gap is the difference between the domestic price and the global price of similar goods in the absence of free trade.

[23] World Bank, "Mid-Term Evaluation of China's 11th Five-Year Plan," December 18, 2008, 7–8.

[24] OECD, *Agricultural Policies in Non-OECD Countries*, 2007, 71.

[25] Lohmar et al., *China's Ongoing Agricultural Modernization*, April 2009, 1–2, 13, 20, 22–23, 35.

[26] During the 20th century, China was periodically unable to feed its population. Between 1958 and 1961, more than 30 million Chinese citizens are estimated to have died from famine as a result of poor government planning during the Communist government's Great Leap Forward. This is often known as the "Great Chinese Famine" or "Three Years of Economic Difficulty." Other famines include an estimated 1 million deaths in 1942–43 during World War II; droughts and other natural disasters that claimed approximately 8 million lives in northern China in 1920–21, 1928–29, and 1936; and famines in east-central China in 1907 and 1911.

[27] During the past 100 years China has also been subject to occupation, political revolutions, and other upheavals, including the Japanese occupation of portions of northern and coastal China between 1931 and 1945, the decades-long armed struggle between the Nationalists and Communists, the Communist takeover of China in 1949, and the Cultural Revolution from 1966 to 1976.

[28] In this case, top-down governance means that policy decisions are typically made by the central government and imposed on local governments and citizens.

[29] CIA, CIA World Factbook: China, 2009.

[30] A discussion of rural and urban income differences can be found in chapter 3.

[31] OECD, *Agricultural Policies in Emerging Economies 2009*, 2009, 80.

[32] Owen, "Standards in China," January–February 2010, 42; Promar International, *The Chinese Potato Industry in Transition*, 2007, 123.

[33] Three vice premiers will sit on this 18-member panel, and the chairman, Li Keqiang, is expected to succeed Wen Jiabao to become China's next premier. *Bloomberg Businessweek*, "China Names Vice Premier Food Safety Commission Head," February 10, 2010.

[34] Lum, "Social Unrest in China," May 8, 2006, 1.

[35] Xiao and Nie, *A Report on the Status of China's Food Security*, 2009, 5–8.

[36] The U.S. Agency for International Development (USAID) and the FAO have similar definitions. USAID, "Policy Determination: Definition of Food Security," April 13, 1992.

[37] More formally, the terms are defined as follows: "food availability" is having sufficient quantities of food available on a consistent basis; "food access" is having sufficient resources to obtain appropriate foods for a nutritious diet; and "food use" is the appropriate use based on knowledge of basic nutrition and care, as well as adequate water and sanitation. Of the three, only availability and access will be addressed in this study. WHO, "Glossary" (accessed July 23, 2009).

[38] Thompson and Metz, *Implications of Economic Policy for Food Safety*, 1999.

[39] Ibid.

[40] The United Nations and other organizations acknowledge that China has significantly increased its prospects for food security through the development of domestic agricultural production, but external trade is also a useful

tool for dealing with food production surpluses and shortfalls. Xiao and Nie, *A Report on the Status of China's Food Security*, 2009, 5–8; Thompson and Metz, *Implications of Economic Policy for Food Safety*, 1999.

[41] Lohmar et al., *China's Ongoing Agricultural Modernization*, April 2009, 20.

[42] Ibid., 19.

[43] USDA, FAS, *China: Bio-fuels*, August 8, 2006, 17.

[44] A full discussion of policy instruments can be found in chapter 4.

[45] China's NTMs are described in chapter 9.

End Notes for Chinese Agricultural Trade

[1] GTIS, Global Trade Atlas database.

[2] GTIS, Global Trade Atlas database; IMF, *International Financial Statistics*.

[3] The sharp rise in the agricultural trade deficit in 2010 was largely because of the rise in commodity prices that year, including the prices of many of China's major agricultural imports, such as soybeans, vegetable oils, hides and skins, and cotton.

[4] For most of the previous three decades, China was a net exporter of agricultural products. Gale and Lohmar, "Who Will China Feed?" 2008.

[5] Industry official, interview by Commission staff, Beijing, China, September 7, 2010.

[6] He, Li, and Polaski, "China's Economic Prospects 2006–2010," April 2007, 17; Anderson, Martin, and Valenzuela, "Long Run Implications of WTO Accession for Agriculture in China," 2007.

[7] Using the WTO definition of agricultural products (see the definition in chapter 1 of this report), in 2010 China became the second leading market for U.S. agricultural exports behind Canada. Using the USDA definition of agricultural products (which is narrower than the WTO's because it excludes certain sugar alcohols, plant-based food additives, and fatty acids), the USDA reported that in 2010 China moved ahead of Canada to become the leading market for U.S. agricultural exports. USDA, "Agriculture Secretary Vilsack's Statement on 2010 U.S. Farm Exports Record," February 11, 2011.

[8] Lohmar et al., *China's Ongoing Agricultural Modernization*, April 2009.

[9] Chinese government policies regarding grain self-sufficiency are described in chapter 4.

[10] USDA, FAS, *National Plan for Expansion of Grain Production Capacity*, February 18, 2010.

[11] Trade data cited here are from the United Nations Trade Statistics, as reported by China and compiled by Global Trade Atlas. The country's imports are likely understated because official government statistics do not reflect "grey market" trade, which enters China through Hong Kong and Vietnam.

[12] Industry official, interview by Commission staff, Beijing, China, September 7, 2010.

[13] During 2005–10, although the value of Chinese soybean imports fluctuated between years, the quantity of imports rose each year during the period.

[14] 2009 is the most recent year for which global soybean export data are available.

[15] USDA, FAS, "Strong Processing Margins Support China's Expanding Soybean Import Market Demand," January 2010, 1.

[16] For example, Chinese imports of soybeans have a duty rate of "Free."

[17] ASA, written submission to the Commission, June 22, 2010.

[18] "Stacks" are multiple genes that have been inserted into a seed using biotechnology in order to convey a special characteristic or trait, like the ability to resist certain insects.

[19] ASA, written submission to the Commission, June 22, 2010.

[20] WTO, TAO.

[21] Chinese apparel exports represent only 44 percent of Chinese production, with the remaining 56 percent consumed in the domestic market. Gereffi and Stacey, "The Global Apparel Value Chain, Trade and the Crisis," April 2010, 8, 10.

[22] USDA, ERS, *Fiber Use for Textiles and China's Cotton Textile Exports,* March 2009.

[23] Cotton is subject to a tariff rate quota. The in-quota tariff is 1 percent while the over-quota tariff is 40 percent. For more information see chapter 7.

[24] USDA, ERS, *Cotton and Wool Situation and Outlook Yearbook*, November 2008, iv.

[25] Ibid., June 11, 2010, 6.

[26] During 2006–09, higher returns available to U.S. farmers from alternative crops, such as soybeans and corn, resulted in less land planted with cotton. USDA, ERS, "Cotton: Background," November 10, 2009.

[27] USDA, ERS, *Cotton and Wool Situation and Outlook Yearbook*, December 13, 2010, 2.

[28] In 2005 India accounted for only 5 percent of China's cotton imports.

[29] USDA, FAS, *India: Cotton and Products Annual, 2010*, April 12, 2010, 3–4. Indian cotton exports are expected to decline due to a law requiring licenses to export cotton. USDA, FAS, *India: Export of Cotton Allowed under License*, June 4, 2010; USDA, ERS, *Cotton and Wool Situation and Outlook Yearbook*, December 13, 2010, 6.

[30] WTO, TAO.

[31] EIU, *China: Food: Sub-Sector Update*, January 2009.

[32] EIU, *China: Food, Beverages and Tobacco Profile*, January 17, 2008.

[33] USDA, FAS, *China: Poultry and Products; Annual*, September 1, 2008, 5.

[34] In August and September 2010, China announced CVD rates of between 5.1 percent and 30.3 percent, and AD rates of between 50.3 percent and 105.4 percent, on imports of the subject U.S. poultry. PRC, MOFCOM, "Notice No. 52 of 2010," August 30, 2010; and PRC, MOFCOM, "Notice No. 51 of 2010," September 26, 2010. See chapter 7 for further discussion of the Chinese CVD and AD investigations on U.S. poultry.

[35] USDA, FAS, *Livestock and Poultry: World Markets and Trade,* April 2008, 5.

[36] Chinese food security policy focuses on grain self-sufficiency, with grains including wheat, corn, rice, and tubers. Trade data presented for grains in this chapter include barley but do not include tubers. USDA, FAS, *China: Grain and Feed; Annual,* March 1, 2010, 15.

[37] USDA, FAS, *China: Grain and Feed; Annual,* March 1, 2010, 3.

[38] USDA, FAS, PSD Online (accessed October 14, 2010).

[39] Gale, Lohmar, and Tuan, *How Tightly Has China Embraced Market Reforms in Agriculture?* April 2009.

[40] USDA, ERS, *China Is Using More Corn for Industrial Products,* December 2009.

[41] Gale, "A Tale of Two Commodities," July 2007.

[42] Some corn gets directed to industrial uses, such as starch and ethanol, in places where domestic end users receive a subsidy for purchasing domestically grown corn. USDA, FAS, *China: Grain and Feed; Annual,* March 1, 2010, 8–10.

[43] USDA, FAS, PSD Online (accessed January 7, 2011)

[44] Starting in September 2008, several Chinese dairies were implicated in a scandal involving milk and infant formula which had been adulterated with melamine, leading to kidney stones and, in some cases, to permanent kidney damage and death of infants. Fry, "Update on the Contamination of Dairy Products in China," November 2008.

[45] EIU, *Business China: Main Report,* March 12, 2007, 4–5.

[46] EIU, "Off the Wagon," March 1997, 8.

[47] Zappone, "How to Sell Wine to China," September 18, 2009.

[48] Oil derived from low erucic acid rapeseed is also called "Canadian oil, low acid" or "canola" oil.

[49] Countries with a free trade agreement with China include Bangladesh, Brunei, Burma, Cambodia, Chile, Hong Kong, India, Indonesia, Laos, Macau, Malaysia, New Zealand, Pakistan, Peru, the Philippines, Singapore, South Korea, Sri Lanka, Thailand, and Vietnam.

[50] USDA, ERS, *China's Rising Fruit and Vegetable Imports,* February 2006, 9. The China-wide antidumping (AD) duty rate is 51.74 percent. However, several Chinese exporters were excluded from the AD orders because they received de minimus margins (margins below 2 percent), and a number of other companies have received de minimus margins in subsequent U.S. Department of Commerce (USDOC) administrative reviews. On November 15, 2010, the USDOC announced that it would revoke the AD duties on U.S. imports of non-frozen apple juice concentrate from China owing to non-participation by the U.S. industry. 75 Fed. Reg. 69628 (November 15, 2010).

[51] California Cling Peach Board, written submission to the Commission, July 28, 2010.

[52] USDA, ERS, *China's Rising Fruit and Vegetable Exports,* February 2006, 11–12.

[53] Industry officials, interviews by Commission staff, Beijing, China, September 2007.

54 Dragon head enterprises are leading-edge companies within the agricultural sector in China, and they are eligible for preferential tax treatment, preferential access to loans, and participation in official delegations. Waldron, Brown, and Longworth, *Rural Development in China,* 2003, 40.

[55] USDA, FAS, *China: Annual; Canned Deciduous Fruit,* November 10, 2009.

[56] Industry official, interview by Commission staff, Washington, DC, August 19, 2010.

[57] Through administrative reviews, several Chinese exporters have been assessed lower duties ranging from zero to 40 percent in recent years. 65 Fed. Reg. 76608 (December 7, 2000); USITC, *Fresh Garlic from China,* September 2006. As of the 13th USDOC administrative review in 2009, AD duties on fresh garlic from China

are assessed on a per unit rate of up to $4.71/kg. U.S. government official, telephone interview by Commission staff, Washington, DC, November 2, 2010.

[58] USITC, *Canned Peaches, Pears, and Mixtures,* 2007.

[59] Shaosheng et al., "Agglomeration Effects and Japanese Food Industry Investment in China," August 2006, 3.

[60] Industry official, interview by Commission staff, Washington, DC, July 12, 2010.

[61] Lohmar et al., *China's Ongoing Agricultural Modernization*, April 2009.

[62] Industry officials, interview by Commission staff, Shandong, China, August 19, 2010.

[63] Industry officials, interview by Commission staff, Shandong, China, September 14, 2010.

[64] Ibid.

[65] Trade statistics bear out the current limited nature of competition in pears. U.S. and Chinese pear exports were $162 million and $220 million, respectively, for 2009, but in only Russia and Canada were more than $10 million worth of pears imported from both countries.

[66] Huang, *Global Trade Patterns in Fruit and Vegetables*, June 2004, 63.

[67] GTIS, Global Trade Atlas database.

[68] Ibid.

[69] GTIS, Global Trade Atlas database.

[70] Interestingly, U.S. exports had a lower unit price than Chinese exports, possibly reflecting exports from China by Japanese-owned farms that produce U.S.-quality fresh vegetables. Shaosheng et al., "Agglomeration Effects and Japanese Food Industry Investment in China," August 2006, 3.

[71] Huang and Gale, "China's Rising Fruit and Vegetable Exports Challenge U.S. Industries," February 2006, 5.

End Notes for Chinese Agricultural Consumption

[1] China accounts for 27 percent of world soybean consumption, 29 percent of world cottonseed consumption, and 41 percent of world peanut consumption. China is also the largest single-country consumer of rapeseed (behind the European Union-27) at 25 percent of world consumption. USDA, FAS, PSD Online (accessed October 18, 2010).

[2] See figure 3.1; see also USDA, ERS, *China Agricultural and Economic Data* (accessed October 5, 2010).

[3] Of Asian countries, only Kazakhstan (3,490 per capita per day) and South Korea (3,073) have a higher caloric intake than China. Japan had the fifth-highest caloric intake, at 2,812. Although Taiwan likely has a high caloric intake among Asian countries, data for Taiwan are not available. FAO, FAOSTAT (accessed May 24, 2010).

[4] USDA, ERS, *China Agricultural and Economic Data* (accessed October 7, 2010).

[5] CIA, *The World Factbook: China,* updated April 26, 2010.

[6] Food outlets include restaurants, fast food locations, supermarkets, and wet markets.

[7] Rozelle, *Vegetables in China,* March 2006, 3.

[8] Gale and Huang, *Demand for Food Quality and Quantity in China,* January 2007, 1, 8.

[9] Latner, *Who Is Feeding China?* February 19, 2010, 13. Wet markets are open-air food markets. Products sold at wet markets, including meat from slaughtered animals are generally stored for short periods of time and are always expected to be fresh.

[10] Chan and Tse, *The Consumer Trap,* May 12, 2007.

[11] Gale and Huang, *Demand for Food Quality and Quantity in China,* January 2007, 24.

[12] The marketing year is a 12-month period, usually beginning with a new harvest, during which the product is marketed. Marketing years differ for each commodity and country.

[13] The data on apparent consumption that are discussed in this section are from the USDA's database PSD Online, where consumption is determined by the sum of beginning stocks, domestic production, and imports, less the sum of ending stocks and exports. Thus consumption includes both domestically produced and imported products. This consumption includes human, livestock, and manufacturing consumption of agricultural products.

[14] Barboza, "Virus Spreading Alarm," August 16, 2007.

[15] Gifford, "Food Fears Persist in China," October 26, 2010.

[16] USITC, *Shifts in U.S. Merchandise Trade,* 2010, CHN-3.

[17] FAO, FAOSTAT (accessed July 23, 2010). Caloric consumption measures daily caloric intake by humans.

[18] Caloric data are from 2007. FAO, FAOSTAT (accessed August 26, 2010).

[19] The actual calories coming from cereals rose from 1,459 calories in 1987 to 1,657 calories in 2007. Calories from tubers, on the other hand, declined in both proportional and absolute terms.

[20] In comparison, the average annual population growth rate between 2005 and 2009 was 0.96 percent for the United States, 1.05 percent for Brazil, and 1.47 percent for India. OECD, *OECD Factbook 2010: Economic, Environmental and Social Statistics*, 2010.

[21] OECD, *OECD Factbook 2010: Economic, Environmental and Social Statistics*, 2010.

[22] Hesketh and Xing, "The Effect of China's One-Child Family Policy after 25 Years," 2005, 1171–76; *The Economist*, "China Policy: Rethinking China's One-Child Policy," August 25, 2010.

[23] EC, DGA, "China: Out of the Dragon's Den?" May 2008, 2.

[24] CIA, *The World Factbook: China*, updated April 26, 2010; CIA, *The World Factbook: India*, updated August 19, 2010; CIA, *The World Factbook: Brazil*, updated August 19, 2010.

[25] OECD, *OECD Factbook 2010: Economic, Environmental and Social Statistics*, 2010.

[26] This data is calculated from the population data in table 3.2 of this report.

[27] CIA, *The World Factbook: China*, updated April 26, 2010.

[28] Gale, *China's Growing Affluence*, June 2003, table 3.2.

[29] Gale, *China's Growing Affluence*, June 2003.

[30] Gale and Huang, *Demand for Food Quality and Quantity in China*, January 2007, 21, 23. While higher-quality foods may be more expensive, higher prices do not necessarily denote a higher-quality product.

[31] Rozelle, *Vegetables in China*, March 2006, 3.

[32] Shapouri and Rose, *Developing Countries*, June 2009, 36–37.

[33] Rozelle, *Vegetables in China*, March 2006, 3.

[34] World Bank, Data: China (accessed July 26, 2010). Measured in 2005 dollars.

[35] EC, DGA, "China: Out of the Dragon's Den?" May 2008, 2.

[36] World Bank, Data: China (accessed July 26, 2010).

[37] *China Daily*, "China's Urban, Rural Income Gap Widens," January 22, 2010.

[38] World Bank, Data: China (accessed November 4, 2010).

[39] *China Daily*, "China's Urban, Rural Income Gap Widens," January 22, 2010.

[40] Unless otherwise noted, the exchange rate used throughout this report is RMB 7.50 per dollar, which is based on the period (2005–09) average calculated from IMF daily rates. IMF, Exchange Rate Database (accessed December 17, 2010).

[41] Latner, "Who Is Feeding China?" February 19, 2010, 5.

[42] EIU, *Country Profile 2008: China*, 2008, 39.

[43] Keidel, "The Causes and Impact of Chinese Regional Inequalities in Income and Well-Being," December 2007.

[44] Wages for a select group of staff and workers as defined in China National Bureau of Statistics, *China Statistical Yearbook 2009*, 2009, 147–148,155–157, and 164, tables 4-22, 4-23, and 4-26. These wages exclude a number of important groups such as those employed by a township, a private enterprise, or themselves.

[45] Twenty-two out of the 31 provinces/municipalities had an average annual wage for select staff and workers of between $3,000 and $4,000. China National Bureau of Statistics, *China Statistical Yearbook 2009*, 2009, 147–148,155–157, tables 4-22, 4-23, and 4-26.

[46] Gale, *China's Growing Affluence*, June 2003.

[47] China National Bureau of Statistics, *China Statistical Yearbook 2009*, 2009.

[48] ISI Emerging Markets, CEIC database. Converted from renminbi to dollars using the 2009 average annual exchange rate. U.S. Federal Reserve System, Foreign Exchange Rates (Annual), January 4, 2010.

[49] Gale and Huang, *Demand for Food Quality and Quantity in China*, January 2007, 23.

[50] Gale and Huang, *Demand for Food Quality and Quantity in China*, January 2007, 21.

[51] USDA, FAS, *China: Oilseeds and Products; Annual 2010*, February 23, 2010, 14.

[52] Ibid.

[53] The following shares represent the amount of each food category consumed at home: grains (80 percent), eggs (75 percent), seafood (73 percent), vegetables (83 percent), fruits (95 percent), dairy (75 percent), drinks (63 percent), and bean products (63 percent). Latner, "Who Is Feeding China?" February 19, 2010, 10.

[54] Latner, "Who Is Feeding China?" February 19, 2010, 10.

[55] Gale and Henneberry, "Markets Adapt to China's Changing Diet," 2009.

[56] Ibid.

[57] Ibid.

[58] *Trading Economics*, "China Inflation Rate Slows in December," January 20, 2011.

[59] USDA, FAS, *China Food Price Inflation in 2010*, November 2, 2010, 2; Foster, "China Feels the Bite of Food Inflation," November 22, 2010.
[60] *Bloomberg Businessweek*, "China May Impose Temporary Price Controls," November 23, 2010.
[61] EIU, *China: Food, Beverages, and Tobacco*, January 5, 2010; EIU, *China: Food, Beverages, and Tobacco*, July 10, 2010.
[62] Ibid.
[63] Animal protein includes all meats, poultry, eggs, and aquatic products.
[64] USDA, FAS, *China: Livestock and Products; Annual 2009*, September 14, 2009, 6.
[65] Fuller, Beghin, and Rozelle, "Consumption of Dairy Products in Urban China," November 29, 2006, 6.
[66] Ibid., 2–3.
[67] Ibid., 2.
[68] Roberts and Anders, *Developments in Chinese Agriculture*, July 2005, 9.
[69] EIU, *China: Food, Beverages, and Tobacco*, January 5, 2010; EIU, *China: Food, Beverages, and Tobacco,* July 10, 2010.
[70] Gale and Huang, *Demand for Food Quality and Quantity in China,* January 2007, 1.
[71] Latner, "Who Is Feeding China?" February 19, 2010, 13.
[72] Gifford, "Food Fears Persist in China," October 26, 2010.
[73] Chan and Tse, *The Consumer Trap*, May 12, 2007.
[74] Ibid.
[75] The BRIC countries are Brazil, Russia, India, and China.
[76] Chan and Tse, *The Consumer Trap*, May 12, 2007.
[77] MacDonald and Whitley, *Fiber Use for Textiles and China's Cotton Textile Exports*, March 2009, 2.
[78] Ibid., 15.
[79] USDA, FAS, *China: Cotton Update,* June 29, 2010, 3.
[80] USDA, FAS, *China: Cotton and Products; Annual,* May 1, 2010, 9–12.
[81] USDA, FAS, *China: Guangdong Market for Leather Weathers the Storm,* November 16, 2009.

End Notes for Chinese Agricultural Production and Policies

[1] CIA, *The World Factbook: China,* May 20, 2010. China's GDP was $4,985 billion in 2009. World Bank, Data: China (accessed October 21, 2010).
[2] Lohmar et al., *China's Ongoing Agricultural Modernization*, April 2009, 6.
[3] CIA, *The World Factbook: China,* May 20, 2010; Lohmar et al., *China's Ongoing Agricultural Modernization*, April 2009, 18.
[4] The Chinese government defines the term "grains" differently for different purposes. When discussing self-sufficiency, "grains" means wheat, rice, corn, and tubers. When discussing government stocks, grains include wheat, corn, rice, and soybeans. In some cases, such as in the *Chinese Statistical Yearbook*, pulses are included with grains as well as cereals, tubers, and soybeans. In this investigation, grains include wheat, corn, rice and other cereals, while tubers (i.e., root crops, such as potatoes, cassavas, and sweet potatoes) are included as horticultural products.
[5] About two-thirds of the country is mountainous. Other parts of China are desert or urban. PRC, "About China: Topography."
[6] Data are from 2008. The figure for sowed land changed less than 1 percent between 1998 and 2008. ISI Emerging Markets, CEIC database.
[7] Roberts and Anders, *Developments in Chinese Agriculture*, July 2005, 7.
[8] EC, DGA, "China: Out of the Dragon's Den?" May 2008, 4.
[9] ISI Emerging Markets, CEIC database; Roberts and Anders, *Developments in Chinese Agriculture*, July 2005, 7; EIU, *Country Profile 2008: China*, 2008, 25; USDA, ERS, "Briefing Rooms: China," 2009. Roberts and Anders caution that the numbers at the higher end of the range may be inflated because of the Chinese practice of double-cropping: land that has been planted and harvested twice in one season may be double-counted in the overall "planted area" total.
[10] During the reform period (1978 to 2008), China's cropland grew by just 4 percent. Lohmar et al., *China's Ongoing Agricultural Modernization,* April 2009, 5–6.
[11] ISI Emerging Markets, CEIC database.

[12] Ito, "Inter-regional Difference of Agricultural Productivity in China," March 21, 2010, 394. Agricultural technologies can generally be categorized as mechanical or biological. Generally, the Chinese adoption of mechanical agricultural technology has been limited because agricultural labor remains a fairly abundant resource.

[13] Borlaug, "Biotechnology and the Green Revolution," interview by ActionBioscience.org, November 2002; Nath, *The FAO Field Programme and Agricultural Development in Asia and the Pacific*, December 1999, 21, 25.

[14] Rozelle et al.,"Rise of China's Horticultural Industry," n.d., 6.

[15] China approved *Bacillus thuringiensis* (Bt) rice, which wards off insects and is estimated to improve yields by 2 to 6 percent, as well as phytase corn, which purportedly increases the absorption of phosphate in animals by increasing their growth rate and reduces phosphate pollution from livestock. Resurreccion, "China 'Bt' Rice OK to Boost Supply," March 2, 2010; Qinghui, "China's GM Gamble: Seeds of Change," May 5, 2010, 15.

[16] Huang and Rozelle, "China's Agricultural," August 2009, 4.

[17] Huang and Rozelle, "China's Agricultural," August 2009, 2; EIU, *Country Profile 2008: China*, 2008, 39; Rozelle, *Vegetables in China*, March 2006, 2. More information on changes in consumer demand can be found in chapter 3 of this report.

[18] ISI Emerging Markets, CEIC database.

[19] The marketing year is a 12-month period, usually beginning with a new harvest, during which the product is marketed. Marketing years differ for each commodity and country.

[20] Chinese production statistics should be seen more as generally indicative of patterns than accurate exact figures. Gale, "China's Statistics: Are They Reliable?" April 2002; industry official, interview by Commission staff, Shanghai, September 13, 2010.

[21] Owing in part to increased demand for feedstuffs, returns to corn production were higher than for soybeans in MY 2008/09, which encouraged farmers to increase corn production. USDA, FAS, *China: Grains and Feed; Annual 2009*, March 3, 2009, 3; South East Farm Press, "China on Track for Good Grain Harvest," October 5, 2010.

[22] Beckman and Junyang, *China: Grain and Feed; Annual 2010*, March 1, 2010, 7.

[23] Resurreccion, "China 'Bt' Rice OK to Boost Supply," March 2, 2010; Qinghui, "China's GM Gamble: Seeds of Change," May 5, 2010, 15.

[24] USDA, FAS, *China: Grains and Feed; Annual 2009*, March 3, 2009, 3.

[25] While wheat production for MY 2009/10 is reported by USDA in PSD Online as 115 million metric tons (mt), USDA officials in China estimate that wheat production for that crop year was only 106 million mt due largely to poor weather. USDA, FAS, *China: Grain and Feed; Annual 2010*, March 1, 2010, 3.

[26] USDA, FAS, *China: Citrus; Annual 2007*, November 29, 2007, 3; USDA, FAS, *China: Grain and Feed; Annual 2010*, March 1, 2010, 2; USDA, FAS, *China: Agricultural Situation; Oilseed Situation Update*, October 23, 2009, 2.

[27] Government support programs for agricultural production in China, including minimum guaranteed purchase prices, direct payments for certain commodities based on cultivated area, and subsidies for inputs such as fertilizers, seeds, and investments in machinery, are discussed in detail below.

[28] USDA, FAS, *China: Grain and Feed; Annual 2010*, March 1, 2010, 2.

[29] USDA, FAS, *China: Grain and Feed; Annual 2010*, March 1, 2010, 3; USDA, FAS, *China's Rising Fruit and Vegetable Exports*, February 2006, 11.

[30] USDA, FAS, *China: Oilseeds and Products; Annual; Part 1 of 2*, March 1, 2008, 4; USDA, FAS, *China: Oilseeds and Products; Annual; Part 1 of 2*, April 15, 2009, 4–5. Official government statistics show a decline of 6 percent in soybean sown area between 2005 and 2008. ISI Emerging Markets, CEIC database.

[31] The Chinese government provides some support for soybean farmers in the form of technical assistance, purchases of soybeans at a floor price for state reserves, and a subsidy for crushers to purchase domestic soybeans.

[32] USDA, FAS, *China: Oilseeds and Products; Annual; Part 1 of 2*, April 15, 2009, 7.

[33] USDA, FAS, PSD Online (accessed November 4, 2010).

[31] Soybean meal and oil are both products produced from crushing soybeans so increased production of one leads to a rise in production of the other. U.S. soybeans are favorable for producing oil because they have a higher oil content, between 19 and 21 percent, than Chinese soybeans, which average 16 percent. Industry official, interview by Commission staff, Beijing, September 6, 2010.

[35] Industry official, interview by Commission staff, Shanghai, September 13, 2010.

[36] For China to grow enough soybeans to replace its imports, it would need to devote 26 million ha, or approximately one-fifth of all crop land, to soybean production. Gale, "China's Future," February 19, 2010.

[37] USDA, FAS, *China: Oilseeds and Products; Annual: Part 1 of 2,* March 1, 2008, 13; USDA, FAS, *China Oilseeds and Products; Annual: Part 1 of 2,* April 15, 2009, 13.

[38] USDA, FAS, *China: Oilseeds and Products; Annual: Part 1 of 2,* April 15, 2009, 14.

[39] USDA, FAS, *China: Citrus; Annual 2007,* November 29, 2007, 3.

[40] USDA, FAS, *China: Citrus; Annual 2007,* November 29, 2007, 3–4; USDA, FAS, *China: Citrus; Annual 2008,* December 1, 2008.

[41] It typically takes about four years for new apple trees to bear fruit. USDA, FAS, *China: Fresh Deciduous Fruit; Annual 2006,* September 22, 2006, 3–4; USDA, FAS, *China Fresh Deciduous Annual,* November 15, 2007, 4–5; USDA, FAS, *China Fresh Deciduous Fruit; Annual,* November 15, 2008, 3–4.

[42] Sow stocks and piglet crops declined an estimated 5 percent and 7 percent, respectively, in 2007 due to the disease outbreak. USDA, FAS, *China: Livestock and Products; Semi-Annual Report 2007,* March 1, 2008, 13.

[43] USDA, FAS, *China: Livestock and Products; Semi-Annual Report 2007,* March 1, 2008, 13.

[44] USDA, FAS, *China: Livestock and Products; Semi-Annual Report 2009,* March 9, 2009, 5; USDA, FAS, *China: Livestock and Products; Semi-Annual Report 2009,* September 15, 2009, 4.

[45] USDA, FAS, *China: Livestock and Products; Semi-Annual Report 2009, September* 14, 2009, 4–5.

[46] USDA, FAS, *China: Poultry and Products Annual,* September 16, 2009; USDA, FAS, *China: Poultry and Products; Semi-Annual 2010,* April 14, 2010.

[47] Fuller, Beghin, and Rozelle, "Consumption of Dairy Products in Urban China," November 29, 2006, 2–3.

[48] Ibid., 2.

[49] USDA, FAS, *China: Dairy and Products; Annual Report 2007,* October 12, 2007, 3–4.

[50] Melamine is an industrial chemical used to make milk appear to have a higher protein content in laboratory tests.

[51] Gifford, "Food Fears Persist In China 2 Years After Milk Scare," October 26, 2010; BBC, "Timeline: China Milk Scandal," January 25, 2010; Wee and Mao, "China Arrests 96 Over Tainted Milk," January 13, 2011.

[52] Industry official, interview by Commission staff, Shanghai, September 14, 2010.

[53] *Dairy Markets,* "Rabobank Warns of Tighter Dairy Markets Next Year," November 25, 2010.

[54] USDA, FAS, *China: Cotton and Products; Annual 2007,* May 1, 2007, 3, 6; USDA, FAS, *China: Cotton and Products Annual 2008,* May 23, 2008, 7; USDA, FAS, *China: Cotton and Products; Annual 2009,* April 15, 2009, 3, 6.

[55] USDA, FAS, *China: Cotton and Products; Annual 2007,* May 1, 2007, 3; USDA, FAS, *China: Cotton and Products; Annual 2009,* April 15, 2009, 3.

[56] USDA, FAS, *China: Cotton and Products; Annual 2009,* April 15, 2009, 3.

[57] USDA, FAS, *China: Cotton and Products; Annual 2009,* May 1, 2007, 5; USDA, FAS, *China Cotton and Products; Annual 2009,* May 23, 2008, 6; USDA, FAS, *China: Cotton and Products; Annual 2009,* April 15, 2009, 5–6.

[58] The Eleventh Five-Year Plan, as translated, describes the rural development goals as (1) developing modern agriculture, (2) increasing farmers' incomes, and (3) improving rural conditions. World Bank, "Mid-Term Evaluation of China's Eleventh Five Year Plan," December 18, 2008, 7–8.

[59] In this report, the term "Chinese government" will denote the Chinese central government, unless otherwise indicated.

[60] As measured by the OECD's Producer Support Estimate (PSE), the amount of support provided to Chinese farmers has gradually risen since the 1990s to 8.6 percent of gross farm receipts in 2007, the latest year for which Chinese data are available. OECD, *Agricultural Policies in Emerging Economies: Highlights,* 2009, 9.

[61] Lieberthal, "China's Governing System and Its Impact on Environmental Policy Implementation," 4–5.

[62] Industry official, interview by Commission staff, Hong Kong, September 20, 2010.

[63] Academics, interviews by Commission staff, Beijing, China, September 7, 2010; Promar International, *The Chinese Potato Industry in Transition,* 123.

[64] OECD, *Agricultural Policies in Non-OECD Countries,* 2007, 71.

[65] Grain is defined in this security plan to include wheat, rice, corn, and tubers such as potatoes, but not soybeans. USDA, FAS, *China: Grain and Feed; Annual 2009,* March 3, 2009, 13.

[66] USDA, FAS, *China: Grain and Feed; Annual 2010,* March 1, 2010, 6.

[67] OECD, *Agricultural Policies in Non-OECD Countries,* 2007, 72.

[68] PSE is defined as the estimated monetary value of transfers from consumers and taxpayers to farmers, expressed as a percentage of gross farm receipts (defined as the value of total farm production at farmgate prices), plus budgetary support. OECD, *Agricultural Policies in Emerging Economics: Highlights,* 2009, 9.

[69] OECD, *Agricultural Policies in Emerging Economics: Highlights,* 2009, 9.

[70] OECD, *Agricultural Policies in Emerging Economics: Highlights,* 2009, 9.

[71] Kwiecinski and von Tongeren, "Quantitative Evaluation of Agricultural Policy Reforms in China," July 8–9, 2007, 215–16.

[72] Ibid.

[73] In 2000, farmers also paid taxes to the township government amounting to RMB 26.8 billion ($3.6 billion) and to the village committee totaling RMB 35.2 billion ($4.8 billion), as well as additional fees estimated at RMB 90 billion ($12.2 billion). OECD, *Agricultural Policies in Non-OECD Countries*, 2007, 73.

[74] Aubert and Xiande, "'Peasant Burden': Taxes and Levies Imposed on Chinese Farmers," 2002, 160–76.

[75] Chen, *A Study on China's County and Township Public Finance,* 2007, 20.

[76] Legislative reforms included the removal of the animal slaughter tax and the special agricultural tax on all products except tobacco. OECD, *Agricultural Policies in Non-OECD Countries*, 2007, 74.

[77] Lardy, *China's Consumption*, July 8–9, 2007, 22.

[78] OECD, *Agricultural Policies in Non-OECD Countries*, 2007, 74. Agricultural taxes are different than China's individual and corporate income taxes, but the latter are rarely paid by farmers. Chinese farms are generally operated by individuals and would not be subject to corporate taxes. In any case, under Article 86 of China's enterprise tax law, effective on January 1, 2008, Chinese farms which are incorporated and engage in crop production, horticulture, livestock rearing, and/or forestry are exempted from corporate income taxes. In practice, few farmers classified as individuals are subject to income taxes either; it only applies to those with income above a high threshold, which most farmers never attain. Government official, e-mail to Commission staff, October 18, 2010; USDA, FAS, *China's Corporate Income Tax Exemption for Agricultural Enterprises 2008*, January 13, 2009, 4.

[79] In 2009, VAT revenues collected from both domestic sources and imports equaled 32.2 percent of total government tax revenues. The second and third largest sources of Chinese government revenue in 2009 were a corporate income tax (21.2 percent of the total) and the domestic consumption tax (13.2 percent). PRC, Ministry of Finance, *2009 Central Financial Revenue Structure.*

[80] The tax is administered by the State Administration of Taxation, and with the exception of VAT collected from imported products, VAT is shared between the central government (75 percent) and local governments (25 percent). Hoffman, "China's VAT System," August 3, 2009.

[81] China's State Council Decree No. 538 of 2008. State Council Decree No. 134 of January 1, 1994, was the first decree to specifically exempt Chinese agricultural producers from payment of the VAT. Subsequent decrees and clarifications have confirmed the policy.

[82] A Chinese tax collector collects the difference between the "sales VAT" and the "purchase VAT" shown on business receipts. USDA, FAS, *China: Trade Policy Monitoring, VAT Protections,* March 19, 2007, 3.

[83] China's central government also refunds VAT to ethanol producers, in addition to other tax exemptions, compensations, and production incentives.

[84] USDA, FAS, *China: Trade Policy Monitoring, VAT Protections,* March 19, 2007, 3.

[85] Agricultural producers are defined as farmers, farm cooperatives, most state trading enterprises (e.g., grain storage facilities) that buy from producers and from such processors as feed or flour mills. In addition, Chinese companies operating large state farms incorporated as companies are exempt from VAT on agricultural products they purchase as raw materials if they are designated by the government as a "dragon head enterprise." As noted earlier in this study, dragon head enterprises are leading-edge companies within the agricultural sector in China, and they are eligible for preferential tax treatment, preferential access to loans, and participation in official delegations. Companies are nominated by local governments and approved at the provincial level; they must have direct contact with farmers and contribute to vertical integration within the sector. Profits, market share, taxes paid, and company growth rates are typical criteria evaluated by provincial government officials in determining which companies attain dragon head status. Government official, e-mail to Commission staff, October 18, 2010; Waldron, Brown, and Longworth, *Rural Development in China*, 2003, 40.

[86] USDA, FAS, *China: Trade Policy Monitoring, VAT Protections,* March 19, 2007, 4.

[87] It appears that labor costs are not to be included as deductible in the VAT calculation. If true, this would lessen the VAT deduction benefit for purchasers of Chinese domestic agricultural goods. See China's State Decree 538, Article 1.

[88] OECD, *Agricultural Policies in Non-OECD Countries*, 2007, 73.

[89] OECD, *Agricultural Policies in Non-OECD Countries*, 2007, 73.

[90] USDA, FAS, *China: Grain and Feed; Annual 2010*, March 1, 2010, 17.

[91] USDA, FAS, *China: Grain and Feed; Annual 2010*, March 1, 2010, 18. Other sources estimate that between 45 and 69 percent of China's wheat production was purchased at minimum support prices during 2006–09. Zhong, "Analysis of Wheat Market in 2009 and Outlook for the Later," 2009, 5–10.

[92] USDA, FAS, *China: Grain and Feed; Annual 2010*, March 1, 2010, 18.

[93] USDA, FAS, *China: Grain and Feed; Annual 2010*, March 1, 2010, 16; USDA, FAS, *China: Grain and Feed Annual 2010*, March 3, 2009, 13.

[94] Lohmar et al., *China's Ongoing Agricultural Modernization*, April 2009, 27.

[95] Gale, Lohmar, and Tuan, "How Tightly Has China Embraced Market Reforms in Agriculture?" 2009, 34.

[96] USDA, FAS, *China: Livestock and Products; Semi-Annual 2010*, March 2, 2010, 4.

[97] OECD, *Agricultural Policies in Non-OECD Countries*, 2007, 73.

[98] To ensure that Chinese farmers benefit, provincial and local governments are required to publicize details about the use of the grain risk fund. Penalties are levied for inappropriate uses of payments. OECD, *Agricultural Policies in Non-OECD Countries*, 2007, 73.

[99] *Farmers Daily*, "Nongye Bu Fabu 2010 Nian Qiang Nong Hui Nong Zhengce (Ministry of Agriculture Announces 2010 Policies Strengthening Agriculture and Benefiting Farmers)," March 11, 2010.

[100] *People's Daily Online*, "China Increase Subsidies to Grain Growers," April 12, 2006.

[101] USDA, FAS, *China: Fertilizer; Annual 2009*, December 14, 2009, 10.

[102] Hudson et al., *Crop Subsidies in Foreign Countries: Different Paths to Common Goals*, April 2009, 34.

[103] Academic, interview by Commission staff, Beijing, China, September 9, 2010.

[104] OECD, *Agricultural Policies in Non-OECD Countries*, 2007, 74.

[105] Academic, interview by Commission staff, Beijing, China, September 9, 2010; industry official, interview by Commission staff, Shandong province, China, September 13, 2010.

[106] Academic, interview by Commission staff, Beijing, China, September 9, 2010.

[107] Wang et al., "Understanding the Water Crisis in Northern China," March 1, 2009, 148.

[108] Lohmar et al., *China's Ongoing Agricultural Modernization,* April 2009, 19.

[109] Academic, interview by Commission staff, Beijing, China, September 7, 2010.

[110] Lohmar et al., *China's Ongoing Agricultural Modernization,* April 2009, 19.

[111] Wang et al., "Understanding the Water Crisis in Northern China," March 1, 2009, 148.

[112] Industry official, interview by Commission staff, Beijing, China, September 7, 2010.

[113] Lohmar et al., *China's Ongoing Agricultural Modernization,* April 2009, 19.

[114] OECD, *Agricultural Policies in Non-OECD Countries*, 2007, 75.

[115] Lohmar et al., *China's Ongoing Agricultural Modernization,* April 2009, 28.

[116] Sheldon, *China's Agricultural Trade*, 2007; Kwiecinski and von Tongeren. *Ch. 10: Quantitative Evaluation of Agricultural Policy Reforms in China: 199--2005,* July 8–9, 2007, 203.

[117] OECD, *Agricultural Policies in Non-OECD Countries*, 2007, 75.

[118] Gale, Lohmar, and Tuan, "How Tightly Has China Embraced Market Reforms in Agriculture?" 2009, 33.

[119] Ibid., 31.

[120] Promar International, *The Chinese Potato Industry in Transition*, 2007, 8.

[121] *Dairy Markets*, "Beijing Increases Financial Aid for Dairies," October 25, 2010.

[122] Gale, Lohmar, and Tuan, "How Tightly Has China Embraced Market Reforms in Agriculture?" 2009, 33.

[123] Li and Liu, "The Effects of Rural Agriculture Infrastructural Development," 2009, 3.

[124] National Bureau of Statistics of China, *China Statistical Yearbook*, 2006.

[125] Ibid.

[126] OECD, *Agricultural Policies in Emerging Economies 2009*, 2009, 79.

[127] Sheldon, *China's Agricultural Trade,* 2007; Kwiecinski and von Tongeren, "Quantitative Evaluation of Agricultural Policy Reforms in China," July 8–9, 2007, 205.

[128] *People's Daily Online*, "China to Further Boost Farm Produce Distribution," October 22, 2010.

[129] Various "special funds" fall under the responsibility of ministries and government institutions such as the National Development and Reform Commission, Ministry of Finance, Ministry of Science and Technology, Ministry of Water Conservancy, Ministry of Agriculture, State Administration of Forestry, China Meteorological Bureau, Ministry of Land Resources, State Council Poverty Alleviation Office, State Office for Preventing Flood and Drought, and the Ministry of Health.

[130] OECD, *Agricultural Policies in Non-OECD Countries*, 2007, 76.

[131] Ibid., 77.

[132] BBC, "Chinese Melamine Scandal Widens," October 31, 2008.

[133] Kwiecinski and von Tongeren, *Quantitative Evaluation of Agricultural Policy Reforms in China,* July 8–9, 2007, 205.
[134] The U.S.-China Business Council, "Recent Developments in China's Food Safety Regime," 2010, 1.
[135] Ibid.
[136] Industry official, interview by Commission staff, Shanghai, China, September 13, 2010.
[137] The U.S.-China Business Council, "Recent Developments in China's Food Safety Regime," 2010, 1.
[138] Ibid., 4.
[139] Industry official, interview by Commission staff, Chengdu, China, September 17, 2010.
[140] Ibid.
[141] Certain provincial and local governments also keep food reserves. USDA, ERS, *Briefing Rooms: China*, updated March 12, 2009.
[142] Another central government policy goal is energy independence. See box 4.6 for a discussion of biofuel policies.
[143] Gale, Lohmar, and Tuan, "How Tightly Has China Embraced Market Reforms in Agriculture?" 2009, 33.
[144] Ibid., 32.
[145] Ibid., 32–33.
[146] Chen, "China Issues New Policies to Attract Foreign Investment," n.d.
[147] According to Chinese government statistics, FDI in agriculture includes farming, forestry, animal husbandry, and fisheries. FDI in food processing includes agricultural nonstaple products processing, grain milling, slaughtering and meat processing, and aqua-products processing. PRC, MOFCOM, "Reference Data on FDI Distribution in Major Sectors." Annual FDI Data.
[148] USDA, FAS, *China Changes Agriculture Investment Restrictions 2007*, November 23, 3007, 2.
[149] Lohmar et al., *China's Ongoing Agricultural Modernization*, April 2009, 45.
[150] Ibid., 22.
[151] FDI in genetically modified plants was banned before the 2007 catalog revisions. USDA, FAS, *China Changes Agriculture Investment Restrictions 2007*, November 23, 2007, 2.
[152] USDA, FAS, *China Changes Agriculture Investment Restrictions 2007*, November 23, 2007, 2–3.

End Notes for Competitive Factors Affecting China's Agricultural Sector

[1] The most efficient mix of productive inputs is known as allocative efficiency, an economic concept in which inputs such as capital, labor, or water can be allocated to produce the optimal level of output. Assuming several possible methods of production, allocative efficiency will largely depend on the cost and availability of necessary inputs. Pearce, *The MIT Dictionary of Modern Economics*, 1997, 13.
[2] In instances where a competitive factor (such as water or land) affects more than one category (i.e., delivered cost and reliability of supply), the factor will be discussed briefly in both sections of the chapter, while attempting to minimize any textual or analytical overlap.
[3] A detailed description of the Commission's analytical framework for competitive factors affecting agriculture is in appendix E.
[4] For a discussion of how exchange rates affect delivered cost, see Gale and Tuan, "China Currency Appreciation Could Boost U.S. Agricultural Exports," August 2007.
[5] Gifford, "Food Fears Persist in China 2 Years after Milk Scare," October 26, 2010; BBC, "Timeline: China Milk Scandal," January 25, 2010.
[6] FAO, "A Competitive Analysis and Strategy," 1992.
[7] For example, Russia imposed export controls on wheat exports, disrupting supplies to its major markets, especially Japan. *The New York Times,* "Russia, Crippled by Drought, Bans Grain Exports," August 5, 2010.
[8] For example, Mali produces and exports high-quality cotton. However, because Mali's production is variable and its internal transportation infrastructure is poorly developed, it is considered an unreliable supplier of cotton by international cotton buyers.
[9] China has a large agricultural labor force relative to the United States. In 2009, about 321 million people (40 percent of the total population) were working in the Chinese agricultural sector, compared to only about 1 million people (less than 1 percent of the total population) in the U.S. agricultural sector. Data for the U.S. agricultural sector include laborers in farming, forestry, and fisheries. CIA, *The World Factbook: China,* May 20, 2010; CIA, *The World Factbook: United States*, October 28, 2010.

[10] Industry officials, interviews by Commission staff, Beijing and Shandong province, China, September 6–16, 2010; Scheineson, "China's Internal Migrants," May 14, 2009.

[11] *China Daily,* "China's Urban, Rural Income Gap Widens," January 22, 2010.

[12] Ibid.

[13] Batson, "Farm Wages Trip Beijing's March Against Inflation," December 23, 2010.

[14] Industry official, interview by Commission staff, Beijing, China, September 6, 2010; USDA, NASS, "Farm Labor," August 19, 2010.

[15] Industry officials, interviews by Commission staff, Beijing and Shandong province, China, September 6–16, 2010.

[16] Lohmar et al., *China's Ongoing Agricultural Modernization*, April 2009, 11, 35.

[17] In China, the percentage of harvested agricultural cropland that is irrigated has been rising in recent years and is now estimated to be approximately 50 percent. *China Daily*, "Poor Irrigation Facilities Threaten Grain Production," March 12, 2009; industry official, interview by Commission staff, Beijing, China, September 9, 2010; Guang-xin and Wei, "Groundwater Crisis and Sustainable Agriculture in Northern China," 2010.

[18] Industry officials, interviews by Commission staff, Beijing and Shandong province, China, September 6–15, 2010.

[19] Powell, "Falling Up! Water Prices Must Meet the True Costs," May 2010, 35.

[20] Government official, interview by Commission staff, August 19, 2010.

[21] Industry officials, interviews by Commission staff, Beijing, China, September 7–9, 2010.

[22] Brown, "Aquifer Depletion," January 23, 2010.

[23] Industry officials, interviews by Commission staff, Beijing and Shandong province, China, September 6–16, 2010.

[24] The use of efficient irrigation technology such as drip irrigation is typically limited to certain high-value cash crops, such as tomatoes and apples, because of the equipment's high costs. Industry official, interview by Commission staff, Beijing, China, September 10, 2010.

[25] In China, only 40 percent of water used in irrigation actually reaches the crops because it leaks from pipes, evaporates or is otherwise lost on the way to the fields, compared to between 70 and 80 percent in most developed countries. Xie, "Addressing China's Water Scarcity," January 1, 2009, 61; Taige, "Drought without End?" March 3, 2009.

[26] A farmer's right to transfer a farmland lease is a recent policy change, and the implementation of this right varies significantly among localities.

[27] Industry official, interview by Commission staff, Beijing, China, September 6, 2010.

[28] Industry officials, interviews by Commission staff, Beijing and Shandong province, China, September 6–15, 2010.

[29] Lohmar et al., *China's Ongoing Agricultural Modernization,* April 2009, 9; industry official, interview by Commission staff, Beijing, China, September 10, 2010.

[30] Industry official, interview by Commission staff, Beijing, China, September 6, 2010; Wang et al., "Producing and Procuring Horticultural Crops with Chinese Characteristics," 2009.

[31] Industry officials, interviews by Commission staff, Beijing and Shandong province, China, September 6–16, 2010.

[32] *China Daily*, "Chinese Farmers Lease Their Farmland," June 16, 2010.

[33] Industry officials, interviews by Commission staff, Beijing and Shandong province, China, September 6–15, 2010.

[34] Industry official, interview by Commission staff, Beijing, China, September 7, 2010. A farmer cooperative is an organization in which the farmers own and operate their individual plots of land but pool resources and centralize administrative functions in order to consolidate product, streamline product distribution, and reduce transaction costs.

[35] Industry officials, interviews by Commission staff, Beijing and Shandong Province, China, September 6–16, 2010.

[36] Industry official, interview by Commission staff, Beijing, China, September 7, 2010.

[37] Industry official, interview by Commission staff, Shanghai, China, September 14, 2010.

[38] Industry officials, interviews by Commission staff, Beijing, China, September 6–10, 2010.

[39] Mullan, Grosjean, and Kontoleon, "Land Tenure Arrangements and Rural-Urban Migration in China," August 26, 2010.

[40] Ford, "China's Land Reform Aims to Revolutionize 750 Million Lives," October 27, 2008.

[41] U.S. Government official, interview by Commission staff, Washington, DC, August 19, 2010.

[42] Recent examples include farm plots of over 3 hectares in Gansu province which are being leased annually for RMB 30,000 ($4,600), with additional sums paid to farmers who continue to work the land for the new leaseholder. *China Daily*, "Chinese Farmers Lease Their Farmland," June 16, 2010.

[43] Industry officials, interviews by Commission staff, Beijing and Shandong province, China, September 6–15, 2010.

[44] USDA, NASS, "Land Values and Cash Rents Summary," August 2010, 18.

[45] In order to lease land either for growing or processing purposes, agricultural firms may need to negotiate the lease terms with local village leaders, even in cases where individual farmers own leases. In other cases, land use rights may be sold by villages in an auction format. Ford, "China's Land Reform Aims to Revolutionize 750 Million Lives," October 27, 2008.

[46] Zhang,"China's Slow-motion Land Reform," February/March 2010.

[47] Fertilizer, pesticide, herbicide, and fungicide costs account for about 14 percent of the cost per *mu* of growing potatoes in northern China.

[48] Schwartz, "Balancing Act in Global Fertilizer Use: Science Report," June 29, 2009.

[49] Zhang, "Improving Fertilizer Use Efficiency through Management Practices: China Experience," n.d.

[50] USDA, FAS, *China: Grain and Feed; Annual*, March 1, 2010, 16.

[51] The transportation subsidy is estimated to be approximately RMB 5 billion per year. Liao, "The Evolution of Fertilizer Subsidies in China," December 2008.

[52] USDA, FAS, *China: Fertilizer*, December 14, 2009.

[53] Southeast Farm Press, "China, India Waking Up to Fertilizer with Huge Subsidies," n.d.

[54] Kahrl et al., "Toward Sustainable Use of Nitrogen Fertilizers in China," November/December 2010, 6.

[55] USDA, FAS, *China: Grain and Feed; Annual, 2010*, March 1, 2010, 16.

[56] USDA, FAS, *China: Grain and Feed; Annual, 2008*, March 1, 2008. The payments per acre vary by product and region, but for cotton it is estimated to be $13 per acre. National Cotton Council, written submission to the Commission, September 15, 2010, 2.

[57] Industry official, interview by Commission staff, September 13, 2010; *Poultex News*, "Half of China's Antibiotics are Fed to Animals," December 1, 2010.

[58] The combined figure for the mechanization of plowing, sowing, and harvesting came to 43 percent of China's agricultural land in 2007, 3 percentage points higher than the previous year. Mechanization for sowing and harvesting reached 78 percent and 79 percent, respectively, of the area planted with wheat; paddy rice mechanization for sowing and harvesting reached 11 percent and 46 percent; and corn mechanization for harvesting came to 7 percent. PRC, MOA, "General Surveys—Agricultural Mechanization," June 25, 2009.

[59] Industry official, interviews by Commission staff, Shandong province, China, September 13–15, 2010.

[60] PRC, MOA, "General Surveys, Agricultural Mechanization," June 25, 2009.

[61] Industry official, interview by Commission staff, Shandong province, China, September 13, 2010.

[62] Industry official, interview by Commission staff, Beijing, China, September 7, 2010.

[63] Academic, interview by Commission staff, Beijing, China, September 9, 2010; Industry official, interview by Commission staff, Shandong province, China, September 13, 2010.

[64] Industry official, interview by Commission staff, Shandong province, China, September 13, 2010.

[65] *Wall Street Journal*, "China's Farmers Could Use Land As Collateral," July 29, 2010.

[66] OECD, "Rural Policy Reviews: China 2009," May 22, 2009, 43.

[67] Stewart and Stewart, written submission to the Commission, June 29, 2010, 4.

[68] Industry officials, interviews by Commission staff, Beijing and Shandong province, China, September 6–15, 2010.

[69] Industry official, interview by Commission staff, Beijing, China, September 6, 2010.

[70] Ibid., September 10, 2010.

[71] *People's Daily Online*, "China's Rural Roads to Cover 3.5 Million Km by 2011," October 28, 2010.

[72] Tolls on expressways in China vary between 7 and 16 cents per mile. *Tollroad News*, "Central Government Struggling to Control Tollroads in China," February 29, 2008.

[73] Terreri, "Go West ... with Caution," November 1, 2010.

[74] Industry official, interview by Commission staff, Beijing, China, September 10, 2010; Gonzalez, Guasch, and Serebrisky, "Latin America: High Logistics Costs and Poor Infrastructure for Merchandise Transportation," 2007.

[75] Xiang, "PE Firms Eye Rich Harvest in Agricultural Logistics," June 11, 2010.

[76] Industry official, interview by Commission staff, Beijing, China, September 10, 2010.

[77] Bolton and Liu, "Creating an Effective China 'Cold Supply Chain,'" 2006.

[78] Industry official, interview by Commission staff, Shanghai, China, September 14, 2010.
[79] Morrison and Labonte, "China's Currency Peg: A Summary of the Economic Issues," April 25, 2005, 1.
[80] *Reuters*, "China No Longer Pegging Yuan to Dollar." July 21, 2005 and Wolverson, "Confronting the China-U.S. Economic Imbalance," October 19, 2010.
[81] Wolverson, "Confronting the China-U.S. Economic Imbalance," October 19, 2010.
[82] U.S. Department of the Treasury, *Report to Congress on International Economic and Exchange Rate Policies,* February 2011, 12.
[83] Morrison and Labonte, "China's Currency Peg: A Summary of the Economic Issues," April 25, 2005, 3.
[84] IMF, Meetings of G-20 Deputies, Seoul, Korea, *Global Economic Prospects and Policy Challenges*, February 27, 2010, 9.
[85] The estimate is the result of purchasing power parity (PPP)-based modeling. Subramanian, "New PPP-Based Estimates of Renminbi Undervaluation and Policy Implications," April 2010, 1, 7.
[86] U.S. Department of Treasury, Office of International Affairs, "Report to Congress on International Economic and Exchange Rate Policies," July 8, 2010, 18.
[87] U.S. Department of the Treasury, *Report to Congress on International Economic and Exchange Rate Policies,* February 2011, 15.
[88] Ulics and Mead, "Current Price Topics: China's Exchange Rate Policy Reflected in U.S. Import Prices," August 2010.
[89] Not all U.S. agricultural commodities will be affected equally by changes in the exchange rate between the U.S. dollar and Chinese renminbi. Other competitive factors must also be taken into account. Gale and Taun, *China Currency Appreciation*, August 2007, 1.
[90] Equilibrium value is often considered to be "purchasing power parity." Gale and Tuan, *China Currency Appreciation,* August 2007, 4.
[91] Gale and Tuan, *China Currency Appreciation,* August 2007, 3.
[92] For more information on industry views, see appendix D.
[93] Industry officials, interviews by Commission staff, Shandong province, China, September 13–15, 2010.
[94] Ibid.
[95] Industry official, interview by Commission staff, Shandong province, China, September 15, 2010.
[96] Ibid., September 13, 2010.
[97] The VAT is assessed differently depending on the type of product; the VATs on raw agricultural products and processed products are 13 and 17 percent, respectively. See chapter 4 for more detailed information on the value-added tax. U.S. Wheat Associates, written submission to the Commission, September 16, 2010, 4.
[98] USDA, FAS, *China: Trade Policy Monitoring; VAT Protections,* March 19, 2007, 1.
[99] For more information, see the discussion on the elimination of certain agricultural taxes in chapter 4.
[100] Industry officials, interviews by Commission staff, Beijing and Shandong province, China, September 6–15, 2010.
[101] USDA, FAS, *China: Grain and Feed; Annual, 2010,* March 1, 2010, 16.
[102] USDA, FAS. *China: Grain and Feed; Annual, 2008*, March 1, 2008.
[103] USDA, FAS, *China: Grain and Feed; Annual, 2010*, March 1, 2010, 16. Industry officials, interviews by Commission staff, Beijing, Shandong Province, Shanghai, Sichuan province, and Hong Kong, China, September 6–21, 2010.
[104] Further discussion of Chinese consumption patterns is found in chapter 3.
[105] Chinese cold storage capacity is estimated to be approximately 7 million cubic meters, compared to 88.8 million cubic meters in the United States. Bolton and Liu, "Creating an Effective China 'Cold Supply Chain,'" 2006.
[106] Industry official, interview by Commission staff, Beijing, China, September 15, 2010; Bolton and Liu, "Creating an Effective China 'Cold Supply Chain,'" 2006.
[107] The cold chain system in China's major cities tends to be of higher quality and more widely available. Imported products also benefit from improvements in cold chain storage because they tend to be marketed in larger, more affluent Chinese cities. Liu, "More Cold Chain Gives More Choice in China," September 2009, 20.
[108] Industry official, interview by Commission staff, Beijing, China, September 10, 2010.
[109] Ibid., September 6, 2010.
[110] Bolton and Liu, "Creating an Effective China 'Cold Supply Chain,'" 2006, 5.
[111] Industry official, interviews by Commission staff, Beijing, China, September 10, 2010.
[112] Industry officials, interviews by Commission staff, Beijing, China, September 6–10, 2010.
[113] Ibid., September 6, 2010.
[114] Lohmar et al., *China's Ongoing Agricultural Modernization*, April 2009, 6.

[115] In China, there is a disproportionate use of fertilizers like urea compared to phosphorous and potash fertilizers. FAO, "Synthesis of Asia and the Pacific Region's Perspective on Nutrient Management and Soil Productivity," n.d.

[116] Lohmar et al., *China's Ongoing Agricultural Modernization*, April 2009, 6.

[117] Industry officials, interviews by Commission staff, Beijing and Shandong province, China, September 6–15, 2010.

[118] Industry official, interview by Commission staff, Beijing, China, September 6, 2010.

[119] See case study on fresh apples in chapter 6.

[120] Industry official, interview by Commission staff, Beijing, China, September 5, 2010.

[121] Industry official, interview by Commission staff, Shanghai, China, September 14, 2010.

[122] Charron, "Agriculture, Food and Beverages Sector Profile: Beijing, China," December 2008.

[123] Li, "Shandong Agricultural Product Exports," March 2009; industry official, interview by Commission staff, Beijing, China, September 6, 2010.

[124] Li, "Shandong Agricultural Product Exports," March 2009; industry official, interview by Commission staff, Beijing, China, September 6, 2010; Charron, "Agriculture, Food and Beverages Sector Profile: Beijing, China," December 2008.

[125] *People's Daily Online*, "China's Rural Roads to Cover 3.5 Million Km by 2011," October 28, 2010; Terreri, "Go West ... with Caution," November 1, 2010.

[126] Bolton and Liu, "Creating an Effective China 'Cold Supply Chain,'" 2006; Xiang, "PE Firms Eye Rich Harvest in Agricultural Logistics," June 11, 2010.

[127] Government official, interview by Commission staff, Washington, DC, August 19, 2010.

[128] Parsai, "India Turns to China for Grain Storage Expertise," June 18, 2010.

[129] Terreri, "Go West ... with Caution," November 1, 2010.

[130] Industry official, interview by Commission staff, Beijing, China, September 10, 2010.

[131] Ibid.

[132] The world average is just over 7,000 cubic meters (m^3) of water per capita, but Chinese resources are slightly over 2,000 m^3 of water per capita. Lohmar et al., *China's Ongoing Agricultural Modernization*, April 2009, 18.

[133] Southern China's average rainfall is just over 2,000 mm/year, compared to the north, which averages only 200–400 mm/year. Xie et al., "Addressing China's Water Scarcity," January 1, 2009, 10.

[134] Lohmar et al., *China's Ongoing Agricultural Modernization*, April 2009, 17.

[135] Roberts, "China Faces a Water Crisis,"April 15, 2009.

[136] *Newsweek*, "Where Rivers Run Dry," April 16, 2007.

[137] Graham-Harrison, "China's Water Pollution Level Higher Than Estimated in 2007," February 10, 2010.

[138] Lohmar et al., *China's Ongoing Agricultural Modernization,* April 2009, 19.

[139] EC, DGA, "China: Out of the Dragon's Den?" May 2008, 4.

[140] Roberts and Anders, *Developments in Chinese Agriculture*, July 2005, 7.

End Notes for Competitive Factor Case Studies

[1] In Shaanxi and Shandong, fertilizer prices increased by 10 and 30 percent, respectively, in 2009, while pesticide prices rose by between 5 and 10 percent. Reportedly, in the Shaanxi province, total production costs—which include fertilizer, pesticide, fruit bagging, and labor—were approximately $2,315 per hectare in 2009, a sharp increase from 2008. On the other hand, farmgate prices for apples in Shaanxi and Shandong increased by between 20 and 30 percent in 2009. USDA, FAS, *China: Fresh Deciduous Fruit; Annual*, November 17, 2009, 3; industry officials, interviews by Commission staff, Beijing, China, September 6, 2010; USDA, FAS, *China: Fresh Deciduous Fruit; Annual, 2010*, November 5, 2010, 4.

[2] USDA, FAS, *China: Fresh Deciduous Fruit; Annual*, November 17, 2009, 3.

[3] In comparison, the leading variety produced in the United States is Red Delicious, which accounts for approximately 30 percent of total production. Industry officials, interviews by Commission staff, Beijing, China, September 6, 2010; USDA, FAS, *China: Fresh Deciduous Fruit; Annual,* November 17, 2009, 3.

[4] Rootstocks are trees bred specifically for the performance of their roots. Different rootstocks have been developed for specific purposes, such as the size of the tree, disease resistance, and certain temperature conditions.

[5] Industry official, interview by Commission staff, Beijing, China, September 6, 2010.

[6] Poland is also a large low-cost supplier of apples, but its exports are destined primarily for two markets: Russia and Ukraine. GTIS, Global Trade Atlas database.

[7] In comparison, in 2009 the United States exported approximately 17 percent of total production. USDA, FAS, PSD Online (accessed November 2, 2010).

[8] Industry official, interview by Commission staff, Beijing, China, September 6, 2010.

[9] Industry official, interview by Commission staff, Beijing, China, September 6, 2010; Honglin et al., "Producing and Procuring Horticultural Crops with Chinese Characteristics," August 7, 2008, 3.

[10] Industry official, interview by Commission staff, Beijing, China, September 6, 2010.

[11] Ibid.

[12] Estimates vary, but labor accounts for anywhere between 35 percent of costs in China to 61 percent of variable costs in Washington state, "Editor Studies the Chinese Apple Industry with IFTA," n.d.; Gallardo and Hinman, "2009 Cost Estimates of Establishing and Producing Gala Apples in Washington," n.d., 3.

[13] Industry official, interview by Commission staff, Beijing, China, September 6, 2010. In China apple workers are typically paid by the day, especially for jobs such as applying pesticides or bagging trees, but some farms may pay workers based on the quantity harvested.

[14] USDA, FAS, *China: Fresh Deciduous Fruit; Annual*, November 17, 2009, 2.

[15] USDA, NASS, "Farm Labor," August 19, 2010, 7.

[16] Zeitner, "China's Apple and Pear Industry: Challenges and Opportunities," January 2006, 10.

[17] Industry official, interview by Commission staff, Beijing, China, September 6, 2010; Honglin et al., "Producing and Procuring Horticultural Crops with Chinese Characteristics," August 7, 2008, 6–7.

[18] Hughes and Ning, "Farmers Slowly Cultivate a New Image," May 31, 2010.

[19] See chapter 5 for more detailed information on land costs. Industry officials, interviews by Commission staff, Beijing, China, September 6, 2010.

[20] Gao and Thornsbury, "Increasing Food-Safety Protection: Fresh Apple Markets in China," March 2008, 3.

[21] Industry official, interview by Commission staff, Beijing, China, September 8, 2010.

[22] Industry officials, interviews by Commission staff, Beijing, China, September 6–10, 2010.

[23] Peng et al., "Fertigation Management in Young Apple Trees in Shandong, China," December 18, 2008.

[24] Industry officials, interviews by Commission staff, Beijing, China, September 10, 2010.

[25] Ibid., September 6, 2010.

[26] USDA, FAS, *China: Fresh Deciduous Fruit; Annual*, November 17, 2009, 3; industry officials, interviews by Commission staff, Beijing, China, September 6, 2010.

[27] USDA, FAS, *China:Fresh Deciduous Fruit; Annual, 2006*, September 22, 2006, 3.

[28] The Chinese Ministry of Agriculture provides cash subsides for the purchase of the bags, which are normally made of paper and cost around $0.01, in export-oriented provinces. Gao, "Regulating Trade with a Systems Approach," 2008, 81; USDA, FAS, *China: Fresh Deciduous Fruit; Annual, 2006*, September 22, 2006, 3.

[29] Despite the improved appearance of high-quality Chinese apples, most have not developed the sweetness and flavor needed to fully compete with imports. Industry official, interview by Commission staff, Beijing, China, September 6, 2010.

[30] Industry officials, interviews by Commission staff, Beijing, China, September 6, 2010.

[31] The United States is the third-largest producer of pork and accounted for approximately 10 percent of global production in 2009.

[32] USDA, FAS, *China: Livestock and Products; Semi-Annual*, March 2, 2010, 4–5. In comparison, there were an estimated 64.9 million swine in the United States, with 5.9 million kept for breeding. USDA, NASS, *Quarterly Hogs and Pigs*, June 25, 2010, 1.

[33] The United States is the world's largest exporter of pork. Most U.S. pork is consumed domestically, but exports account for a generally increasing share of U.S. pork production—approximately 18 percent in 2009.

[34] Somwaru, Xiaohui, and Tuan, *China's Hog Production Structure and Efficiency*, 2003, 6.

[35] USDA, FAS, *China: Livestock and Products; Annual*, September 14, 2010, 5. However, some industry sources estimate that the share of China's pork production from farms with less than 50 hogs was still at least 30 percent in 2009; industry official, interview by Commission staff, May 6, 2010.

[36] Gale and Huang, *Demand for Food Quantity and Quality in China*, 2007, 21–22; Latner, *Who is Feeding China?* 2010, 10.

[37] The cold chain includes both cold storage and transportation.

[38] USDA, FAS, *China: Livestock and Products; Annual*, September 14, 2009, 6.

[39] Ibid., September 25, 2007, 1.

[40] USDA, FAS, *China: Livestock and Products; Semi-Annual*, March 2, 2010, 4–5.

[41] Ibid., 5.
[42] USDA, FAS, *China: Chinese Government Purchases Pork to Support Local Prices*, April 12, 2010.
[43] Wang and Xiao, "Development of the Hog Industry and Its Integration in China," 2007, 14.
[44] A feeder pig is a young hog, typically 40-50 pounds in weight. Feeder-to-finish production includes raising swine from the feeder stage through finishing for slaughter.
[45] USDA, ERS, *U.S. hog production costs and returns per hundredweight gain*, 2009.
[46] Farrow-to-finish swine production encompasses the entire life cycle of swine raised for slaughter, from birth (farrowing) through finishing for slaughter.
[47] For producers that purchase feeder pigs, the cost of feeder pigs is also a major component of cost.
[48] Industry official, interview by Commission staff, Shandong province, China, September 13, 2010.
[49] Imported soybeans are crushed to produce soybean oil for human consumption and soybean meal for animal feed. Soybeans yield approximately 20 percent oil and 80 percent meal.
[50] China's imports of corn have also increased. China imported over 1.5 million mt of corn in 2010 compared to 65,000 mt in 2006. GTIS, Global Trade Atlas database (February 14, 2011).
[51] USDA, FAS, *China: Livestock and Products; Semi-Annual*, March 2, 2010, 4.
[52] Industry officials, interviewed by Commission staff, Shandong province, China, September 16 and 18, 2010.
[53] Ortega, Wang, and Wu, "Food Safety and Demand," 2009, 53.
[54] Processed foods can be defined as agricultural products that undergo some form of further manufacturing that alters their original state. Processed products are typically found in chapters 6, 7, and 17– 22 of the Harmonized Tariff Schedule of the United States.
[55] New Zealand-China Trade Association, "Processed Food for Thought: China's Food Processing Market," n.d.
[56] Data are from China's National Bureau of Statistics reported by CEIC Database. Estimates vary significantly, depending on the source and what type of food processors are included. Some reports show that if all facilities are counted individually, including those with 10 or less employees, there may be as many as 400,000. Baker, "Food and Agricultural Imports from China," September 26, 2008, 14.
[57] Industry officials, interviews by Commission staff, Shandong province, China, September 13–15, 2010; industry officials, interviews by Commission staff, Chengdu, China, September 17, 2010.
[58] See chapter 4 for more detailed information on Chinese farmer cooperatives.
[59] Just 30 percent of food in China is processed, compared to approximately 80 percent in Western countries. USDOC, U.S. Commercial Service, "China's Emerging Markets: Opportunities in the Food Processing Industry," n.d., 1.
[60] USDA, FAS, *China: Potato and Potato Products; Annual, 2010*, August 16, 2010, 9; *Meat Trade News Daily*, "China: McDonald's Leading the Way," August 10, 2010.
[61] Promar International, "The Chinese Potato Industry in Transition," July 2007, 112.
[62] The United States uses a much higher level of mechanization in its processing plants. Industry official, interview by Commission staff, Shandong province, China, September 13, 2010.
[63] Promar International, "The Chinese Potato Industry in Transition," July 2007, 122; industry officials, interviews by Commission staff, Beijing, China, September 6, 2010.
[64] Industry officials, interview by Commission staff, Shanghai, China, September 13, 2010.
[65] Industry officials, interview by Commission staff, Beijing, China, September 7, 2010.
[66] Promar International, "The Chinese Potato Industry in Transition," July 2007, 112.
[67] Industry officials, interview by Commission staff, Beijing, China, September 7, 2010.
[68] Ibid.
[69] Ibid.
[70] Ibid.
[71] In China, the most efficient FFF processors use just under 2 pounds of fresh potatoes to produce 1 pound of FFF, but this ratio varies widely in Chinese processing facilities, depending on quality of the raw inputs, and can occasionally reach 2.5 pounds of fresh potatoes. In comparison, the average in the United States is 1.7 pounds of fresh per pound of FFF output. This disparity makes processing costlier in China than in the United States. Industry officials, interview by Commission staff, Beijing, China, September 7, 2010; industry officials, telephone interview by Commission staff, November 10, 2010.
[72] Industry official, interview by Commission staff, Shanghai, China, September 13, 2010; industry official, interview by Commission staff, Shandong province, China, September 13, 2010.
[73] Industry official, interview by Commission staff, Shandong province, China, September 13, 2010; USDA, FAS, *China: Poultry and Products; Annual, 2010*, September 30, 2010, 2.
[74] Industry officials, interview by Commission staff, Shandong province, China, September 14, 2010.

[75] Ibid., September 15, 2010.

[76] Industry officials, interviews by Commission staff, Beijing and Shandong province, China, September 6–16, 2010.

[77] In comparison, in the United States certain varieties of horticultural products, such as cling peaches and garlic, are grown specifically for processing and are therefore not suitable for the fresh market.

[78] For example, processors may test soil, give financial support, identify suitable land, and advise on good agricultural practices. Industry officials, interview by Commission staff, Shandong province, China, September 15, 2010.

[79] Industry officials, interview by Commission staff, Beijing and Shandong province, China, September 6–16, 2010.

[80] Liu, "Researches on the Effect of Foreign Direct Investment," n.d., 3.

[81] The United States uses far more mechanization in its processing plants.

[82] Industry officials, interview by Commission staff, Beijing, China, September 7, 2010.

[83] Industry official, telephone interview by Commission staff, September 1, 2010; industry official, interview by Commission staff, Beijing, China, September 6, 2010.

[84] Industry official, telephone interview by Commission staff, September 1, 2010.

[85] Industry official, telephone interview by Commission staff, September 1, 2010; industry official, interview by Commission staff, Beijing, China, September 6, 2010.

[86] Industry officials, interviews by Commission staff, Beijing and Shandong province, China, September 6–16, 2010.

[87] Industry official, interview by Commission staff, Shandong province, China, September 13, 2010.

[88] Domestic consumers are still wary of China's food supply. Gifford, "Food Fears Persist in China 2 Years after Milk Scare," October 26, 2010; BBC, "Timeline: China Milk Scandal," January 25, 2010.

[89] Industry official, telephone interview by Commission staff, November 15, 2010; Baker, "Food and Agricultural Imports from China," September 26, 2008, 14.

[90] Industry officials, interviews by Commission staff, Beijing and Shandong province, China, September 6–16, 2010.

[91] Industry official, interview by Commission staff, Shandong province, China, September 15, 2010.

[92] Industry official, interview by Commission staff, Shandong province, China, September 15, 2010; industry official, interview by Commission staff, Beijing, China, September 7, 2010.

[93] Industry official, interview by Commission staff, Shandong province, China, September 15, 2010.

[94] Industry official, interview by Commission staff, Shandong province, China, September 13, 2010; Lohmar and Gale, "Who Will Feed China?" June 2008, 15; *Poutlex News,* "Half of China's Antibiotics Are Fed to Animals," December 1, 2010.

[95] Industry official, interview by Commission staff, Shanghai, China, September 13, 2010; industry official, interview by Commission staff, Shandong province, China, September 13, 2010; Calvin et al., "Food Safety Improvements Underway in China," November 2006. Medical costs are approximately RMB 2 per bird in China, compared with only about RMB 0.5 in the United States.

[96] The National Grain Plan issued in November 2009 includes plans to improve yields measures and mechanized technology for farm use. The focus of the No. 1 Documents is increasing farmers' incomes. For further description of such polices see chapter 4.

[97] China is the world's leading single-country producer of wheat.

[98] USDA, ERS, *China Provincial Data,* July 1, 2009.

[99] USDA, FAS, *China: Grain and Feed; Annual,* March 1, 2010, 3.

[100] World Grain, "Changes in Chinese Flour Milling," September 1, 2010.

[101] USDA, PSD Online (accessed January 14, 2011).

[102] NBSC, *China Statistical Yearbook 2009,* September 2009, table 12-12.

[103] Zhang, "Chinese Wheat: Current Situation and Prospects," 2007, 7.

[104] World Grain, "Changes in Chinese Flour Milling," September 1, 2010.

[105] PRC, NDRC, SAG, "Message from the Administrator of SAG," 2002.

[106] Sinograin, "Company Overview," accessed July 28, 2010.

[107] COFCO, "Wheat Division," accessed November 10, 2010.

[108] COFCO, "Agricultural Primary Products Storage and Logistics," accessed November 10, 2010.

[109] USDA, FAS, *China: National Plan for Expansion of Grain Production Capacity,* February 18, 2010, 7.

[110] PRC, NDRC, *National Agricultural Product Cost and Profit Materials Compilation,* 2003–08. Costs are on a per *mu* basis.

[111] Land costs in some parts of China are similar to those in parts of the United States. See the discussion on land costs in chapter 5.
[112] USDA, FAS, *China: National Plan for Expansion of Grain Production Capacity*, February 18, 2010, 17, 22, 33.
[113] PRC, NDRC, *National Agricultural Product Cost and Profit Materials Compilation*, 2003–08.
[114] USDA, FAS, *China: National Plan for Expansion of Grain Production Capacity*, February 18, 2010, 13.
[115] PRC, NDRC, *National Agricultural Product Cost and Profit Materials Compilation*, 2003–08. Data represent national averages.
[116] PRC, NDRC, *National Agricultural Product Cost and Profit Materials Compilation*, 2003–08. Measured in terms of days per *mu*.
[117] Ibid.
[118] In 2007, nearly 80 percent of the total wheat area was sown and harvested by machine. PRC, MOA, "General Surveys: Agricultural Mechanization," June 25, 2009. A greater proportion of wheat is mechanically harvest compared to other crops because of the nature of the plant and the large contiguous planted areas (figure 6.2).
[119] USDA, FAS, *China: National Plan for Expansion of Grain Production Capacity*, February 18, 2010, 9.
[120] USDA, FAS, *China: Grain and Feed; Annual*, March 1, 2010, 4.
[121] USDA, FAS, *China: National Plan for Expansion of Grain Production Capacity*, February 18, 2010, 39.
[122] USDA, FAS, *China: Grain and Feed; Annual*, March 1, 2010, 4.
[123] USDA, FAS, *China: National Plan for Expansion of Grain Production Capacity*, February 18, 2010, 5, 15, 39.
[124] Zhang, "Chinese Wheat: Current Situation and Prospects," 2007, 73.
[125] Chinese imports of wheat are relatively insignificant, as the TRQ volume accounts for less than 10 percent of domestic consumption.
[126] USDA, FAS, *China: National Plan for Expansion of Grain Production Capacity*, February 18, 2010, 1.
[127] USDA, FAS, *China: 2010 Agricultural Policy Directive*, February 18, 2010, 4.
[128] USDA, FAS, *China: Grain and Feed; Annual*, March 1, 2010, 16. The data are on a per hectare basis.
[129] USDA, FAS, *China: Grain and Feed; Annual*, March 1, 2010, 18.
[130] USDA, FAS, *China: National Plan for Expansion of Grain Production Capacity*, February 18, 2010, 28.

End Notes for Chinese Agricultural Tariff Measures

[1] Value-added taxes (VATs) may also affect U.S. exports of agricultural products. The impact of VATs on U.S. agricultural exports are described in chapter 9.
[2] WTO, "Members and Observers," (accessed June 8, 2010).
[3] China's free trade agreements and their impact on U.S. agricultural exports are described in greater detail in chapter 8.
[4] The simple average tariff is the average of tariffs not weighted by trade volume.
[5] Rosen, Rozelle, and Huang, "China and the WTO Agriculture Agenda," 2004, 8.
[6] WTO, "WTO Successfully Concludes Negotiations on China's Entry," September 17, 2001.
[7] WTO, "Schedule CLII: People's Republic of China," October 1, 2001.
[8] The trade-weighted average tariff is the average of all tariffs weighted by trade volume.
[9] WTO, "China Tariff Profile," 2009. China's trade-weighted average for all products (agricultural and non-agricultural) declined from 40.6 percent ad valorem in 1992 to 9.1 percent ad valorem in 2001, or by 78 percent. In 2002, this tariff fell by another 30 percent, to 6.4 percent ad valorem, after China joined the WTO. Rumbaugh and Blancher, "China: International Trade and WTO Accession," 2004, 18. In 2007, China's trade-weighted average tariff for all products was 4.5 percent.
[10] PRC, Customs Tariff Commission of the State Council, *Tariff Execution Plan 2010*, December 8, 2009.
[11] PRC, SAG, "Message from the Administrator of SAG," 2002.
[12] PRC, MOFCOM, "Decree of the State Council of the People's Republic of China (No.539)," November 30, 2008.
[13] WTO, "China and the WTO." Applied and bound tariffs are similar for these items.
[14] USDA, FAS, *China: Cotton and Products; Annual, 2007*, May 1, 2007, 11.
[15] Blue Diamond Growers, written submission to the Commission, June 3, 2010, 4.
[16] These items were identified by the U.S. government. The White House Office of Public Liason, "Summary of U.S.-China Bilateral WTO Agreement," November 1999.
[17] U.S. Wheat Associates, written submission to the Commission, September 16, 2010, 1–2.

[18] National Cotton Council, written submission to the Commission, September 15, 2010, 6.
[19] WTO, "Schedule CLII: People's Republic of China," October 1, 2001. The WTO prohibits absolute quotas.
[20] PRC, MOFCOM, *Interim Measures for the Administration of Import Tariff Quotas of Agricultural Products*, September 27, 2003.
[21] There is little annual change in TRQ quantities and conditions, except for cotton.
[22] WTO, *Responses to Questionnaire on Import Licensing Procedures*, October 20, 2009.
[23] PRC, MOFCOM, "Announcement No. 93, 2005," December 7, 2005.
[24] WTO, *Trade Policy Review China*, April 26, 2010, 34.
[25] Calculated by Commision staff based on data from USDA, FAS, PSD Online.
[26] China usually increases its WTO minimum cotton TRQ annually by announcing a supplemental quantity that can enter at a reduced duty rate, or "sliding duty."
[27] WTO, *Responses to Questionnaire on Import Licensing Procedure,* September 30, 2005, 2.
[28] Ibid.
[29] WTO, *Responses to Questionnaire on Import Licensing Procedure,* September 30, 2004, 4.
[30] Unused allocations must be returned by September 15 of the quota year. PRC, MOFCOM, *Interim Measures for the Administration of Import Tariff Quotas*, September 27, 2003.
[31] PRC, MOFCOM, "Notice No. 64 of 2010," September 30, 2010.
[32] PRC, MOFCOM, "Notice No. 65 of 2010," September 30, 2010.
[33] NDRC, "Notice No. 27 of 2010," September 25, 2010.
[34] WTO, *Trade Policy Review China*, April 26, 2010, 32.
[35] Zhou and Kang, "Uncover the Causes of China's Low Utilisation of Grain TRQ," 2009, 41.
[36] WTO, "Schedule CLII: People's Republic of China," October 1, 2001.
[37] U.S. Wheat Associates, written submission to the Commission, September 16, 2010, 1–2; National Cotton Council, written submission to the USITC, September 15, 2010, 6; USTR, *2009 Report to Congress on China's WTO Compliance*, December 2009, 72.
[38] WTO, *Trade Policy Review China*, April 26, 2010, 12.
[39] PRC, MOFCOM, "Notice No. 52 of 2010," August 30, 2010; and PRC, MOFCOM, "Notice No. 51 of 2010," September 26, 2010.
[40] USTR, *2010 Report to Congress on China's WTO Compliance,* December 2010, 34.
[41] USDA, FAS, *China: Poultry and Products; Annual, 2010*, September 30, 2010, 2.
[42] The Commission's simulation for tariff and TRQ removal were performed with a framework that links a partial equilibrium trade model to an economy-wide trade model, the Global Trade Analysis Project model. The simulation framework is described in appendix F.
[43] The ad valorem equivalent tariffs for products subject to TRQs capture the restrictiveness of the quota volume level and may be equal to the in-quota tariff, the over-quota tariff, or some level in between.
[44] Commission model results for the 131 commodity sectors under consideration in this report are presented in appendix G.
[45] As table 7.5 and appendix table G.1 show, China's tariff rates vary considerably from product to product. The trade-weighted average import tariff levied on U.S. products is 6 percent because a large share of U.S. exports (e.g., soybeans and cotton) are levied at relatively low tariffs (approximately 1 and 6 percent for soybeans and cotton, respectively).
[46] The Commission's estimate seems reasonable because a simple calculation involving tariff removal, an average trade elasticity, and the share of imports of Chinese consumption for food and agricultural products results in an increase in U.S. agricultural exports to China of approximately 19 percent.
[47] A range of simulated effects was obtained by varying the magnitude of trade elasticities to account for the degree of statistical uncertainty in the econometric estimates of the elasticities. Low and high values for the elasticities of demand (ESM) were computed as ESM + standard deviation. Hertel et al. (2007) provide estimated values for ESM and standard deviations. Hertel et al., "How Confident Can We Be of CGE-Based Assessments?" 2007, 611–635.
[48] The expansion in wheat trade is larger than that for other products because the AVE for wheat is relatively high and the import demand elasticities for wheat are large due to the relatively homogeneous nature of the commodity.
[49] Further discussion of the mechanics of the general equilibrium model used in this simulation can be found in appendix F.

End Notes for China's Trade Agreements

[1] In this report, the term "free trade agreement" applies to all of China's trade agreements.

[2] Shulong, "China's Approach to the Free-Trade Area," October 4, 2006, 3–4.

[3] SARs are highly autonomous regions in China which fall under the principal of "one country, two systems." SARs may have different social, economic, political, and legal systems from mainland China, while their foreign and defense affairs are controlled by China's central government. Under the current laws, Hong Kong and Macau will retain their SAR status until 2047 and 2049, respectively.

[4] Shulong, "China's Approach to the Free-Trade Area," October 4, 2006, 15. It is notable that rice, produced extensively in China and ASEAN countries, was excluded from the China-ASEAN FTA.

[5] China's agricultural trade, including that with FTA partners, is discussed in detail in chapter 2.

[6] All of China's trade arrangements are discussed in this chapter.

[7] The China-Costa Rica FTA has been signed but not ratified.

[8] The China-ASEAN FTA does allow relatively sensitive products to be excluded by certain reduction formulas.

[9] Taiwan, MEA, "Ministry of Economic Affairs Announced ECFA Online English Text," September 21, 2010.

[10] HKSAR, TID, *MHKCEPA Annexes*, September 29, 2003, annex 1, 1.

[11] Ibid., annex 1, table 1, 1.

[12] Ibid., annex 1, 2.

[13] HKSAR, TID, *Goods Entitled to CEPA Zero Tariff Preference: Mainland 2010 Tariff Codes; Product Description and Rules of Origin*, May 31, 2010.

[14] HKSAR, TID, *MHKCEPA Annexes*, September 29, 2003, annex 2, 1.

[15] Ibid., annex 2, 2.

[16] Ibid., annex 2, 3.

[17] HKSAR, TID, *MHKCEPA Annexes*, September 29, 2003, annex 1, table 1.

[18] Ibid., annex 6, 4.

[19] Ibid., annex 6, 1.

[20] Ibid., September 29, 2010, annex 6, 2.

[21] Macau is a SAR of China with a similar status to that of Hong Kong.

[22] DSE, "CEPA (October 17, 2003)," 2007.

[23] ASEAN, *Agreement on Trade in Goods*, November 29, 2004.

24 See, for example, Yue, *ASEAN–China Free Trade Area*, April 2004, 14; Wang, *The Logic of China-ASEAN Free Trade Agreement*, August 2007, 18; UACT, *ASEAN-China Free Trade Agreement: A Primer*, 8.

[25] ASEAN, "ASEAN-China Free Trade Area."

[26] Brunei, Indonesia, Malaysia, the Philippines, Singapore, and Thailand.

[27] WTO, *Communication from the Parties to the Agreement,* October 16, 2007, 3.

[28] ASEAN, *Framework Agreement,* article 6, November 4, 2002.

[29] ASEAN, *Protocol to Amend the Framework Agreement,* October 6, 2003, article 6, 6. The EHPspecific product list is available at http://www.aseansec.org/22199.pdf, and the EHP product exclusion list is available at http://www.aseansec.org/22198.pdf.

[30] The Philippines entered the EHP after the other FTA members and submitted its EHP concessions at a later date. RPTC, *Executive Order No. 485*, December 29, 2005. The list of products for which the Philippines granted concessions to all ACFTA members is included in annex A, and additional products for which the Philippines granted concessions to China are listed in annex B.

[31] ASEAN, *Agreement on Trade in Goods*, November 29, 2004, annex 1. Each country determined the products it placed in the normal track.

[32] ASEAN, *Agreement on Trade in Goods*, November 29, 2004, annex 1, appendix 1.

[33] RPTC, *Executive Order No. 487,* January 12, 2006, annex A.

[34] It is possible for a product to maintain its pre-ACFTA tariff. For example, under the highly sensitive list criteria, a certain number of items may remain at a tariff rate of 50 percent ad valorem. China designated raw and refined sugar, which had an existing most-favored-nation (MFN) rate of 50 percent ad valorem, as highly sensitive products, effectively resulting in no tariff reductions. Also, products may be excluded under general and security exceptions set forth in article 12 and article 13 of the Agreement on Trade in Goods.

[35] ASEAN, *Agreement on Trade in Goods*, November 29, 2004.

[36] Ibid., annex 2, appendix 1.

[37] Ibid., appendix 2.

[38] Ibid., annex 2.

[39] WTO, *Communication from the Parties to the Agreement*, October 16, 2007, 11–12.
[40] ASEAN, *Agreement on Trade in Goods*, November 29, 2004, Annex 3.
[41] The criteria for wholly obtained or produced is similar those in the MHKCEPA.
[42] ASEAN-China Free Trade Area Business Portal, "Rules of Origin (ROO)," December 23, 2009.
[43] ASEAN, *Agreement on Trade in Goods*, November 29, 2004, Article 8.2.
[44] Ibid., Article 12(b).
[45] ASEAN, *Memorandum of Understanding*, November 20, 2007.
[46] ASEAN, "ASEAN Joint Statement," October 25, 2010.
[47] ASEAN, *Agreement on Investment*, August 15, 2009.
[48] The Commission's simulation of the implementation of bilateral tariff and TRQ provisions in China's FTAs was performed with a framework that links a partial equilibrium trade model to an economy-wide general equilibrium trade model, the Global Trade Analysis Project model. The simulation framework is described in appendix F.
[49] Commission model results for the 131 commodity sectors under consideration in this report are presented in appendix G.
[50] A range of simulated effects was obtained by varying the magnitude of trade elasticities to account for the degree of statistical uncertainty in the econometric estimates of the elasticities. Low and high values for the elasticities of demand (ESM) were computed as ESM + standard deviation. Hertel et al. (2007) provide estimated values for ESM and standard deviations. Hertel et al., "How Confident Can We Be of CGE-Based Assessments?" 2007, 611–635.
[51] The former group of products includes soybeans, cotton, poultry, and bovine hides, skins, and miscellaneous products. The latter group of products includes palm oil, other vegetables, and other fresh and dried fruits.
[52] Further discussion of the mechanics of the general equilibrium model used in this simulation can be found in appendix F.

End Notes for Chinese Agricultural Nontariff Measures

[1] WTO Secretariat, "Data Day at the WTO," May 18–19, 2009, 25.
[2] Ibid.
[3] The VAT on imports is assessed on the import's cost plus duties paid.
[4] WTO, SPS Agreement, art. 3, para. 3.
[5] WTO, SPS Agreement, para. 4.
[6] The organization changed its name to the World Organization for Animal Health in 2003, but is still widely known by its former French acronym, OIE.
[7] WTO, SPS Agreement, art. 5, para. 1.
[8] Ibid., para. 2.
[9] For instance, imports of strawberries from the United States were allowed into China in advance of the Beijing Olympic Games, apparently because of the increased demand, but since the Games ended exporters have been denied clearance for subsequent shipments. USTR, *2010 Report on Sanitary and Phytosanitary Measures*, 2010, 36–37.
[10] Industry official, interview by Commission staff, Hong Kong, September 22, 2010.
[11] International Plant Protection Convention, art. VI.1b, X.4, and II.1.
[12] China has allowed imports of two varieties of apples (Red Delicious and Golden Delicious) from Idaho, Oregon, and Washington since 1995, but has not permitted imports of additional varieties or apples from other states. USTR, *2010 Report on Sanitary and Phytosanitary Measures*, 2010, 36.
[13] USTR, *2010 Report on Sanitary and Phytosanitary Measures*, 2010, 36.
[14] Ibid.
[15] National Potato Council, "U.S. Fresh Potato Market Access to China," June 22, 2010.
[16] The U.S. first officially sought access to the Chinese market for potatoes from the Pacific Northwest in 2000. USTR, *2010 Report on Sanitary and Phytosanitary Measures*, 2010, 36.
[17] Industry official, interview by Commission staff, Hong Kong, September 22, 2010.
[18] USTR, "Agreement on U.S.-China Agricultural Cooperation," 1999, 5–6.
[19] U.S. Wheat Associates, written submission to the Commission, September 16, 2010. In one case involving a U.S. shipment of both spring wheat and soft white wheat to a northern port, spring wheat (which does not have a

TCK risk) was allowed to discharge, but the soft white wheat (which has a TCK risk) was redirected to Guangdong to be treated. Since then, U.S. exporters have been unwilling to risk shipping soft white wheat to ports in northern China. Industry official, interview by Commission staff, Hong Kong, September 22, 2010.

[20] Industry official, interview by Commission staff, Hong Kong, September 22, 2010.

[21] OIE, *Terrestrial Animal Health Code*, 2010, art. 5.1.2.

[22] USTR, *National Trade Estimate Report*, 2009, 95.

[23] USTR, *National Trade Estimate Report*, 2009, 95; USDA, FSIS, *Export Requirements for People's Republic of China*, December 23, 2010; USTR, *2010 Report on Sanitary and Phytosanitary Measures*, 2010, 35.

[24] OIE, *Terrestrial Animal Health Code*, 2010, art. 10.4.20. A compartment is a subpopulation that can be defined by management practices related to biosecurity.

[25] BSE is a fatal neurological disease affecting adult cattle. It is believed to be caused by a modified form of a prion, a type of protein that is found in the central nervous system of infected cattle and that is not destroyed by measures commonly used to control pathogens. USDA, FSIS, *Bovine Spongiform Encephalopathy*, March 2005. The infective agent can infect humans and is associated with variant Creutzfeldt-Jacob Disease (vCJD), a fatal neurological disease in humans.

[26] BSE is rarely observed in cattle less than 30 months of age. In a series of experiments designed to determine the infectivity of different tissues from cattle that had been fed the infected material, the infective agent was first observed outside the digestive system of infected cattle at 32 months after inoculation. Wells, et al., "Preliminary observations," 1998, 1. Consequently, many regulations related to BSE distinguish between cattle over or under 30 months.

[27] Specified risk materials (SRMs) are tissues in which the infective agent has been found. OIE guidelines allow trade in beef from cattle of any age as long as SRMs are properly removed.

[28] Tallow is a rendered form of fat.

[29] OIE, *Terrestrial Animal Health Code,* 2010, article 11.5.1.

[30] USTR, *2010 Report on Sanitary and Phytosanitary Measures*, 2010, 34–35.

[31] Pet Food Institute, "Dossier on Chinese Barriers to U.S. Pet Food Products," December 2008, 3; USTR, *National Trade Estimate Report*, 2009, 83; USTR, *National Trade Estimate Report*, March 2010, 64.

[32] Similarly, China's Ministry of Agriculture (MOA) also reportedly requires that U.S. seed companies seeking safety certificates for biotechnology products provide confidential information beyond that needed for a safety assessment. USTR, *2009 National Trade Estimate Report on Foreign Trade Barriers*, 2009, 96.

[33] Pet Food Institute, "Dossier on Chinese Barriers to U.S. Pet Food Products," December 2008, 4–5.

[34] USTR, *National Trade Estimate Report*, 2009, 95.

[35] USTR, *2010 Report on Sanitary and Phytosanitary Measures*, 2010, 33. This strain of influenza, which can infect both humans and swine, has been referred to as "swine flu."

[36] OIE, "Joint WTO/OIE/WHO/FAO Statement on A/H1N1 Influenza," May 2, 2009.

[37] USDA, FAS, *China: Livestock; Annual*, June 24, 1999, 19; industry official, e-mail to Commission staff, May 27, 2010. Scrapie is relatively uncommon in the United States. Between 1947, when the disease was first reported in a U.S. flock, and October 2003, there were approximately 2,350 cases reported in sheep. USDA, "Scrapie Factsheet," August 2004, 2. An ongoing surveillance and eradication program reported 78 cases in U.S. sheep in FY 2009. USDA, APHIS, "National Scrapie Eradication Program: Fiscal Year 2009 Report," January 15, 2010, 15–16.

[38] OIE, *Terrestrial Animal Health Code*, 2010, art. 14.9.1.

[39] Industry official, e-mail message to Commission staff, November 30, 2010.

[40] WTO Committee on Sanitary and Phytosanitary Measures, "Notification: G/SPS/N/CHN/203," January 19, 2010.

[41] The OIE World Animal Health Information database reports that in the most recent period (January– June 2010), anthrax was present in certain zones or regions of dairy-producing countries such as Argentina, Australia, China, and the United States; paratuberculosis was present in Argentina and the United States, and was restricted to certain zones or regions of Australia; and bovine tuberculosis was present in Argentina, restricted to certain zones or regions of the United States, and not present in Australia.

[42] Australia, Department of Agriculture, Fisheries, and Forestry, Quarantine and Inspection Service, Health Certificate, accessed November 24, 2010.

[43] OIE, "Old Classification of Diseases Notifiable to the OIE," January 25, 2005.

[44] Republic of Argentina, Ministry of Agriculture, Livestock and Fisheries, National Service of Agrifood Health and Quality, Health Certificate, accessed November 24, 2010.

[45] U.S. industry officials, e-mail message to Commission staff, October 14, 2010.

[46] U.S. industry officials, e-mail messages to Commission staff, October 14, 2010, and November 17, 2010.

[47] The JECFA recommended an MRL of 0.01mg/kg in muscle and fat for both pork and beef; JECFA, "Summary of Evaluations Performed by the Joint FAO/WHO Expert Committee," January 28, 2006.

[48] Codex Commission, "Report on the Nineteenth Session of the Codex Committee," September 2010, 8; industry officials, telephone interview by Commission staff, October 20, 2010; JECFA, "Residue Evaluation of Certain Veterinary Drugs," 2010, 38. If China were to establish an MRL for ractopamine in lung tissue, it would have no effect on its imports, as lung tissue is not certified as fit for human consumption.

[49] USTR, *2010 Report on Sanitary and Phytosanitary Measures*, 2010, 33–34. With the data submitted in 2010, China did provide the JECFA with an evaluation of the recommended MRL ("Evaluation Report on the Establishment of MRL for Ractopamine in China"), but the JECFA found that the data submitted did not support China's position.

[50] USTR, *2010 Report on Sanitary and Phytosanitary Measures*, 2010, 24.

[51] In many ways, pork trade with China is a good fit for U.S. producers, as cuts that are highly prized in China are in lower demand in the United States.

[52] WTO, *SPS Agreement Training Module,* (accessed October 26, 2010). The Codex, the Commission on Phytosanitary Measures, and the OIE have all begun work on establishing, but have not completed, standards for trade in GMOs.

[53] USTR, *2010 Report on Technical Barriers to Trade*, 2010, 52.

[54] USTR, *2010 Report on Sanitary and Phytosanitary Measures*, 2010, 37.

[55] Reuters, "China Quarantine Bureau Rejects U.S. Corn Cargo," November 2, 2010; industry official, telephone interview by Commission staff, November 8, 2010; Biotechnology Industrial Organization, Commercial Status of Certain Agricultural Biotechnology Products database; Center for Environmental Risk Assessment, GM Crop database.

[56] American Soybean Association, written submission to the Commission, June 22, 2010, 3.

[57] WTO, "Accession of the People's Republic of China," November 23, 2001, 4; USDA, FAS, *Food and Agricultural Import Regulations and Standards*, June 13, 2007, 3.

[58] USTR, *National Trade Estimate Report*, March 2010, 63; industry official, interview by Commission staff, Hong Kong, September 20, 2010.

[59] Industry official, interview by Commission staff, Hong Kong, September 20, 2010.

[60] For instance, licenses for poultry imports were reportedly delayed following U.S. imposition of duties on Chinese tire imports in 2009. Industry official, interview by Commission staff, Hong Kong, September 20, 2010.

[61] USDA, FAS, *China: Challenges to Increasing U.S. Sales of Chicken Paws to China*, December 15, 2006, 11; industry official, interview by Commission staff, Shanghai, September 13, 2010.

[62] USTR, *National Trade Estimate Report*, March 2010, 63.

[63] Industry official, interview by Commission staff, Hong Kong, September 20, 2010.

[64] International Trade Centre, "Nontariff Requirements in the Domestic Market."

[65] Government official, interview by Commission staff, Washington, DC, May 24, 2010.

[66] USDA, FAS, *Food and Agricultural Import Regulations and Standards*, June 13, 2007, 1.

[67] USTR, *National Trade Estimate Report*, March 2010, 64.

[68] Ibid., 63.

[69] The VAT is 13 percent on raw agricultural products and 17 percent on processed products. PRC, MOFCOM, Decree 538, November 2008, art. 2.

[70] PRC, MOFCOM, Decree 538, November 2008, art. 14.

[71] U.S. Wheat Associates, written submission to the Commission, September 16, 2010, 3; industry officials, interviews by Commission staff, Washington, DC, May 6, 2010.

[72] PRC, MOFCOM, Decree 538, November 2008, art. 8.

[73] Ministry of Finance and State Administration of Taxes, "Circular on VAT Exemption for Staple Crops Enterprises," 1999.

[74] OECD, "Analysis of Non-Tariff Measures: The Case of Labeling," November 13, 2003, 6.

[75] Ibid., 10.

[76] U.S. Department of State, "U.S. Concerned That EU's Biotech Labeling Could Be Burdensome," July 4, 2003; Mattson, Won, and Taylor, "Non-Tariff Trade Barriers in Agriculture," March 2004, 9.

[77] OECD, "Analysis of Non-Tariff Measures: The Case of Labeling," November 13, 2003, 13.

[78] Government official, interview by Commission staff, Washington, DC, May 24, 2010.

[79] Ibid.

[80] Industry official, interview by Commission staff, Shanghai, China, September 13, 2010.

[81] Ibid., September 14, 2010.

[82] Ibid.
[83] PRC, AQSIQ, "Mission." Conformity assessment procedures set the testing methods used to determine that products conform to standards.
[84] USTR, *National Trade Estimate Report*, March 2009, 87.
[85] WTO, *Trade Policy Review; China*, April 26, 2010, viii.
[86] When enacted, these sectoral or local standards must be registered with SAC. Standards Portal, "Standardization Organizations in China: Standardization Administration of China."
[87] Industry officials, interviews by Commission staff, Chengdu, China, September 17, 2010.
[88] Industry officials, interviews by Commission staff, Beijing, China, September 7, 2010; industry officials, interviews by Commission staff, Shanghai, China, September 13, 2010.
[89] Industry official, interview by Commission staff, Washington, DC, July 26, 2010.
[90] Industry official, interview by Commission staff, Shanghai, China, September 13, 2010.
[91] Industry officials, interviews by Commission staff, Shanghai, China, September 13–14, 2010.
[92] USTR, *2010 Report to Congress on China's WTO Compliance*, December 2010, 28–29; industry officials, interviews by Commission staff, Shanghai, China, September 13–14 , 2010; industry officials, interviews by Commission staff, Hong Kong, September 20, 2010.
[93] USDA, FAS, *China's Customs Valuation Operation, 2008*, October 6, 2008, 2.
[94] WTO, *Trade Policy Review; China*, April 26, 2010, S/230, 26.
[95] USTR, *2010 Report to Congress on China's WTO Compliance*, December 2010, 29; USTR, *National Trade Estimate Report*, March 2010, 61; industry officials, interviews by Commission staff, Beijing, China, September 6, 2010; industry officials, interviews by Commission staff, Shanghai, China, September 14, 2010; industry officials, interviews by Commission staff, Hong Kong, September 21, 2010.
[96] Industry official, interview by Commission staff, Hong Kong, September 20, 2010.
[97] Industry official, interview by Commission staff, Shanghai, China, September 13, 2010.
[98] Industry officials, interviews by Commission staff, Shanghai, China, September 13–14, 2010. There are provisions to appeal a valuation, but it may take up to nine months to receive a refund of the overpaid amount.
[99] Industry official, interview by Commission staff, Shanghai, China, September 13, 2010.
[100] WTO, *Trade Policy Review: China*, April 26, 2010, 25.
[101] Industry officials, interviews by Commission staff, Shanghai, September 13–14, 2010; industry officials, interviews by Commission staff, Hong Kong, September 23, 2010; USTR, *National Trade Estimate Report*, March 2010, 61.
[102] Industry official, interview by Commission staff, Hong Kong, September 20, 2010.
[103] Ibid., September 21, 2010.
[104] U.S. Wheat Associates, written submission to the Commission, September 16, 2010, 1–2.
[105] Industry official, interview by Commission staff, Hong Kong, September 22.
[106] USDA, FAS, *Cotton and Products Annual*, May 1, 2010, 20–21.
[107] Industry official, telephone interview by Commission staff, November 3, 2010; Cotton International, *China Sets Limit on Cotton Imports for 2011*, December 30, 2010.
[108] WTO, *Trade Policy Review: China*, April 26, 2010, 11.
[109] Government official, interview by Commission staff, Washington, DC, May 24, 2010.
[110] WTO, SPS Agreement, Annex B, para. 5.
[111] WTO, TBT Agreement, art. 2, para. 2.9.2 and Annex 3, para. L.
[112] USTR, *2010 Report on Technical Barriers to Trade*, 2010, 74; USTR, *National Trade Estimate Report*, March 2010, 91.
[113] WTO, SPS Agreement, Annex B, para. 7 and 8; WTO, TBT Agreement, Article 10, para. 10.5.
[114] USTR, *National Trade Estimate Report*, March 2010, 91.
[115] USTR, *2010 Report Congress on China WTO Compliance*, December 2010, 8.
[116] Government official, interview by Commission staff, Washington, DC, May 24, 2010; industry official, interview by Commission staff, Beijing, China, September 10, 2010.
[117] Industry official, interview by Commission staff, Hong Kong, September 20, 2010. The fact that import licenses depend on the status of political tensions is not unique to exports from the United States. Exports of soybean oil from Argentina to China were disrupted because of trade disputes over products unrelated to soybeans or soybean oil. *China Daily*, "China Said to Seek Brazil, U.S. Soybean Oil Supply," May 5, 2010. After relations between the two countries eased, restrictions were apparently lifted. *Bloomberg News*, "China Agrees to Reopen Market to Argentine Soybean Oil Imports, People Say," October 11, 2010.
[118] National Potato Council, "U.S. Fresh Potato Market Access to China," June 22, 2010.

[119] Industry officials, interviews by Commission staff, Shanghai, China, September 13–14, 2010; industry officials, interviews by Commission staff, Hong Kong, September 21 and 23, 2010.

[120] The Commission's simulation for NTM removal was performed with a framework that links a partial equilibrium trade model to an economy-wide trade model, the Global Trade Analysis Project (GTAP) model. The simulation framework is described in appendix F.

[121] Quantity gaps were converted into price gaps, which were then incorporated into the model simulation and removed.

[122] For further explanation of the methodology used to identify tariff equivalents for simulating the effects of NTMs, see appendix F. Price gaps for the 12 product groups are presented in table F.4.

[123] Ranges for the NTM effects were obtained by employing low and high values in the estimated price gaps. Low and high values were computed as average price gap x (1 + 0.10).

[124] The U.S. Meat Export Federation has estimated the value of U.S. beef sales in the first year U.S. exporters gain access to the Chinese market at $200 million. Industry official, telephone interview by Commission staff, February 3, 2011.

[125] The Commission's tariff model simulation estimates that Chinese tariffs and TRQs on all 131 product groups had effects on U.S. exports ranging from $1.3–2.1 billion.

End Notes for Appendix D

[1] The Almond Board, hearing statement to Commission, September 22, 2010.

[2] The Almond Board stated that almonds in China are mainly consumed as snacks and are roasted and flavored in China.

[3] American Dehydrated Onion and Garlic Association, written submission to the USITC, September 8, 2010.

[4] American Soybean Association, hearing statement to the Commission, June 22, 2010.

[5] Blue Diamond Growers, written submission to the Commission, June 3, 2010.

[6] California Cling Peach Board, written submission to the Commission, July 28, 2010.

[7] National Cotton Council, written submission to the Commission, September 15, 2010.

[8] National Milk Producers Federation and U.S. Dairy Export Council, written submission to the Commission, September 15, 2010.

[9] National Potato Council, written submission to the USITC, June 22, 2010, and National Potato Council, written submission to the USITC, June 24, 2010.

[10] Northwest Horticultural Council, written submission to the USITC, June 14, 2010.

[11] Ranchers-Cattlemen Action Legal Fund, United Stockgrowers of America, written submission to the Commission, June 9, 2010.

[12] R-CALF USA asserted that the United States is hurting itself and its trade by importing Canadian cattle of any age when Canada has been struggling with BSE outbreaks.

[13] Stewart and Stewart, written submission to the Commission, June 3, 2010, and June 29, 2010.

[14] U.S. Wheat Associates, written submission to the Commission, September 16, 2010.

[15] Western Pistachio Association, written submission to the Commission, September 15, 2010.

[16] Wine Institute, California Association of Wine Grape Growers, and Wine America, written submission to the USITC, September 13, 2010.

End Notes for Appendix E

[1] Examples include USITC, *Conditions of Competition for Milk Protein Products in the U.S. Market*, 2004; USITC, *Conditions of Competition for Certain Oranges and Lemons in the U.S. Fresh Market*, 2006; USITC, *Canned Peaches, Pears, and Mixtures: Conditions of Competition between U.S. and Principal Foreign Supplier Industries*, 2007.

[2] For example, in recent Commission investigations, factors affecting competitiveness in the canned fruit and citrus fruit industries were identified as natural resource endowments, production costs, technology, market size, industry concentration, government involvement, and exchange rates, and in the U.S. milk protein industry,

competitiveness factors included costs of production, government programs, production technology, transportation costs, and exchange rates.

[3] Industry representatives, interviews with Commission staff, China, September 2010; written hearing testimony and submissions submitted to the U.S. International Trade Commission in connection with inv. nos. 332-518, *China's Agricultural Trade: Competitive Conditions and Effects on U.S. Exports*, June 2010.

[4] USITC, *Guidelines for Developing an Economic Framework for an USITC Study,* 2008.

[5] Other definitions of competitiveness are (1) "the ability of a nation, national industry, or firm to produce goods and services that consumers choose over competing alternative." President's Commission on Industrial Competitiveness, *Global Competition—New Reality* (vol. 1), January 1985, 6; and (2) "the ability of producers to sell goods in foreign markets at price, quality and timeliness comparable to competing foreign products." USITC, *Sub-Saharan Africa: Effects of Infrastructure Conditions on Export Competitiveness*, 2009.

[6] Porter, Michael E., *Competitive Strategy*, 1980.

[7] Porter, Michael E., *Competitive Advantage*, 1985, 13.

[8] Porter provides an example of a focus strategy used by Martin-Brower, once the third largest food distributor in the United States. "Martin-Brower has reduced its customer list to just eight leading fast food chains. Its entire strategy is based on meeting the specialized needs of the customers, stocking only their narrow product lines, order taking procedures geared to their purchasing cycles, locating warehouses based on their locations, and intensely controlling and computerizing record-keeping." Porter, Michael E., *Competitive Advantage*, 1985, 40.

[9] Klein, "A Critique of Competitive Advantage," July 2001, 1–2.

[10] Aktouf, "The False Expectations of Michael Porter's Strategic Management Framework," January/June 2005, 76.

[11] Downes, "Beyond Porter," December 1997.

[12] Sharp and Dawes, "Is differentiation optional? - a critique of Porter's competitive strategy typology," 1996, 269–271.

[13] Klein, "A Critique of Competitive Advantage," July 2001, 6.

[14] O'Shaugnessy, "Michael Porter's Competitive Advantage Revisited," 2006, 13.

[15] Porter's five forces are bargaining power of customers, buying powers of suppliers, rivalry between existing players, threat of substitutes, and threat of new entrants. Recklies, "Beyond Porter: A critique of the critique of Porter," 2001.

[16] Speed, "Oh Mr. Porter! A Reappraisal of Competitive Strategy," May/June 1989, 10.

[17] Porter, *The Competitive Advantage of Nations*, 1990.

[18] The focus strategy discussed by Porter can be viewed as a type of product differentiation in which sellers adapt their products to fit the particular desires of a narrow segment of consumers. This is particularly true for branded items such as wine produced in small lots and hand-crafted cheese. For this reason, further discussion of the focus strategy is dropped from this analytical framework.

[19] The assertion that reliability of supply is a major factor in importer purchasing decisions is based on Commission staff experience in dealing with agricultural industry exporters and importers.

[20] USITC, *India: Effects of Tariffs and Nontariff Measures*, 2009.

[21] This does not mean that the United States is not competitive in all markets. In other markets, buyers may base their purchasing decisions less on delivered cost and more on product quality and the ability of U.S. exporters to meet the desired product specifications of customers. In such markets, the U.S. product may be more competitive than that of Brazil and Argentina.

[22] In this investigation, China's agricultural sector is defined using the HS subheadings identified by the WTO. These are agricultural products in HS chapters 1–24 (excluding chapter 3), 41, 51, and 52 (plus certain industrial/chemical products from other chapters). Agricultural products form a continuum based on the level of processing. As an illustration, broad categorizations are unprocessed products for further processing (wheat, corn, soybeans); minimally processed products for immediate consumption (meat, fresh horticultural products); semiprocessed products for the food manufacturing sector (soybean oil, flour, raw sugar); semiprocessed products for the textile, apparel, and home furnishings sector (cotton, wool, hides and skins); semiprocessed products for immediate consumption (dairy, processed horticultural products); highly processed generic products for immediate consumption (crackers, infant formula); and highly processed branded products for immediate consumption (Oreo Cookies, Philadelphia cream cheese).

End Notes for Appendix F

[1] For more information about the GTAP model, see Hertel, *Global Trade Analysis: Modeling and Applications*, 1997; Narayanan and Walmsley, *Global Trade, Assistance, and Production: The GTAP 7 Data Base*, 2008.

[2] The GE model is based on version 7 GTAP data.

[3] ASEAN (the Association of Southeast Asian Nations) is an economic organization of 10 countries: Brunei, Cambodia, Indonesia, Laos, Malaysia, Burma, the Philippines, Singapore, Thailand, and Vietnam. The GTAP database version 7 does not indentify Brunei because of the lack of certain economic statistics. Thus Brunei has not been included in either the GE model or the PE model in this report.

4 Other works applying a PE-GE approach are Grant, Hertel, and Rutherford, "Extending General Equilibrium to the Tariff Line," June 2007; Narayanan, Hertel, and Horridge, "A Nested PE/GE Model for GTAP," June 2008; Jansson et al., "Getting the Best of Both Worlds?" June 2008; USITC, *Global Beef Trade*, September 2008; USITC, *India: Effects of Tariffs and Nontariff Measures*, November 2009.

[5] A CET production possibilities frontier is characterized by a constant percentage change in a product-mix ratio to changes in the marginal rate of transformation. Powell and Gruen, "The Constant Elasticity of Transformation Production Frontier," 1968, 315–28.

[6] A CES function is characterized by a constant percentage change in product proportions because of a percentage change in the marginal rate of technical substitution. Arrow et al., "Capital-Labor Substitution and Economic Efficiency," 1961, 225–50.

[7] Armington, "A Theory of Demand for Products," 1969, 159–76.

[8] United Nations Commodity Trade Statistics Database, statistics downloaded from WITS (World Integrated Trade Solution) on July 1, 2010.

[9] FAO statistics for production, consumption, and trade for various products and product groups were downloaded from the FAOSTAT online database.

[10] TRAINS statistics were downloaded from WITS on July 1, 2010. For documentation of the MAcMapHS6 database, see Boumellassa, Laborde, and Mitaritonna, "A Consistent Picture of the Protection," 2009.

[11] This rule was proposed by Jomini et al., *The SALTER Model of the World Economy*, 1994. Another study failed to reject this rule statistically. Lui, Arndt, and Hertel, "Parameter Estimation and Measures of Goodness of Fit," 2004, 626–49.

[12] Examples of real-world complexities that are difficult to reflect in the model include the changing relative growth of different economies; politically motivated, export-oriented investment; relationships between multinational subsidiaries that influence trade patterns; and such events as catastrophic weather or violence, which are inherently unpredictable (at least in their details).

[13] For the foundations of the method for estimating price gaps for NTMs at the Commission, see Linkins and Arce, "Estimating Tariff Equivalents of Non-Tariff Barriers," August 2002. For further descriptions of the price-gap method, as well as literature reviews, see Deardorff and Stern, *Measurement of Non-Tariff Barriers*, 1998; Ferrantino, "Quantifying the Trade and Economic Effects of Nontariff Measures," January 2006.

[14] In general, it is not feasible to correct for all possible quality differences while estimating NTM price gaps because some of these differences are unobservable. Certain countries consistently export products at higher unit values than other countries, however, suggesting a quality difference, particularly for relatively homogeneous goods. The methods used in this study exploit the observed quality differences arising from differences in exporter-specific unit values.

[15] This estimation procedure is similar to that used in a recent Commission study. See USITC, *India: Effects of Tariff and Nontariff Measures,* November 2009.

[16] U.S. export statistics were compared with U.S. official data as reported on the Commission's DataWeb system and revised as necessary.

[17] This method uses the relation (reference export unit value)*(1 + transport margin + NTM price gap) = (reference import unit value). Thus, implied NTM rents = (NTM price gap/(1 + transport margin + NTM price gap))*(value of c.i.f. imports). The usual method of trade weighting yields unrealistic results in this case because, for price gaps exceeding 100 percent, the implied tariff revenues are larger than the value of c.i.f. imports and not a realistic estimate of the value of NTM rents, which must be a share of the value of imports.

[18] These difficulties included, inter alia, nonstandard units of measurement and thinly traded products exported from small countries, for which a reference price could not be established. The standard unit of measurement for almost all agricultural products is the kilogram or the metric ton. Beverages are typically measured in liters.

In: Agricultural Trade in China
Editor: Mark T. Devine

ISBN: 978-1-62100-602-2

Chapter 2

Chinese Agricultural Exports Provide Growing Competition*

United States Department of Agriculture

Summary

With China becoming the second largest U.S. market in fiscal year 2010, China's emergence as a major agricultural importer is well-known. While China is a large net food importer, it is also becoming a formidable competitor in the export market with shipments nearly tripling over the past 10 years and market share increasing. While some of China's exports are bound for the United States, others directly compete with U.S. products in foreign markets. Exports of consumer-oriented high-value products (HVPs) have shown particular growth, especially to nearby markets in Japan and Southeast Asia. Although China faces production constraints and booming domestic consumption, future exports, particularly of high value products, have room for expansion.

Exports Rise along with Market Share

Chinese agricultural exports began to surge after 1999, with shipments increasing in value from $10.3 billion in 1999 to an estimated $28 billion in 2010. This $18 billion increase is impressive, but as global agricultural trade was also on the rise over this period, perhaps more important was the increase in market share. Chinese exports accounted for 4.5 percent of global agricultural trade in 1999, but climbed to 5 percent in 2009. Meanwhile, over the same period, U.S. export share fell from 22 percent to 18 percent. Although other exporters, particularly Brazil and Argentina, played a larger role in the fall of U.S. share, the growth of Chinese exports likely contributed to the drop, particularly in certain markets for consumer-oriented HVPs.

* This is an edited, reformatted and augmented version of a United States Department of Agriculture Foreign Agricultural Service Report, updated February 9, 2011

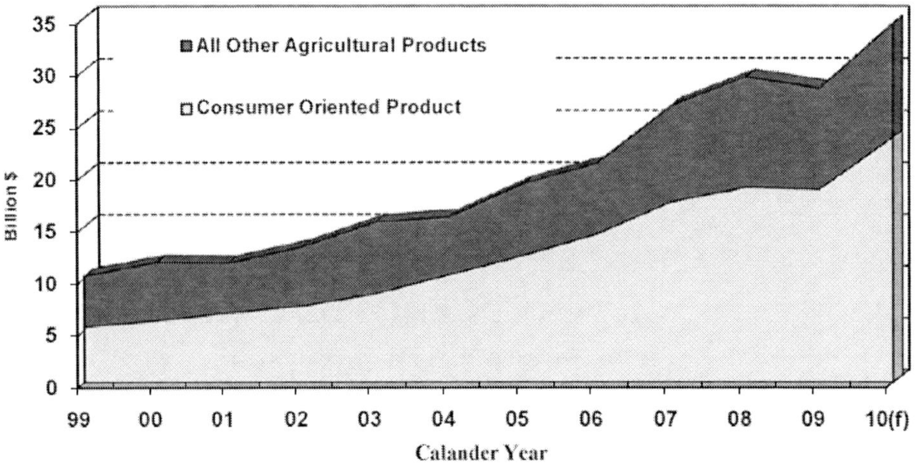

Consumer-Oriented Products Push Exports.

Consumer-Oriented HVP Exports on the Rise

As seen in the chart above, much of the growth in Chinese exports over the past decade was due to greater shipments of consumer-oriented products, which increased by more than $18 billion compared to under $6 billion for all other products. While the category consists of over 300 items, top products include garlic, tomato paste, fresh apples and juice and processed poultry. Compared to other major agricultural exporters, consumer-oriented products make up a large part of the overall Chinese export basket, accounting for two-thirds of exports in 2009. Meanwhile, the category accounts for about one-third of U.S. exports, 39 percent for Canada, and 30 percent of Brazilian exports.

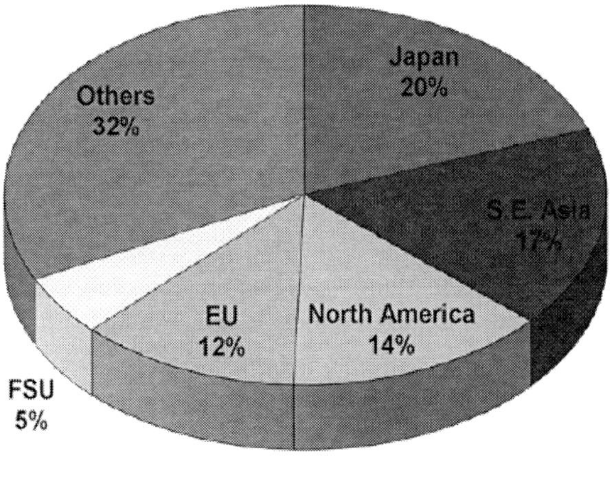

Chinese Consumer Oriented Export Markets; (2009).

CHINESE EXPORTS HEAT UP IN BATTLEGROUND MARKETS

As seen in the chart above, Japan and Southeast Asia[1] have been the largest markets for Chinese consumer-oriented products, in part due to a freight cost advantage and shorter delivery times compared to U.S. and European suppliers. The Japanese market has been particularly important for Chinese exports, some of which compete with U.S. products, including meats and processed fruits and vegetables. Chinese exports of consumer-oriented products to Japan in 2009 were $3.5 billion compared with U.S. exports of $4.4 billion. However, Chinese exports have grown over the past decade (1999 vs. 2009) by $1.3 billion while U.S. consumer- oriented exports fell by $244 million. Chinese consumer- oriented exports also compete directly with U.S. products in several other markets including Southeast Asia, Europe, South Korea and Canada. China's trade with Southeast Asia was boosted by the "early harvest" provision in its new free trade agreement with the Association of Southeast Asian Nations (ASEAN) that reduced tariffs on agricultural products. Although Chinese exports to Canada were only $321 million last year, shipments of low-priced apple juice concentrate, pasta and fruits provide competition for U.S. products. Chinese exports to the European Union, another top market for U.S. consumer-oriented products, have also performed well and doubled over the past five years from $1.1 billion in 2004 to $2.2 billion in 2009.

APPLE MARKET DEMONSTRATES EMERGENCE OF GROWING COMPETITION

Perhaps no commodity better illustrates the emergence of China as an export competitor than apples. China's share of global apple exports in 1999 and 2000 (July/June) was just 10 percent, but reached 23 percent in 2009 and 2010. Meanwhile, over the same time frame, the U.S. share of global apple exports fell from 21 percent to 16 percent. While Chinese apples appear to be a growing competition to U.S. apples around the world, nowhere is this more pronounced than in Southeast Asia. Though U.S. apple exports to the region have grown over the past decade, they have grown at a slower rate than Chinese exports. As seen in the chart below, Chinese exports to the region have increased dramatically from only $57 million in 1999 to $315 million in 2009. Chinese apples are generally lower-priced and compete in a separate quality category from U.S. apples. However, whether creating new demand based on price or competing directly with U.S. apples, China has clearly captured the growth in Southeast Asian demand for apples.

Nevertheless, U.S. apples continue to perform well in many markets, including Mexico, due to higher U.S. quality and safety standards.

As with many of China's consumer- oriented products, the future looks bright for increasing apple production and exports. With abundant cheap labor, China has a natural

[1] For this paper, Southeast Asia includes Malaysia, Thailand, Vietnam, Singapore, and Indonesia.

advantage in labor-intensive crops. Though land area available for expansion is limited, yield gains for high value crops could provide future production growth as has been the case with apples. Area dedicated to apples in traditional growing regions of Eastern China has stagnated, but new orchards are being planted in western provinces. Moreover, yields have improved. Government programs are encouraging dissemination of new apple varieties, providing subsidies to bag apples in the field after harvest, thus improving quality, assisting producers in adopting "green food," good agricultural practices and other quality management systems, setting up centers to test for toxic pesticide residues, and sponsoring trade fairs. Quality has also improved due to better orchard management. As with many of China's consumer-oriented products, however, price should continue to be the principal driver behind increased exports.

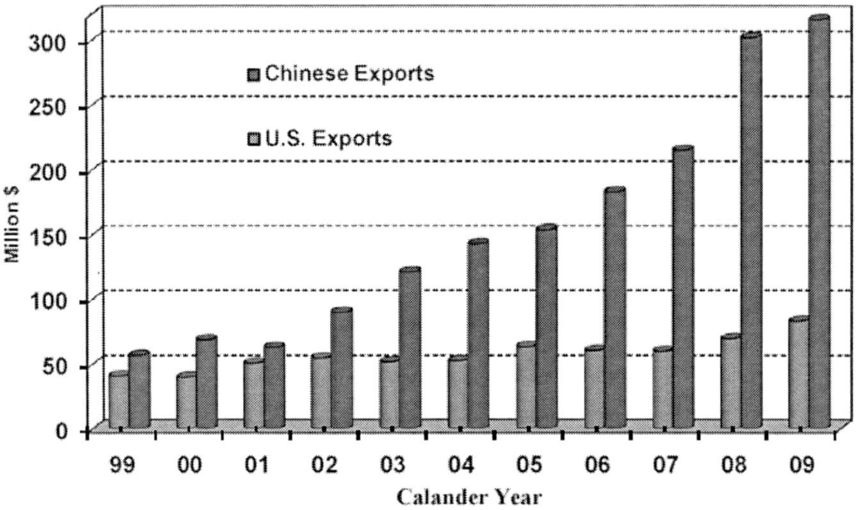

Date Source: GTA.

Chinese Apple Exports to S.E. Asia Soar Past Competition.

FUTURE EXPORT GROWTH FACES CHALLENGES

Rising costs and concerns about commodity price inflation may constrain China's export growth. Farm labor costs are rising as much as 10 percent annually as rural laborers migrate to cities while the competition for scarce land is raising rents. Rapid urbanization is consuming suburban vegetable plots and Chinese food manufacturers are pushing farther into China's interior to search for supplies of fruits, vegetables and aquaculture products as raw materials for domestic and exported products. Rapidly-rising prices of garlic and other vegetables during 2010 raised concerns about domestic vegetable supplies. Meanwhile, China's State Council directed municipal authorities to set up vegetable farms and establish vegetable reserve systems.

Continued growth in both overall exports and global market share will depend on several factors including 1) production growth given rising domestic consumption and land constraints; 2) overall food safety conditions in agricultural production; and 3) labor and other production costs. Though China enjoys a freight advantage to East and Southeast Asian markets, relatively high internal transportation costs can mitigate some of the price competition of Chinese products. Despite the rising cost structure on Chinese farms, robust investment in new processing facilities (encouraged by subsidized loans and tax breaks) tends to push down prices of final products. Relatively low production costs and the push of exporters into western regions may enable China to maintain its status as a low-cost supplier for many products.

In: Agricultural Trade in China
Editor: Mark T. Devine

ISBN: 978-1-62100-602-2

Chapter 3

CHINA EMERGES AS THE SECOND LARGEST U. S. AGRICULTURAL EXPORT MARKET*

United States Department of Agriculture

SUMMARY

Agricultural exports to China finished fiscal year 2010 at $15.1 billion as the country surpassed Mexico to become the number two market for U.S. agricultural exports. Record soybean shipments along with strong Chinese demand for cotton and distillers dried grains (DDG's) contributed to the overall record in U.S. exports to China. Moving forward, exports in FY 2011 are forecast to reach another record at $17.5 billion.

CHINA GROWS FASTER THAN NAFTA PARTNERS

Although agricultural exports to Mexico and Canada have grown impressively over the past decade, sales to China have grown faster, and finished the fiscal year up $4 billion over FY 2009's record exports of $11.1 billion. Total exports to China in FY 2010 hit $15.1 billion due primarily to record soybean shipments of $9.3 billion. Strong demand throughout the year combined with less competition from South American supplies early in the year, led to sustained soybean shipments. Increased demand from the pork and poultry industries spurred exports of soybeans along with some late-season corn shipments. Additionally, rising demand and high Chinese corn prices led to an increase in U.S. DDG shipments, which soared from a previous record of $35 million in FY 2009 to $447 million in FY 2010.

* This is an edited, reformatted and augmented version of a Foreign Agricultural Service, International Agricultural Trade Report, dated December 20, 2010.

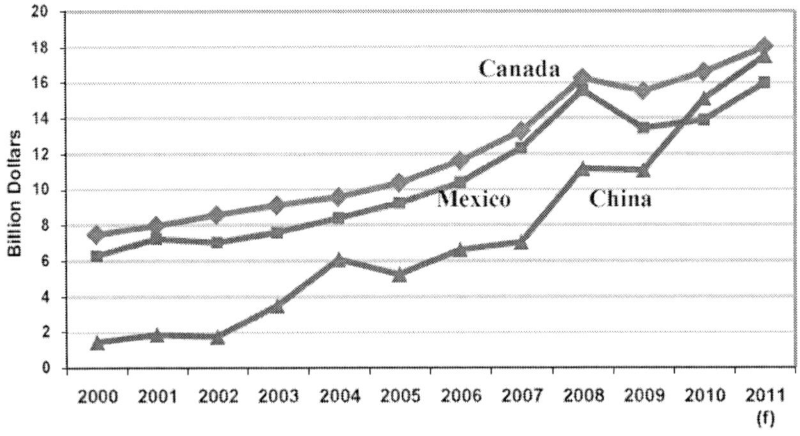

U.S. Agricultural Exports to Canada, China, & Mexico.

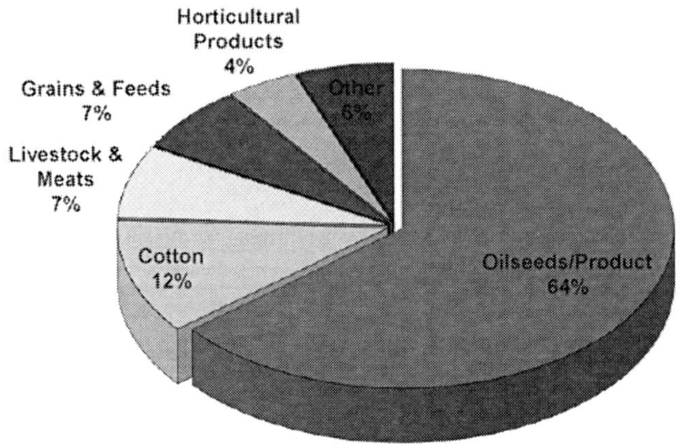

U.S. Exports to China by Product; (FY 2010).

While exports to China were soaring in early 2010, shipments to Mexico grew only slightly as punitive duties on horticultural products, slow economic growth, and a strong U.S. dollar against the Mexican Peso kept a lid on U.S. shipments. Exports increased only 5 percent over the previous year, compared to 36 percent in China and 13 percent growth in exports to all markets. This higher growth in China compared to Mexico is expected to continue in FY 2011 and China should narrow the gap with the top market of Canada.

TOTAL CHINESE AGRICULTURAL IMPORTS GROW

The United States was not the only exporter to benefit from increasing Chinese demand for agricultural products this past year. The total agricultural import market is estimated to have grown from $48 billion in FY 2009 to $64 billion in FY 2010 with the U.S. share steady at 25 percent. Most of the increase of $16 billion came from greater purchases of soybeans, cotton, rubber, and palm and rapeseed oil. Regional competing suppliers such as Thailand and India increased market share as did Uzbekistan, due to increased cotton exports. Meanwhile, market share for South America (Brazil and Argentina) fell from 28 percent to 23 percent though the actual value of exports from the region increased due to strong demand and higher prices for oilseeds.

FY 2011 Exports Forecast to Reach Another Record

Exports from the United States to China are forecast at a record $17.5 billion in FY 2011, which is just $500 million behind the largest export market of Canada. High prices for cotton, soybeans, and other products exported to China support the record forecast. Furthermore, early season export sales of soybeans are ahead of last year's record pace, indicating that soybeans should lead the push for another year of record U.S. exports to China.

In: Agricultural Trade in China
Editor: Mark T. Devine
ISBN: 978-1-62100-602-2

Chapter 4

India and China: Divergent Markets for U.S. Agricultural Exports[*]

United States Department of Agriculture

India and China share many similarities. The world's only two population billionaires are both among the global leaders in GDP growth over the past decade and both face similar challenges of producing enough food to meet the demands of an increasingly affluent populace. However, when it comes to agricultural trade, China and India appear to have taken divergent paths. China has emerged as a major market for agricultural products whereas India (though somewhat liberalizing trade) remains a largely closed market. This is particularly true for U.S. agricultural exports, as China is the fourth largest and a fast growing market, while India is not among the top 20 markets. However, both markets hold significant potential for U.S. agricultural exports, albeit with a different mix of products.

Trade Policy

India has long been a largely closed market to agricultural imports. Imports are generally viewed negatively, partially due to the fact that roughly 65 percent of the population depends on agriculture for their livelihoods and agricultural imports are seen as providing unwanted competition. Additionally, failed monsoons and reliance on large scale grain imports and food aid during the 1960s and 1970s has led the government to focus on food self sufficiency. In many respects, imports are seen mostly as an unnecessary evil.

---

[*] This is an edited, reformatted and augmented version of a Foreign Agricultural Service, International Agricultural Trade Report, dated February 24, 2010.

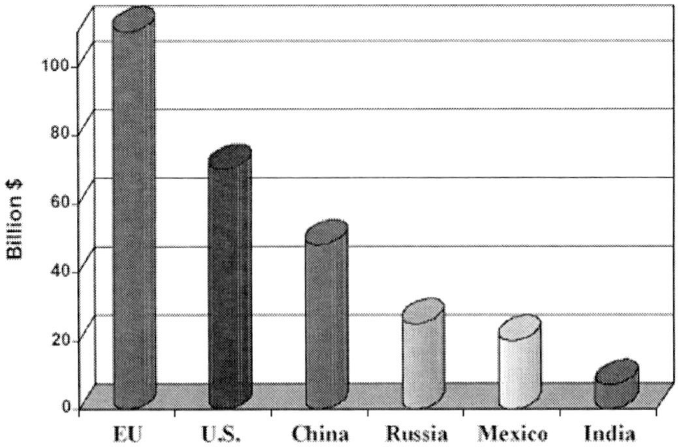

Source: GTA.

Chart 1. Total Agricultural Imports (2008).

Total Indian agricultural imports in 2008 (chart 1) were one-third that of Mexico and only a small fraction of import levels seen in China, the European Union, and United States. So how has the second most populous country, with a rapidly growing middle class, remained relatively closed to agricultural imports? While Indian import duties have come down in recent years, the average bound rate exceeds 100 percent, while the average applied rate is about 35 percent. High duties are meant to protect domestic industries, as in the case of apples where U.S. products face a 50 percent duty. However, duties also act as a government revenue source, as in the case of almonds and various "luxury" products, where there is little domestic production to protect. Along with high duties, the government also uses non-tariff barriers to block imports. Sanitary and phytosanitary barriers are employed to keep out U.S. dairy products, meats, grains, oilseeds, and nearly all biotech products.

In contrast, China has adopted trade policies that have helped it to become more integrated with the world economy. Throughout the 1990s, the government lowered tariffs and other trade barriers to many agricultural products, and repealed state-trading companies' monopoly on imports and exports[1]. Since joining the WTO in December 2001, China has rapidly become a large agricultural importer, and now is ranked as the fourth largest U.S. market (chart 2).

China's liberalized trade policy was borne out of the overall economic reforms that began in the late 1970s. Reforms in the agricultural sector changed collective production to a household responsibility system where farmers had autonomy to make production decisions. Marketing also shifted to the private sector, facilitating the creation of independent markets and fostering greater inter-regional trade. Although self-sufficiency, particularly for strategic commodities like food grains, is still paramount to the Chinese government, the reforms have unleashed an inexorable tide of market liberalization. Both imports and exports have boomed under the reforms. With scarce arable land and relatively abundant and inexpensive labor, the country's agricultural economy has benefited greatly by importing land-intensive bulk and intermediate commodities and exporting labor-intensive consumer-oriented products. Cotton

---

[1] Lohmar, B., F. Gale, F. Tuan, and J. Hansen, 2009, "China's Ongoing Agricultural Modernization: Challenges Remain After 30 Years of Reform."

and textiles are a prime example of this development model. China's export-driven textile and apparel industry is the world's largest, which depends on imported cotton, hides, and skins to fuel its burgeoning growth. The government has provided support to this industry by, among other measures, allowing cotton imports well above China's Tariff-Rate Quota under its WTO obligations without imposing the prohibitive 40 percent out-of-quota tariff.[2] Since its WTO accession, growth of China's agricultural imports has outstripped that of exports, resulting in a widening trade deficit.

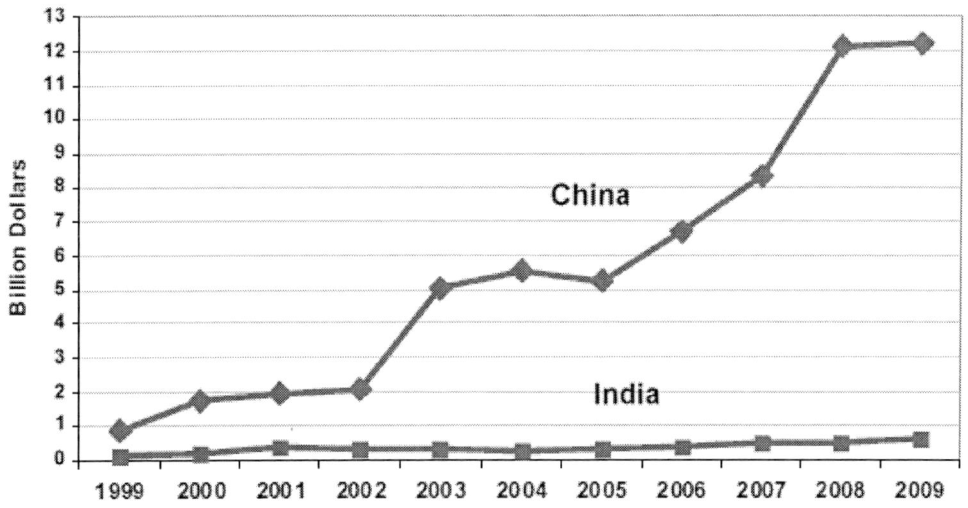

Source: U.S. Census Bureau.

Chart 2. U.S. Ag Exports to China and India.

Consumption Patterns

Higher meat consumption may be the most important factor explaining China's greater level of agricultural imports. Rising meat consumption since the 1980s drew more of its land into production of feed crops and created robust demand for imported soybeans and fishmeal that add protein to feed for poultry, hogs, and fish. China also imports chicken paws, wing tips, and meat offal. Meanwhile, Indian per capita consumption of beef, pork and poultry combined is less than 10 percent that of China (chart 3). Over the past decade Indian consumption increased only 1.6 kg/person, compared to 8.2 kg/person in China. India is largely a Hindu nation, with approximately 40 percent of the population estimated to be vegetarians for either religious or economic reasons. Moreover, India trails China in economic development, with Indian GDP per capita less than half that of China, while unemployment is nearly double. Most Indian consumers simply do not have the purchasing power of their Chinese neighbors.

Not only has the very low level of meat consumption meant that India has not been a market for meat, but India has also not been a market for feed ingredients such as soybeans, feed grains, distillers grains, and meals. These commodities destined for the livestock industry have, however, been important drivers of China's impressive import growth.

Between 1990 and 2009, production of meat and poultry nearly tripled, spurred by urbanization and income growth, which also drove up demand for vegetable oil. As a result, soybean demand (mostly for oil and meal) expanded over 460 percent during the last 20 years, while production only grew 32 percent. To bridge the rapidly widening gap between domestic production and soaring consumption, imports grew from virtually nothing in 1990 to 41 million metric tons in 2009—three-fourths of the country's domestic demand and over 50 percent of total world trade (chart 4). The United States has been the dominant supplier, with 60 percent of all U.S. agricultural exports to China comprised of soybeans.

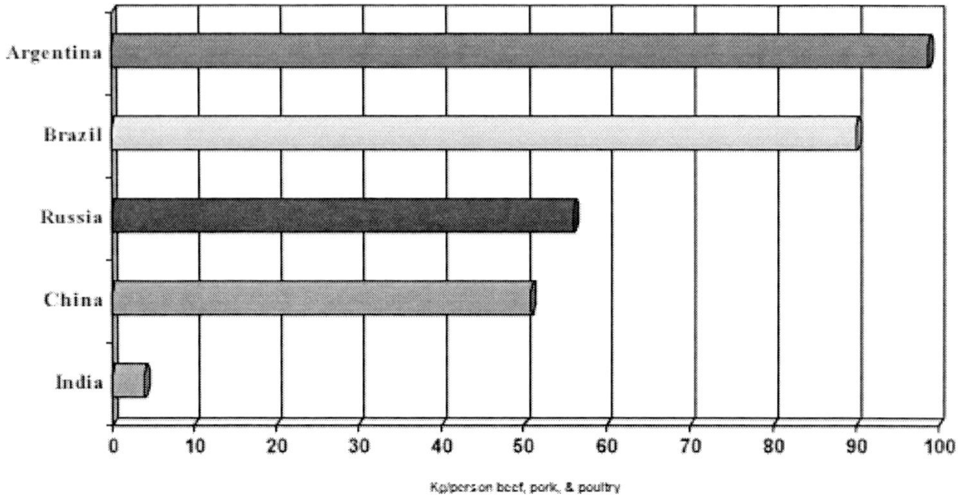

Source: USDA/FAS.

Chart 3. Per Capita Meat Consumption.

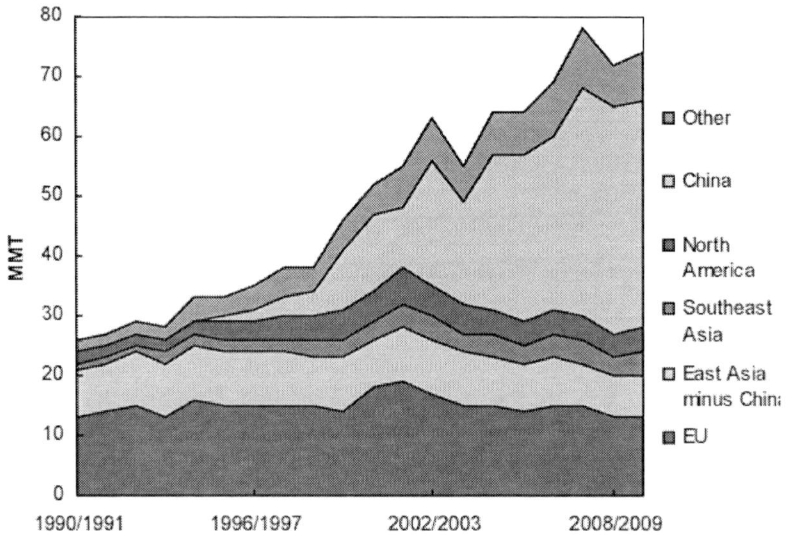

Source: GTA.

Chart 4. Global Soybean Imports.

In addition to soybeans, China also imports large quantities of cotton and hides. The textiles and apparel industry, famous for its booming exports, is increasingly driven by surging domestic use as a result of rising disposable income and population growth. According to China's National Statistics Bureau, urban per capita expenditure on clothing more than doubled between 1997 and 2007. Cotton imports have surged dramatically since 2001, accounting for over a quarter of total world trade. In contrast, India's textile industry is much smaller and its cotton imports have fallen dramatically as domestic use has not kept pace with production. As a result, India is a relatively small market for U.S. cotton while China is the top U.S. market (chart 5).

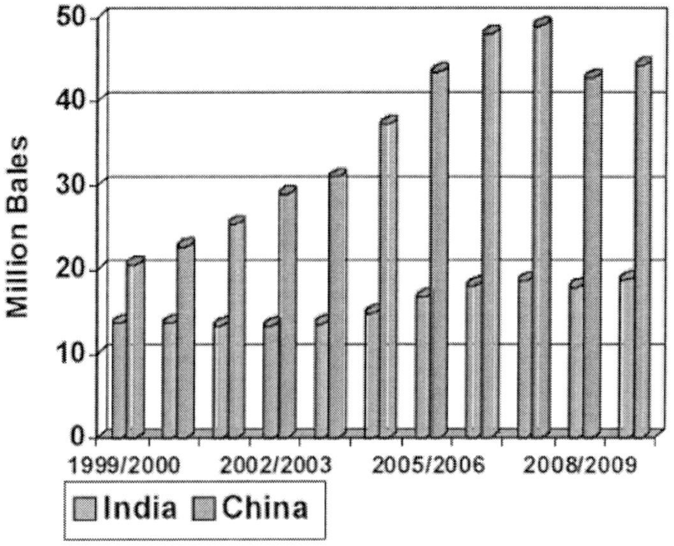

Source: USDA PS&D.

Chart 5. Cotton Use.

PROSPECTS

Looking ahead, the prospects for Chinese agricultural imports are promising. China's per capita arable land is 40 percent less than the world average and production of land-intensive bulk and intermediate commodities, such as soybeans, cotton, and hides simply cannot keep up with surging demand. It has relied heavily on U.S. supplies and will continue to do so. Imports of consumer-oriented products, such as poultry, dairy, processed fruits and vegetables, and tree nuts, will also likely continue their robust growth, although their amounts are still small compared to bulk imports.

U.S. agricultural exports to India, though much lower than China, are expected to continue the impressive growth seen over the past few years, particularly in the high value categories. While Chinese imports are mostly driven by bulk commodities, India's growth is likely to continue in fruits, nuts, and processed products due to a booming middle class, emergence of organized retail outlets, and the positive image of U.S. products. Additionally, some growth is expected in pulses and to a lesser extent cotton. A major driver behind increased demand for agricultural imports in both countries is the growing middle class. Projections from Global Insight's consumer markets data indicate that India and China together will account for nearly 75 percent of the global growth of middle class households over the next ten years.

While India and China share many similarities, approaches toward agricultural trade, and imports in particular, have taken opposite courses. China has emerged as a top U.S. market while India remains a minor market with restrictive import policies and low meat consumption. However, the future is encouraging in both markets for U.S. agricultural exports.

INDEX

A

D

E

F

G

H

I

J

K

L

M

N

O

P

Q

R

T

U

V

W

Y